PROCEEDINGS OF THE SIXTH BERKELEY SYMPOSIUM

VOLUME V

PROCEEDINGS *of the* SIXTH BERKELEY SYMPOSIUM ON MATHEMATICAL STATISTICS AND PROBABILITY

Held at the Statistical Laboratory
University of California
June 21–July 18, 1970
April 9–12, 1971
June 16–21, 1971
July 19–22, 1971

with the support of
University of California
National Science Foundation
National Institutes of Health
Air Force Office of Scientific Research
Army Research Office
Office of Naval Research

VOLUME V

DARWINIAN, NEO-DARWINIAN, AND
NON-DARWINIAN EVOLUTION

April 9–12, 1971

EDITED BY LUCIEN M. LE CAM,
JERZY NEYMAN, AND ELIZABETH L. SCOTT

UNIVERSITY OF CALIFORNIA PRESS
BERKELEY AND LOS ANGELES
1972

UNIVERSITY OF CALIFORNIA PRESS
BERKELEY AND LOS ANGELES
CALIFORNIA

CAMBRIDGE UNIVERSITY PRESS
LONDON, ENGLAND

ISBN: 0-520-02188-6

LIBRARY OF CONGRESS CATALOG CARD NUMBER: 49-8189

PRINTED IN THE UNITED STATES OF AMERICA

CONTENTS OF PROCEEDINGS VOLUMES I, II, III, IV, V, AND VI

Volume I—Theory of Statistics

Multivariate Analysis

Volume II—Probability Theory

Introduction

Measure Theory

Inequalities

Combinatorial Analysis

Ergodic Theory

Gaussian Processes

Central Limit Theorem

Volume III—Probability Theory

Passage Problems

Markov Processes—Potential Theory

Markov Processes—Trajectories—Functionals

Point Processes, Branching Processes

Information and Control

Volume IV—Biology and Health

Clinical Trials and Sequential Procedures

Population Studies and Branching Processes

Biostatistics

Cellular Phenomena and Carcinogenesis

Psychological Aspects of Observational Studies

Volume V—Darwinian, Neo-Darwinian, and Non-Darwinian Evolution

Darwinian and Non-Darwinian Evolution Theory

DNA, RNA, Amino acid Sequences

Population Studies and Evolution

F. AYALA, Darwinian *versus* non-Darwinian evolution in natural populations of *Drosophila*. R. W. ALLARD and A. L. KAHLER, Patterns of molecular variation in plant populations. W. BODMER and L. CAVALLI-SFORZA, Variation in fitness and molecular evolution.

Role of Theory in Evolutionary Studies

L. L. GATLIN, Evolutionary indices. T. A. REICHERT, The amount of information stored in proteins and other short biological code sequences. R. BELLMAN, Hierarchies of control processes and the evolution of consciousness. R. HOLMQUIST, Theoretical foundations of paleogenetics. W. J. EWENS, Statistical aspects of the non-Darwinian theory.

Volume VI—Effects of Pollution on Health

Programs and Studies of Governmental Agencies

J. R. GOLDSMITH, Statistical problems and strategies in environmental epidemiology. J. V. BEHAR, Application of computer simulation techniques to problems in air pollution. J. R. TOTTER, Research programs of the Atomic Energy Commission's Division of Biology and Medicine relevant to problems of health and pollution. J. F. FINKLEA, M. F. CRANMER, D. I. HAMMER, L. J. McCABE, V. A. NEWILL, and C. M. SHY, Health intelligence for environmental protection: a demanding challenge. W. B. RIGGAN, D. I. HAMMER, J. F. FINKLEA, V. HASSELBLAD, C. R. SHARP, R. M. BURTON, and C. M. SHY, CHESS, a community health and environmental surveillance system. W. C. NELSON, V. HASSELBLAD, and G. R. LOWRIMORE, Statistical aspects of a community health and environmental surveillance system. M. G. SIRKEN, Survey strategies for estimating rare health attributes.

Radiation and Health Controversy

E. J. STERNGLASS, Environmental radiation and human health. M. H. DEGROOT, Statistical studies of the effect of low level radiation from nuclear reactors on human health. J. W. GOFMAN and A. R. TAMPLIN, Epidemiologic studies of carcinogenesis by ionizing radiation. E. A. TOMPKINS, P. M. HAMILTON, and D. A. HOFFMAN, Infant mortality around three nuclear power reactors. V. L. SAILOR, Population exposure to radiation: natural and man-made. H. W. PATTERSON and R. H. THOMAS, Radiation and risk—the source data. E. LANDAU, Radiation and infant mortality—some hazards of methodology.

Problems of Monitoring

E. B. HOOK, Monitoring human birth defects: methods and strategies. G. D. FRIEDMAN and M. F. COLLEN, A method for monitoring adverse drug reactions. A. D. KEITH, Chemical induction of mutagenesis and carcinogenesis. M. C. CLARK, D. GOODMAN, and A. C. WILSON, The biochemical approach to mutation monitoring in man. F. S. GOULDING, X-ray fluorescence—an improved analytical tool for trace element studies. R. E. BARLOW, Averaging time and maxima for air pollution concentrations.

Pollutants in Food Chains

R. W. RISEBROUGH, Effects of environmental pollutants upon animals other than man. H. L. ROSENTHAL, Implications of environmental strontium 90 accumulation in teeth and bone of children. T. D. STERLING, Problems in determining if a commonly used herbicide (2,4,5-T) has an effect on human health. B. E. VAUGHAN, Ecological and environmental problems in the application of biomathematics.

Ecological Studies

Skeletal Plans for a Comprehensive Health-Pollution Study, Discussion and Epilogue

PREFACE

Berkeley Symposia on Mathematical Statistics and Probability have been held at five year intervals since 1945, with the Sixth Symposium marking a quarter of a century of this activity. The purpose of the Symposia is to promote research and to record in the *Proceedings* the contemporary trends in thought and effort. The subjects covered in the Berkeley statistical Symposia range from pure theory of probability through theory of statistics to a variety of fields of applications of these two mathematical disciplines. The fields selected are those that appear especially important either as a source of novel statistical and probabilistic problems or because of their broad interdisciplinary character combined with particular significance to the society at large. A wide field of application traditionally represented at the Berkeley Symposia is the field of biology and health problems. Physical sciences, including astronomy, physics, and meteorology are also frequently represented. Volume 5 of the *Proceedings* of the Fifth Symposium was entirely given to weather modification.

With the help of advisory committees and of particular scholars, the participants of the Berkeley Symposia are recruited from all countries of the world, hopefully to include representatives of all significant schools of thought. In order to stimulate fruitful crossfertilization of ideas, efforts are made for the symposia to last somewhat longer than ordinary scholarly meetings, up to six weeks during which days with scholarly sessions are combined with excursions to the mountains and other social events. The record shows that, not infrequently, novel ideas are born at just such occasions.

According to the original plans, the entire Sixth Berkeley Symposium was to be held during the summer of 1970, with the generous support of the University of California, through an allocation from the Russell S. Springer Memorial Foundation, of the National Science Foundation, of the National Institutes of Health, of the Office of Naval Research, of the Army Research Office, and of the Air Force Office of Scientific Research. This help is most gratefully acknowledged. Certain circumstances prevented the Biology-Health Section from being held in 1970 and the meeting held in that year, from June 21 to July 18, was concerned with mathematical domains of probability and statistics. The papers presented at that time, and also some that were sent in by the individuals who were not able to attend personally, fill the first three volumes of these *Proceedings.* Volume 1 is given to theory of statistics and Volumes 2 and 3 to the rapidly developing theory of probability.

The Biology-Health Section of the Sixth Symposium had to be postponed to 1971. Every postponement of a scholarly meeting involves a disruption of the plans and all kinds of difficulties. Such disruption and difficulties certainly occurred in the present case. As originally planned, the Biology-Health Section of the Sixth Symposium was to be comparable to that of the Fifth, the *Proceedings* of which extended close to 1,000 pages in print. This is much larger than

Volume 4 of the present *Proceedings* that summarizes the Biology-Health Section held from June 16 to 21, 1971. However, the losses suffered in some respects have been compensated by gains in others. Those gains are reflected in Volumes 5 and 6 of these *Proceedings*.

During the fall of 1970 we became much impressed by the development and rapid growth of a new field of biological studies which includes the areas known as "non-Darwinian" and "neo-Darwinian" studies of evolution. These are studies based on the structure of macromolecules present in many now living species and performing in them similar functions. One example is the hemoglobin molecule, carried by all mammals as well as by fish. The differences among the homologous macromolecules in different species are usually ascribed to mutations that are in some sense inconsequential, and are supposed to occur more or less at a uniform rate. The number of differences between any two species is indicative of the time that elapsed from the moment of separation from the presumed common ancestor. The probabilistic-statistical problems involved in such studies include the estimation of philogenetic trees of several species and, in particular, the estimation of the time since two species separated from their ancestor.

It was found that, with only a few exceptions, mathematical statisticians are not familiar with the new domain and that, at the same time, a great many biologists make strong efforts to treat the statistical problems themselves. A joint meeting of biologists and statisticians was clearly indicated and a separate conference, especially given to novel studies of evolution, was held from April 9 to 12, as part of the Biology-Health Section of the Sixth Berkeley Symposium. It is summarized in Volume 5 of these *Proceedings*. Somewhat unexpectedly, it appeared that the new field of studies of evolution involves controversies that are just as sharp as those that occasionally enliven the meetings of mathematical statisticians . . .

We were introduced to problems of evolution treated on the level of macromolecules by Professor T. H. Jukes, V. N. Sarich, and A. C. Wilson. Their very interesting seminar talks and later their advice on the organization of the conference on evolution are highly appreciated.

While studies of evolution involve observational research, particularly that concerned with the relation between classical population genetics and novel findings on the level of molecular biology, the whole domain is clearly conceptual. Contrary to this, the third part of the Biology-Health Section of the Sixth Symposium was totally given to observational studies in a domain of great importance to society at large and of great public interest.

The domain in question, a highly controversial domain, is that of the relation between environmental pollution and human health. The growing population in the United States and in other countries needs more electric power, more automobiles, and other products. The relevant industries are eager to satisfy these needs. However, the expanded industrial activity, unavoidably conducted with an eye on costs, leads to pollution of the environment. The controversies at public hearings, in the daily press, and in scholarly publications center around the question whether the currently adopted standards of safety are sufficient or not. The volume of research, largely statistical, surrounding this question is immense. The intention that the *Proceedings* provide a cross section of contemporary statistical work dictated the organization of a special conference entirely given to the problem of health and pollution. This conference, held from July 19 to 22, is summarized in Volume 6 of these *Proceedings*. In organizing the conference we benefitted greatly from the advice of Dr. S. W. Greenhouse of the National Institutes of Health, of Professor B. Greenberg of Chapel Hill, North Carolina, and of Drs. J. M. Hollander and H. W. Patterson of the Berkeley Lawrence Laboratory.

The first purpose of the Health-Pollution Conference was to take stock of the studies already performed. The second and the ultimate purpose was to see whether a novel statistical study is called for, hopefully more comprehensive and more reliable than those already completed. With this in mind, invitations to the conference were issued to Federal and State governmental agencies concerned with health and pollution, to authoritative scholarly institutions, and to a number of particular individual scholars known to have worked on one or another aspect of this problem.

As a special stimulus for thought on the entire problem of pollution and health, its present state and the future, the invitations to the conference were formulated to include a call for submission of skeletal plans for a fresh comprehensive statistical study, capable of separating the effects of particular pollutants. Four such plans were submitted and they are published in Volume 6.

All the participants had complete freedom of expression, both in their prepared papers and in their contributions to the discussion. Thus it is likely that the goal of providing a realistic cross section of contemporary statistical research on the problem is reasonably approached. Also it is not unlikely that the present state of knowledge on human health and pollution, and the scholarly level of the substantive studies prepared are fairly reflected in these *Proceedings*.

In addition to funds provided by the University of California and the National Institutes of Health, the Health-Pollution Conference was organized using a grant from the Atomic Energy Commission, Division of Biology and Medicine. This help is gratefully acknowledged.

The organization and the running of three distinct scholarly meetings, one in April, another in June, and the third in July 1971, each attended by some 100 to more than 300 participants, would not have been possible without the willing, efficient, and cheerful help and cooperation of the staff of the Department of Statistics and the Statistical Laboratory. Our most hearty thanks go to our successive "ministers of finance," Mrs. Barbara Gaugl and Mrs. Freddie Ruhl, who watched the sinking balances and surveyed the legality of proposed expenditures, some appropriate under one grant and not under another, etc. In addition to financial matters, Mrs. Gaugl supervised the local arrangements for scholarly sessions, for several social events and for servicing the participants. In this she was efficiently helped by Mrs. Dominique Cooke, by Miss Judy Whipple and by a number of volunteers from among the graduate students in the Department. Mrs. Cooke and Miss Whipple had their own very important domain of activities: to keep straight the correspondence and the files. Coming in addition to the ordinary university business, this was no mean job and the performance of the two ladies is highly appreciated.

All the above refers to the early part of the year 1971 and up to the end of the conferences. Then the manuscripts of the papers to be published in the *Proceedings* started to arrive, totalling 1849 typewritten pages, not counting figures and numerical tables. This marked a new phase of the job in which we enjoyed the cooperation of another group of persons, who prepared the material for the printers. At the time, the team of editors, Miss Carol Conti, Mrs. Margaret Darland, and Miss Jean Kettler, under the able guidance of Mrs. Virginia Thompson and supervised by Professor LeCam, Chairman of the Organizing Committee, worked assiduously on proofs of papers in Volumes 1, 2 and 3. The arrival of the material for Volumes 4, 5, and 6, unavoidably involving some correspondence with the authors and conferences at the University Press, created heavy burden. We are very grateful to the four ladies whose cooperation has been inspiring to us.

Last but not least, our hearty thanks to the University of California Press, Mr. August Frugé and his colleagues for their help, cooperation and also their patience when confronted with piles of manuscripts which we hoped to see published both excellently as in the past quarter of a century and "right away, yesterday!"

J. NEYMAN E. L. SCOTT L. LE CAM (CHM.)

CONTENTS

Darwinian and Non-Darwinian Evolution Theory

DNA, RNA, Amino-acid Sequences

Population Studies and Evolution

Role of Theory in Evolutionary Studies

DARWINIAN AND NON-DARWINIAN EVOLUTION

JAMES F. CROW
UNIVERSITY OF WISCONSIN

1. Introduction

Evolution by natural selection, by survival and differential reproduction of the fittest, is about as firmly established as any broadly general scientific theory could imaginably be. Why then should it be challenged by a rival theory in 1971? The answer is that it is not, for the proponents of non-Darwinian evolution are not questioning that evolution of form and function has occurred in the orthodox neo-Darwinian manner.

So let me first say what non-Darwinian evolution is not. It is not orthogenesis, emergent evolution, inheritance of acquired characters, catastrophism, vitalism, inherent directiveness, or telefinalism. It is not associated with names such as Lamarck, Osborn, or Teilhard de Chardin. Rather it is evolution by random drift of mutants whose effects are so minute as to render them essentially neutral, and a more appropriate name to mention is Sewall Wright.

Random drift is not a new idea. It was considered quite thoroughly by R. A. Fisher [10] and discounted by him as a factor of any great interest in evolution. He regarded it as a calculable amount of random uncertainty that could cause disorderly fluctuations, but would not alter to any great extent either the direction or the rate of evolution, except in very small populations. To Sewall Wright [47], [48], [50], on the other hand, random gene frequency fluctuations became an important part of his shifting balance theory of evolution. Random fluctuations may enable a population to pass to the other side of an unstable equilibrium, or in a structured population permit a particularly favorable gene combination to arise locally and spread through the entire population. In Wright's view, random drift caused by near neutrality, small population size, and fluctuating selective values is part of a basic mechanism that enhances the probability of evolutionary novelty.

Random drift in the present context is different in emphasis. The idea put forth as non-Darwinian evolution is that most DNA changes and most amino acid substitutions in evolution have been so nearly neutral that their fate was determined mainly by random processes. In this view the chief cause of observed molecular evolution is random fixation of neutral mutations. The effect of all this on fitness is regarded as negligible.

Paper Number 1506 from the Genetics Laboratory, University of Wisconsin.

How similar must a mutant be to the gene from which it arose to be regarded as neutral? For its fate to be determined largely by chance, its selective advantage or disadvantage must be smaller than the reciprocal of the effective population number; so the operational definition of a neutral gene is one for which $|s| \ll 1/N_e$, where s is the selective advantage and N_e is the effective population number [47].

The principal reason for not accepting non-Darwinian evolution, I believe, is an unwillingness to believe that any mutational change can be so slight as to have no effect on fitness when considered over the enormous geological times involved. Another reason, perhaps, is that a random theory may discourage a search for other explanations and thus may be intellectually stultifying. Thirdly, a biologist may well say that if these changes are so nearly neutral as to be governed by chance in large populations and over long periods of time they are not really of much interest. He is more interested in processes that affect the organism's ability to survive and reproduce, and which have brought about such exquisite adaptations to diverse environments. To many biologists the evolution of amino acid changes is rather dull compared to that of the elephant's trunk, the bird's wing, the web spinning skills of a spider, the protective resemblance of mimetic butterflies, the vertebrate eye, or the human brain.

On the other hand, the neutral theory leads to a different formulation with new ideas and with quantitative predictions. It is directly concerned with the gene itself, or its immediate product, so that the well-developed theories of population genetics become available. It produces testable theories about the rates of evolution. I have commented elsewhere [4] on the great enrichment to population genetics that has come through molecular biology, which at last makes it possible to apply population genetics theory to those quantities (that is, gene frequencies) for which it was developed.

The original plan of the Symposium was to have two introductory papers, one on Darwinian evolution and one on non-Darwinian evolution. Due to illness this has not been possible, so I am discussing both subjects. This means that there will be many places with the equivocal "on the one hand . . . but on the other," as I endeavor to present arguments that have been given for both views. I shall probably slight the Darwinian arguments somewhat; they are already too well known to need further elaboration.

2. Some recent history

The neutral evolution hypothesis in its present molecular context was foreshadowed by the work of Sueoka [42] and Freese [15]. Both were concerned with the diversity of base content in bacteria of different species despite rather similar amino acid makeup and suggested that this might depend on mutation rates of individual nucleotides with negligible differences in selective values.

The real beginning of the subject was Kimura's daring challenge to evolutionary orthodoxy, published in 1968 [19]. His argument was based on the difficulty of explaining the enormous number of gene substitutions that would occur if all the DNA were evolving by selection at the same rate as that observed for such proteins as hemoglobin. Because of this difficulty, based on Haldane's [17] idea of the cost of natural selection, he argued that most of the changes are in fact not selective, but the result of random fixation.

Kimura's idea was strongly supported in the influential paper by King and Jukes [28], which gave the name to this Symposium, and in a more tentative way by me [4]. King and Jukes presented several more arguments, chief among which were the now familiar ones based on the constancy of amino acid substitution rates, the predictability of amino acid composition from nucleotide frequencies and the genetic code, and the great difficulty of interpreting the apparent indifference of one *Escherichia coli* strain to an inordinately high mutation rate.

3. The continuum of fitness values of new mutants

It has been observed since the beginnings of modern genetics that mutations that have effects conspicuous enough to be noticed are almost invariably harmful. It is of course to be expected from natural selection theory that the great majority of newly arising mutants would be deleterious, or at best neutral, in the environment where the existing genes evolved. When geneticists look for examples of beneficial mutants they customarily think of mutants that are adapted to a new environment, such as mutations for drug resistance in bacteria, DDT detoxification in houseflies, or industrial melanism in moths.

Fisher [10] argued that mutations with large effects should almost always be harmful, but that as the effect of the mutant gene becomes less the probability of its being favorable increases until near the limit of zero effect the probability of being deleterious approaches ½. Muller [38] emphasized another, related point. He noted that mutants with minor effects were more frequent than those with more drastic effects. In particular it was shown experimentally in *Drosophila* that recessive mutants causing a small decrease in viability are some two to three times as frequent as those causing a lethal effect [18], [45]. These findings have been confirmed and extended by Mukai and his collaborators [36], [37] using methods of greater sensitivity by which smaller differences could be detected. From these experiments the mutation rate of genes causing minor effects on viability is estimated to be at least ten times that for lethals, and perhaps considerably higher since the experiment permits only a minimum estimate. In absolute frequency, this amounts to at least 0.15 per gamete. For 10,000 loci this is a rate of 1.5×10^{-5} per locus.

Although this experiment does not detect neutral mutations, the increasing frequency of mutants as the sensitivity of the experiment increases suggests a continuum of fitness values. Presumably mutations range from severely deleterious, through neutrality, to mildly beneficial. The situation is illustrated in

Figure 1. The part that can be substantiated by direct measurement is shown in the solid line. Whether the extrapolation through zero effect is more like A or like B is unknown.

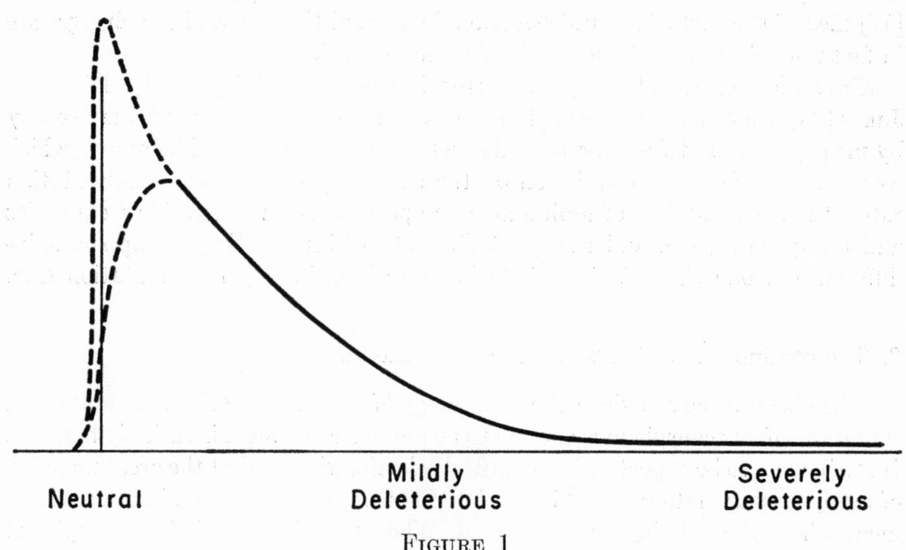

FIGURE 1

Distribution of viabilities of new mutations.
Solid line: from data. Dotted lines: extrapolations.
Farther to the right would be a hump caused by the grouping together of all lethal mutations, regardless of the time of death.

I believe that a debate over whether Darwinian or non-Darwinian evolution is more important is largely fruitless. We know that selection occurs and that some loci are strongly selected. We know further that the main direction of phenotypic evolution is determined by selection, within the limits set by mutational possibility. On the other hand, we know that some loci are so weakly selected that random drift is a major factor in determining their frequency. An elegant laboratory demonstration of random drift of modifying genes affecting the relative viability of inversion types was given by Dobzhansky and Pavlovsky [7]. I suggest that, just as the concept of heritability replaced a meaningless discussion of whether heredity or environment is all important, the right statistical formulation can assign the proper allocation to selection, to mutation, and to random drift as determinants of the evolutionary process.

The subject of this Symposium is mainly molecular evolution rather than evolution of overt traits and processes. Therefore, we are concerned with individual nucleotide changes and their consequences, amino acid replacements. So we ask what fraction of the observed evolution at this level is caused by drift and what fraction by selection.

4. Contrasts between classical and molecular evolution

In selection dominated evolution, as traditionally viewed, the increase or decrease in frequency of a phenotype depends on its survival and fertility relative to competing types. The genetic basis is typically polygenic and essentially continuous. Concealed within a relatively uniform phenotype is a large amount of genetic variability. The amount of variability in the population is very large compared to that which arises in a single generation of mutation, so selection is mainly utilizing variability that is already in the population; in other words, mutation is hardly ever the rate limiting factor. The pattern and direction of evolution are determined by ecological opportunity, diversity of habitat, stability of the environment, and the nature of competing species. Population structure and migration are likely to be important.

Although the general direction of selection is highly deterministic, there may be a large stochastic element in the individual genes involved. There are typically many genetic ways of accomplishing the same phenotypic change, so that the particular genes that increase or decrease in a particular population are largely a matter of chance. It is occasionally true that genetic variability is limiting, as when an insect population that happens to have a mutant gene producing a detoxifying enzyme survives the application of a new insecticide, but this is not thought to be typical.

Neutral molecular evolution, as viewed by its proponents, has quite different kinetics. The stability of the environment, the ecological situation, competing species, population size and such factors are largely irrelevant. There is little effect of the manner of reproduction or of population structure. Species that superficially are evolving very rapidly, like Darwin's finches on the Galapagos Islands, should show no more rapidity of change for neutral amino acid changes than slowly evolving forms. The rate determining factor becomes the rate of mutation of neutral alleles.

The best analogies for non-Darwinian evolution may come from simple asexual systems. For example, there may be phenomena similar to periodic selection in bacteria [40]. As Morton [34] has suggested the amount of polymorphism may be related to the time since a favorable mutant swept through the population, or since a size bottleneck occurred.

Although, in my view, the true situation is an essentially continuous range of fitness values and a range of types of genetic determination from oligogenic to polygenic, it is convenient for classification and discussion to contrast the extreme situations. The evolutionary process can be dichotomized two ways:

- (i) on the basis of phenotype:
  - (A) morphological and physiological traits,
  - (B) molecular changes;
- (ii) on the basis of selection:
  - (1) selected,
  - (2) neutral.

It is likely that many of the component genes in a polygenic system (affecting body size, for example) may be very nearly neutral on the average in a population that is near equilibrium for this trait. For size, as for almost any quantitative trait, the optimum is intermediate rather than at an extreme. Therefore, a gene that increases size is favorable in an individual that is below the optimum and unfavorable in an individual that is too large; the net effect of the gene is neutral.

But this is not the subject of this Symposium. We are here concerned with whether class B2 exists, and if so, what fraction of DNA and amino acid substitutions in evolution are of this type.

5. Arguments for and against neutral evolution

I should like now to list and discuss some of the major arguments that have been put forth for and against the neutral hypothesis. Many of these are discussed in more detail elsewhere in this Symposium.

5.1. *The cost of natural selection.* Haldane [17] first showed that the total amount of selective mortality or differential fertility required for a gene substitution is largely independent of the intensity of selection and depends mainly on the initial frequency. Thus, for a certain excess of reproductive capacity that can be devoted to natural selection, there is a limit to the number of independent gene substitutions that can occur in a given time interval without reduction of the population size.

That a limitation on the rate of gene substitution is inherent in a given pattern of variability in birth and death rates is, I think, generally accepted. But whether the Haldane cost principle provides the most appropriate measure has been seriously questioned. The meaning of a substitution load for an advantageous mutant in a nondeteriorating environment is not clear. Another limitation is the inherent assumption that gene substitutions are independently inherited; linkage may alter this. It is also assumed that the genes being substituted are independent in their effects on fitness. If the genes interact strongly, the principle may be grossly misleading. An extreme model assumes that above a certain level of fitness there is no distinction. By properly adjusting the parameters in such a threshold model, one can demonstrate a system in which a much larger number of gene substitutions can be carried out with the same amount of selection per generation [43], [32], [49].

There are a number of reasons for questioning a strict threshold, or truncation model. For one thing, truncation selection applies to a trait for which there is some underlying variable on which the genes act cumulatively and then selection retains all that are above a certain level on this scale and rejects those that are below. Although selection for yield or performance in livestock and plant breeding approximates this procedure, I doubt that strict truncation applies to much of natural selection. Furthermore, the heritability of fitness must be exceedingly low, which has the effect of blunting the sharpness of the truncation. I suspect that the truth lies somewhere between a strict application of the Haldane principle and a truncation model.

Using his principle, Haldane [17] suggested that a reasonable rate of evolution when the population can devote about ten per cent of its reproductive excess to gene substitutions is about one substitution every 300 generations. Kimura [19] pointed out that if all the mammalian DNA is evolving at the same rate as that observed for hemoglobin and cytochrome c, this is equivalent to a gene substitution every year or two, far faster than would be possible if the Haldane limitation applies. Kimura suggested that this contradiction can best be reconciled by the assumption that most molecular evolution is selectively neutral.

There are two other ways around this dilemma. One is the assumption of truncation selection mentioned above. The other is to doubt that the number of genes is as large as direct DNA measurements would suggest. If the number of genes is 10^4 there is no problem with even the strictest interpretation of Haldane's principle. For example, 10^4 loci evolving at a total rate of one substitution every 300 generations would mean a substitution per locus of one in 3×10^6 generations; if there are 300 codons per locus, the per codon rate would be about one substitution per 10^9 generations, a value of the same order of magnitude as the observed rate.

I shall return to a discussion of gene numbers.

5.2. *The remarkable constancy of molecular evolution rates.* Another argument that has been advanced for the neutral hypothesis is the constancy of evolutionary rates in different proteins and in the same protein in different lineages. One example, elaborated by Kimura in this Symposium and elsewhere [21], is hemoglobin. A rate of about 10^{-9} per codon per year is found in several diverse ancestral sequences. Particularly striking, since it does not depend on an estimate of the time involved, is a comparison of β and α hemoglobins following the duplication which started them on separate evolutionary courses. Whereas human β and carp α differ by 77 of 139 amino acid sites, human α and β differ by 75. Furthermore, the human β differs from the α of mouse, rabbit, horse, and cattle by 75 to 79 amino acids. The constancy is for the total number of changes, not the individual changes themselves which are often at different sites or involve different amino acids at the same site. Since the time of the original duplication, the amount of divergence of the chains in the same organism, man, is almost exactly the same as that between two chains, one in man and one in a fish, despite the fact that the lines of descent of man and fish have been separate for most of the time. Despite the enormous differences in evolution of form and function between fish and mammals, some timing mechanism has kept the hemoglobins evolving at the same rate.

The rate constancy is equally impressive if we compare different proteins. Fitch and Markowitz [14] have classified the amino acid sites into constant and variable. The former presumably contain amino acids that are essential for the proper functioning of the molecule and cannot be changed without damage.

The latter are free to evolve, since they can be changed without seriously affecting the function. By estimating the rate of evolution of those codons that are variable at a given time (concomitantly variable codons, or covarions) Fitch [11] has shown that the number of substitutions per variable codon is 0.50, 0.44, 0.80, and 0.72 for cytochrome c, α hemoglobin, β hemoglobin, and fibrinopeptide A, respectively, in the two lines of descent since the pig and horse diverged from a common ancestor. There is reason to think that the β hemoglobin estimate is too high. It is remarkable that these widely diverse proteins with proportions of covarions ranging from ten per cent or less in cytochrome c to 95 per cent in fibrinopeptide differ in their evolution by amounts no greater than might be expected from errors in the estimating procedures.

Finally, from the data of Kohne [29] the rate of evolution of nonrepetitive DNA, based on thermostability of hybrid DNA between new and old world primates, is estimated as about 2×10^{-9} per nucleotide per year. The rate of 6×10^{-9} per three nucleotide codon is roughly the same as that for the most rapidly evolving protein (fibrinopeptide A, with 18 of its 19 amino acids variable) and for the variable parts of other proteins.

It thus appears that, to a first approximation based on limited data and necessarily involving a number of uncertainties, DNA and the variable codons are evolving at roughly the same rate.

On a selection hypothesis there is no obvious reason to expect this rate constancy. Different proteins would be expected to evolve at different rates depending on their functions and their environments. The same protein might also differ in rate in different phylogenies.

On the other hand, with the neutral hypothesis the rate of gene substitution is equal to the neutral mutation rate and quite independent of other factors [19], [4]. An evolutionary rate of 10^{-9} per codon per year would imply, for a 500-amino-acid gene and a five year average age of reproduction, a mutation rate of $500 \times 5 \times 10^{-9} = 2.5 \times 10^{-6}$. Since this is about ten per cent of the usually accepted mutation rate per locus, this implies that if one tenth of mutants were selectively neutral this would be sufficient to account for the observed rate of molecular evolution.

A difficulty with the neutral interpretation is that the amino acid substitution rate seems to be constant per year, not per generation. This is unexpected from classical knowledge of mutation rates, which have been regarded as being more related to generation time than to calendar time. Human, mouse, and *Drosophila* mutation rates for single loci with conspicuous phenotypes are rather similar when the measure is per generation, but widely different when measured in absolute time units, as discussed by King in this Symposium.

Furthermore, calendar equality of rates can be ruled out for some cases. Consider a comparison of *Drosophila* and man. The spontaneous rate of occurrence of recessive lethal mutations in *Drosophila* is about 0.015 per gamete per generation and these persist in the population long enough to reach an equilibrium frequency of about 0.5 per gamete. The human reproduction cycle is about

1000 times that of *Drosophila*, so if lethals were to arise at the same absolute rate in man there would be at least 15 lethals per gamete per generation, making no allowance for the possibly greater gene number in mammals. If these were to accumulate to anything like the extent that they do in *Drosophila*, each of us would carry several hundred recessive lethal genes. This means that the child of a cousin marriage would never survive! There must have been some adjustment of the lethal mutation rate to correspond to the life cycle. Furthermore, the same argument can be applied to mildly deleterious genes having an effect on viability of less than five per cent. In *Drosophila* these equilibrate at a frequency of about 0.25 lethal equivalents [35] per gamete [44]. If these occurred in man with a frequency 1000 times as high, we would be riddled with them and again consanguineous marriages would inevitably lead to lethality.

We must conclude that for genes having deleterious effects on viability, whether mild or lethal, the mutation rate is much more nearly constant per generation than per year. What does this mean for the neutral hypothesis for evolution of amino acids?

There are two ways out of the dilemma. One is to postulate that DNA changes leading to neutral mutations are a different class from those producing deleterious changes. Perhaps the latter are reduced by repair mechanisms that are somehow adjusted to the generation length. But I find it unappealing to assume that there is a fundamental difference in the mutation process between those amino acid substitutions that are nearly neutral and those that are severely deleterious.

The second way is to question the accuracy of the rate measures. It should be mentioned that the best data are for organisms whose life cycles are not greatly different. Comparison between widely divergent organisms, like mammals and wheat, involve so many differences that correction for multiple changes in the same amino acid site become important, and these are subject to error. It may be that when all the data are in there may be a correlation of evolution rate and life cycle. This is suggested by some of the DNA data [30], [29]. See also King's discussion in this Symposium.

5.3. *Amino acid frequencies and the code.* Kimura [20] and King and Jukes [28] noted that the frequencies of amino acids, averaged over a large number of proteins, agree rather well with random expectations based on the frequency of the nucleotides in these proteins and the genetic code. King and Jukes used this as one of their major arguments for neutral evolution. The methods have been refined since that time and more data have become available. The agreement is remarkably good, with the exception of arginine which is used much less often than would be expected from the number of ways that it can be encoded. I shall discuss this only briefly, since it is considered in other papers in this Symposium.

It is obvious that on the neutral hypothesis the amino acid composition of proteins should be predictable from nucleotide frequencies and the code. There is also a selectionist interpretation, however. Suppose that, perhaps because of a change in internal physiology or environment, a particular protein would function better if its structure were altered. Suppose also that there are several ways in which this improvement could occur. The first mutant to occur that is of suitable type has the best chance of success. The more likely a particular amino acid is within the restriction of the nucleotide frequencies and the code, the more likely it is that the first mutant is one encoding this particular amino acid. In the long run, those amino acids whose codons occur most often will be most frequently incorporated.

The same argument applies when the selection is among pre-existing mutants. On the average, those mutants with the highest initial frequencies have the best chance and therefore those amino acids that occur with the greatest frequency in the coding system will be most likely to prevail. For these reasons, I think the argument is equivocal and the observation is consistent with either hypothesis.

5.4. *The functional equivalence of homologous proteins from different species.* Another argument for the neutral hypothesis is the apparent physiological equivalence of proteins from diverse sources. For example, bovine and yeast cytochrome c appear to function equally well with bovine cytochrome oxidase, despite a large number of amino acid differences. Furthermore, enzymes in species hybrids seem to function properly even though they differ in several amino acids. This is given as an argument for neutrality. Yet, there is an obvious selectionist answer: functional differences far too small to be detected in any such manner could still create selective differences large enough to be effective in large populations and over the enormous periods of time involved.

5.5. *The Treffers mutant in Escherichia coli.* Another argument for selective neutrality of many DNA changes comes from the mutator gene, studied by Cox and Yanofsky [2]. This produces an enormous number of AT → CG transversions throughout the genome. Despite a number of DNA changes equivalent to half a dozen per cell division, this strain had no obvious deterioration in viability after hundreds of generations—enough time that the DNA base change could actually be measured. Furthermore, these produce mutations by purine-pyrimidine interchanges and therefore a smaller fraction are synonymous than if they were purine-purine or pyrimidine-pyrimidine substitutions. The conclusion that the cells are not greatly harmed by these mutations is strengthened by chemostat experiments in which the mutable strain competed effectively with a normal strain; in fact it seemed to do better, perhaps because of being better able to adapt to chemostat conditions [16].

Unless there is some sort of Maxwell's demon that guides all the half dozen mutant genes into the same daughter cell at each division and thus eliminates them from the population in clusters, they must surely accumulate, as in fact shown by direct chemical analysis of DNA. The great majority of these mutants must therefore be very nearly neutral.

5.6. *Correlation between similarity of amino acids and replacement rate.* Clarke [1] has pointed out that there is a correlation between the frequency with which an amino acid substitution occurs in evolution and the smallness of the difference in the two amino acids, as measured by their structural and chemical properties. He argues from this that amino acid substitutions are selective, since those that have the smallest effect are most likely to be beneficial. But this argument can easily be turned around. As Clarke himself notes, the smaller the difference between two amino acids, the more likely the change is to be selectively neutral. His analysis does imply that only a minority of amino acid changes are neutral, but as I said before, the rate of neutral evolution does not demand that most amino acid changing mutations be neutral; a small fraction is sufficient to account for the observed rate of amino acid substitution.

5.7. *Successive substitutions.* Another argument has been advanced by Fitch [12]. He notes that in the history of cytochrome evolution most double changes have followed in close succession during the relatively short period while the particular codon was variable. About 30 per cent of the changes are double. In the selectionist view, this must mean that the best substitution was often two steps removed, but it also means that the first step was also an improvement (although the second step made things still better). It would seem surprising that if the best mutant were two steps away, the intermediate step would also be beneficial in such a large proportion of cases. Furthermore, as Fitch notes, the genetic code seems to have the property that individual nucleotide substitutions on the average lead to more similar amino acids than multiple changes do. Similar changes are more likely to be beneficial. Why then, he asks, should the best substitution so often be two steps away? This would seem to argue for neutrality.

However, as King has also noted, there is a selectionist interpretation. If, because of an environmental change, the existing amino acid at some site is no longer optimum, it is likely that it can be improved to varying extents by more than one type of replacement. The first to occur is not necessarily the best; hence the way is open for successive steps.

These arguments, when viewed collectively, make a substantial case for non-Darwinian evolution. In my opinion it is a very strong case for DNA as a whole and a case strong enough to be taken seriously as a working hypothesis for amino acids at concomitantly varying codons. The hypothesis raises a number of new questions and makes a number of quantitative predictions that can guide further experimental and observational inquiry. For this reason alone it merits further consideration.

6. How many genes are there?

The amount of DNA in a mammalian cell amounts to about 3 to 4×10^9 nucleotides. If this is all divided into genes of several hundred nucleotides each there are millions of genes. This is hard to square with observed deleterious mutation rates of 10^{-5} per locus, or higher. It also raises problems with the substitution load unless one postulates truncation selection or neutrality as ways out of "Haldane's dilemma."

Another possibility is that most of the DNA is not genic in the sense of carrying information for protein synthesis or for specific RNA sequences. We have no basis for estimation of gene number in mammals, but there is good evidence in *Drosophila*. The left end of the X chromosome around the white eye locus has been studied exhaustively by Judd and his associates [41]. The best analyzed region includes 16 salivary gland chromosome bands. Lethal and visible mutants within this region can be unambiguously located and classified for identity by a complementation test. The region now appears to be exhausted in that no new mutants have been found for some time that do not fall into one of the 16 complementation units. Thus, there seems to be a perfect correspondence between salivary chromosome band number and the number of complementation units. Similar data for another region of the chromosome give results that are consistent with this idea, although the study is not so exhaustive.

There are a few loci scattered throughout the genome that are known to produce visible mutants but not lethals; that is to say, the normal gene (or genes) at these loci is not absolutely necessary. But such loci appear to be a small minority. That there is not a large class of loci that produce no harmful or lethal mutants is indicated by the fact that any deletion of more than about 20 to 30 salivary chromosome bands has highly deleterious heterozygous effects, usually lethal.

This all suggests that the number of complementation units (genes?) in *Drosophila* is commensurate with the number of salivary chromosome bands. This number in *Drosophila melanogaster* is about 6000. No corresponding information is available for mammals although the chromomere count in some amphibia seems to give about the same number. The amount of DNA in mammals is an order of magnitude greater than that in *Drosophila,* but there is no reason from this fact alone to think that there are more genes. Some of the organisms with the largest amounts of DNA, such as lungfish, are not any more complex or advanced by other criteria. The absence of correlation between DNA amount and any other property is also true of plants.

The DNA in *Drosophila* is enough for several hundred thousand genes, far too much for the 6000 estimated from the salivary chromosome bands (assuming the propriety of defining the gene by a complementation criterion). What is all this DNA doing? Even if we allow for duplication of ribosomal DNA, satellite DNA, and other forms or repetitive DNA there is still far too much.

I would like to join the group who believes, or at least suspects, that the gene number is not large and that most of the DNA has some function other than coding for proteins. It may be purely structural or mechanical. It may be regulatory. It may once have been informational, but have deteriorated after duplication [39]. It may still have a transcribing function, for it is known that some RNA that is produced by transcription stays in the nucleus and does not participate in protein synthesis. Perhaps this has a timing function, as Watson has suggested [46].

If one were looking for an intracellular structural material that had the desirable properties of replicating itself in synchrony with the cell division process so as to maintain a constant amount, that had a mechanism already existing in the cell for doing this, that had a regular structure of constant shape and rigidity, and (perhaps most important) that maintained its structural integrity and replicative capacity regardless of random chemical alterations in its own composition, he would find that DNA has exactly these properties.

Noninformational DNA, as I would like to designate all DNA whose cellular function does not depend on its exact nucleotide sequence, would have very little mutation load. Its function would depend on average properties, such as the overall AT:GC ratio, but not on the sequence. Mutations increasing the number of AT pairs and those increasing the number of GC pairs would be largely cancelling in their effects. A mutator gene, or simply the ravages of time, could cause a change in overall composition with much less change in function, perhaps none at all. In other words, this kind of DNA would evolve mainly by mutation and random drift. Such changes would show up as differences in DNA hybridization studies (with perhaps no overall change in base ratios), but need not imply any change in function. Note that noninformational DNA need not be repetitive; it can be as varied in sequence as genic DNA.

Thus, the hypothesis of non-Darwinian evolution, or evolution by random drift, can be broken into two parts:

(1) DNA that is noninformational evolves by random drift, or mainly so;

(2) observed changes in amino acid sequences are mainly the result of random drift.

A more general statement of Kimura's original hypothesis is given in terms of total DNA rather than just that part which encodes proteins [39]. In particular, the argument of his 1968 paper [19] based on "Haldane's dilemma," is more convincing for the totality of DNA than for the probably small part of this that codes for proteins.

7. Neutrality versus near neutrality

I should like to return to my original contention, that there is a continuum of fitness values ranging from strongly deleterious through neutrality to slightly beneficial and ask about the rate of substitution of mutants whose advantage or disadvantage is very close to zero.

As I mentioned before, the average number of neutral mutant genes substituted per unit time is equal to the neutral mutation rate. This is independent of the ecological conditions, and of the population structure and size. I should note, however, that the definition of neutrality is dependent on the effective population number. A gene is effectively neutral if its selective advantage or disadvantage is small relative to the reciprocal of the effective population number. This means, then, that a gene that is effectively neutral in a small population may not be in a large population. A slightly harmful gene has a better chance in a small population than in a larger one; a slightly beneficial mutant has a better chance in a large population.

The probability of fixation of a gene with a small selective advantage s in the heterozygote and $2s$ in the homozygote is given by Kimura's formula:

$$u(p) = \frac{1 - \exp\{-4N_e sp\}}{1 - \exp\{-4N_e s\}}, \tag{1}$$

where N_e is the variance effective population number (see [3] and [5], p. 352) and p is the initial frequency of the mutant (see [5], p. 425). Usually the new mutant is present only once in the population, so $p = 1/2N$ in a diploid population of size N. When $p = 1/2N$ and s is small equation (1) becomes

$$u = \frac{2s}{1 - \exp\{-4N_e s\}}, \qquad s \text{ small}, \tag{2}$$

$$u = \frac{1}{2N}, \qquad s = 0, \tag{3}$$

as given by Wright [47]. This is correct even when s is negative. If the actual and effective population numbers are greatly different, the right side of (2) should be multiplied by N_e/N.

Figure 2 is a graph from Kimura and Ohta [27] and shows the probability of fixation $u(p)$ as a function of $4N_e s$. As expected, when $s = 0$ this has a value of $1/2N$. It is smaller when s is negative and greater when s is positive. The point of interest is that there is an appreciable chance of fixation of a slightly deleterious gene as long as $4N_e s$ is greater than -2. Whatever the exact shape of the distribution in Figure 1, it is certain that there are more deleterious than beneficial mutants. Since the prior probability is thus greater for being deleterious than beneficial, there are more mutants to the left of neutrality where the curve deviates less from $1/2N$ than on the right. The result is that the average fixation probability can be rather close to $1/2N$ (or the substitution rate close to the mutation rate) for mutations some distance on either side of $s = 0$. That is to say, the evolution rate for near neutral genes is also equal to the mutation rate, as a rough approximation. For a further discussion of this point, see King's paper in this Symposium.

The gene substitution rate may therefore be somewhat enhanced in a small population for mutations that are slightly deleterious. However, this more rapid substitution of deleterious mutants is at the price of decreased fitness and any such effect in evolutionary time may well be neutralized by the extinction of small populations accumulating too many such mutants.

One other point merits mention in this context. The value of s can hardly be constant, if for no other reason than that even a neutral gene is linked to other genes on a chromosome and somewhere on the chromosome will be one or more genes with selective differences. If s is highly variable, this can have somewhat the same effect as if N_e is small. The probability of fixation of a slightly harmful gene is on the average enhanced whereas that for a favorable gene is slightly depressed. A mathematical treatment of this has been worked out by Ohta (personal communication).

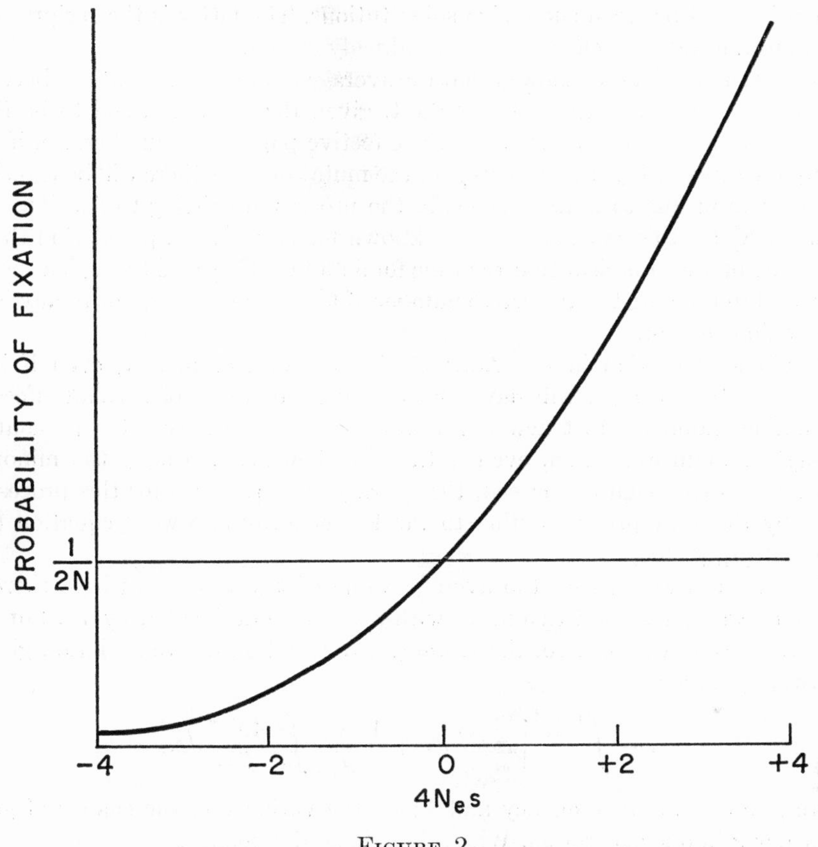

FIGURE 2

Probability of fixation of a new mutant
as a function of effective population number N_e and the selective advantage s,
where N is the actual number of individuals in the population.

Mutants with very slight effects must surely be of great importance in evolution. Evolutionary fine adjustment depends on having a virtually continous range of differences on many scales. The adjustment of such fine differences is the essence of neo-Darwinian evolution. Surely many genes, if not completely neutral, are near enough to neutrality that their individual chances are very much influenced by random factors. For this reason a comprehensive theory of evolution has to consider both deterministic and random processes.

8. Polymorphism

If any appreciable part of amino acid substitution is by random drift, then at any one time there should be some genes in the process of being substituted at that time provided that the time required for such a substitution is large relative to the interval between successive substitutions. The latter is the reciprocal of the mutation rate of neutral alleles, as already stated.

Kimura and Ohta [26] showed that the average number of generations between the occurrence and fixation of a mutant, given that it is destined to be fixed rather than lost, is $4N_e$ where N_e is the effective population number. So, if $4N_e$ is large relative to $1/\mu$, the reciprocal of the mutation rate, there will be transient polymorphism due to mutant genes in the process of drifting to fixation. The value of N_e for this calculation is not known for any natural population that I am aware of, but it is clear that the time for a mutant to spread through a species is related to the long term effective number of the entire species, not to any local subdivision thereof.

Random fluctuation in the value of the selection coefficient, even if it is neutral on the average, will have effects similar to those of a small effective population number. Most new mutants are quickly lost from the population through random extinction, even if they are beneficial. Among the minority that are lucky enough to succeed, the average time required for this process is given by the appropriate solution to the Kolmogorov backward equation (see [26] and [5], p. 403).

Consider the case where the average value of s is zero but where there is random variation around this mean with a variance designated by V_s. For the case where the mean is zero, the Kolmogorov equation has the solution giving the average time as

$$\bar{t} = \int_p^1 \frac{2x(1-x)}{V_{\delta x}}\,dx + \frac{1-p}{p}\int_0^p \frac{4Nx^2}{V_{\delta x}}\,dx, \tag{4}$$

where p is the initial frequency and $V_{\delta x}$ is the variance in the change of gene frequency x in one generation. When there is no dominance:

$$V_{\delta x} = \frac{x(1-x)}{2N_e} + x^2(1-x)^2V_s. \tag{5}$$

If the initial frequency is very small, we can let p approach zero and integrate from 0 to 1. The second term becomes negligible, and the solution is

$$\bar{t} = \frac{8N_e}{C}\log_e\frac{C+K}{C-K}, \tag{6}$$

where $K = 2N_eV_s$ and $C = [K(K+4)]^{1/2}$. When $V_s = 0$, $\bar{t} = 4N_e$, in agreement with the case for random drift of a neutral gene in a population of effective size N_e [26]. The value of $\bar{t}$ in terms of N_e and as a function of $2N_eV_s$ is shown in Figure 3.

A gene can be neutral on the average, but fluctuate in its s value from time to time for at least two reasons. One is that the environment or the background genotype changes so that the gene is sometimes favored and sometimes not in such a way that its average value is neutral; whether such a gene should be classified as neutral is open to debate. On the other hand, a gene that is truly

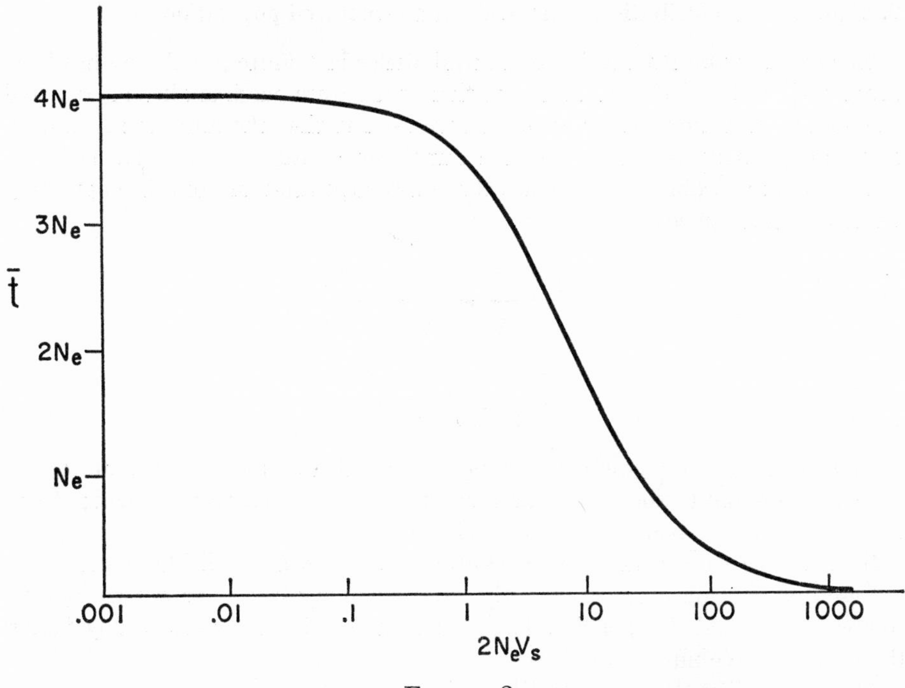

FIGURE 3

Average number of generations until fixation $\bar{t}$ of a new mutant destined to become fixed, where N_e is the effective population number and V_s is the variance in the selective advantage of the mutant.

neutral does not occur in isolation. It has a chromosomal location and therefore is influenced by the selective value of genes linked to it. If a new mutant happens to be on a chromosome that is favored at the time, it has a head start and an increased chance of becoming fixed. This is exactly balanced by the probability of being on a deleterious chromosome, so the mean probability for fixation does not change. The average time until fixation of those that are fixed is shortened however. Equation (6) probably underestimates the influence of this effect since it does not allow for the autocorrelation from generation to generation that is brought about by linkage, for linked combinations persist many generations if the linkage is tight. Equation (6) therefore tends to overestimate the time.

Doolittle, Chen, Glasgow, Mross, and Weinstein [8] noted that no variation at all was found among 125 persons whose fibrinopeptides were analyzed. Fitch and Margoliash [13] noted that if the effective population number is large there should be some polymorphism and the size of the sample should have been large enough to detect this. Perhaps the answer lies along the lines that I have been discussing; random fluctuations around neutrality have the same consequences as lowering the effective population number.

9. Equilibrium distribution of neutrals in a structured population

The equilibrium distribution of neutral alleles in a finite population has been given earlier [24], [20], [22]. Assume that there are k possible allelic states and the neutral mutation rate is μ; that is, we assume that the mutation rate from any state to a particular one of the remaining states is $\mu/(k-1)$. Then, at equilibrium the probability that two alleles drawn at random from the population are identical is approximately

$$f = \frac{4N_e\mu \frac{1}{k-1} + 1}{4N_e\mu \frac{k}{k-1} + 1} \approx \frac{1}{4N_e\mu + 1} \tag{7}$$

when k is large. The reciprocal of f may be regarded as the effective number of alleles. It is equal to the actual number when they are equally frequent; otherwise it is less. The average heterozygosity is $1 - f$.

If the alleles are neutral, the distribution is strongly skewed. Many alleles are represented only once or twice in the population while one or a few drift to comparatively high frequencies. The actual number is then considerably larger than the effective number.

The probability that a neutral locus is polymorphic at equilibrium is determined by whether the mutation rate is larger or smaller than the reciprocal of the effective population number. If $4N_e\mu$ is much larger than 1, the population is mainly heterozygous; if it is much less than 1, the population is mainly homozygous. It should be recalled that if the population size fluctuates, N_e is influenced very much by the smaller values, since it is the harmonic mean of the value at various times.

The equilibrium neutral hypothesis can be tested by seeing how well the distribution of allele frequencies fits the theoretical distribution which is given by a diffusion approximation as

$$\phi(x) = 4N_e\mu(1 - x)^{4N_e\mu - 1}x^{-1}. \tag{8}$$

The sampling distribution of the number of alleles has been worked out by Ewens [9] and this can be used to test whether a sampled population is in agreement with this expectation.

If the population is geographically structured, the probability of identity of two alleles in individuals a specified distance apart is a very complicated function of the mutation rate, the total size of the population, and the structure of the population. One approach to the problem has been given by Morton [34] who has tried various distributions with actual human data. Maruyama [31] has studied theoretical models, including several patterns of migration between partially isolated colonies and also a population of continuous structure with random migration.

One equilibrium relation appears in all these models regardless of the structure of the population or the number of dimensions [6]. It is the extension of equation (7) to a structured population. The relationship is

$$\bar{f} = \frac{1 - f_0}{4N_e \mu} \tag{9}$$

for a large number of potential alleles, with a slight modification if the number of possible alleles is small. In this formula, f_0 is the probability of identity for two alleles drawn from the same locality or from the same individual, $\bar{f}$ is the probability for alleles drawn at random from the entire population, and N_e is the effective population number not taking structure into account.

One conclusion from this, also noted by Maynard Smith [33], is that two alleles drawn from a pair of individuals widely separated from each other geographically should rarely be identical, regardless of the structure of the population. If the population has very little migration, then each part will come to have its own alleles. If there is free migration, the same alleles will be maintained throughout the population, but there will be many of them with individually low frequencies and the probability that any two will be identical is small. This assumes that the total effective number is large relative to the reciprocal of the mutation rate and that the number of potential neutral alleles is large. For a discussion see Kimura and Maruyama [25].

Finally, the global effective number taking the structure into account (N_{es}) is related to the effective number not taking this into account (N_e) by the relation

$$N_{es} = N_e \frac{1 - \bar{f}}{1 - f_0} \tag{10}$$

regardless of the number of alleles. This shows that with a structure of any sort the total effective number is enhanced.

10. Conclusions

I have tried to present the main arguments for and against the hypothesis of evolution by random drift of neutral mutations, or non-Darwinian evolution. I have devoted most of the discussion to non-Darwinian evolution rather than Darwinian since the latter is so well known. The theory of natural selection needs no further description or defense from me.

Mutants range from severely harmful, through neutral, to rare beneficial types. A proper theory would treat the entire range of values with an appropriately greater emphasis on stochastic elements near neutrality.

Despite this continuum of values, it is convenient for discussion to consider the possibility of a distinct class of mutants whose effect is so slight that their fate is mainly determined by random processes. Operationally, this means that the selective advantage or disadvantage is small relative to the effective population number. We then ask whether any substantial fraction of DNA and amino acid changes in evolution, or of polymorphisms, have this explanation.

I suggest that the great majority of DNA is noninformational in that it does not code for proteins or for unique sequence RNA, and that this DNA changes for the most part by mutation and random drift. The possibility that amino acid substitutions observed in evolutionary lineages have this explanation seems promising enough to deserve the exploration that it is clearly getting. At a minimum it has heuristic value, for it lends itself to theoretical developments, quantitative predictions, and testable hypotheses that will surely lead to a deeper understanding of evolution, whatever the outcome of this particular question. Whether any appreciable fraction of molecular polymorphism is neutral is an open question.

REFERENCES

[1] B. Clarke, "Selective constraints on amino-acid substitutions during the evolution of proteins," *Nature,* Vol. 228 (1970), pp. 159–160.

[2] E. C. Cox and C. Yanofsky, "Altered base ratios in the DNA of an *Escherichia coli* mutator strain," *Proc. Nat. Acad. Sci. U.S.A.*, Vol. 58 (1967), pp. 1895–1902.

[3] J. F. Crow, "Breeding structure of populations. II. Effective population number," *Statistics and Mathematics in Biology,* Ames, Iowa State College Press, 1954, pp. 543–556.

[4] ———, "Molecular genetics and population genetics," *Proceedings of the Twelfth International Congress on Genetics,* Idengaku Fukyukai, Mishima, Shizuoka-ken, Japan, 1969, Vol. 3, pp. 105–113.

[5] J. F. Crow and M. Kimura, *An Introduction to Population Genetics Theory,* New York, Harper and Row, 1970.

[6] J. F. Crow and T. Maruyama, "The number of neutral alleles maintained in a finite, geographically structured population," *Theor. Pop. Biol.,* Vol. 2 (1971), pp. 437–453.

[7] Th. Dobzhansky and O. Pavlovsky, "An experimental study of interaction between genetic drift and natural selection," *Evolution,* Vol. 11 (1957), pp. 311–319.

[8] R. F. Doolittle, R. Chen, C. Glasgow, G. Mross, and M. Weinstein, "The molecular constancy of fibrinopeptides A and B from 125 individual humans," *Hum. Genet.,* Vol. 10, (1970), pp. 15–29.

[9] W. J. Ewens, "The sampling theory of selectively neutral alleles," *Theor. Pop. Biol.,* Vol. 3 (1972), pp. 87–112.

[10] R. A. Fisher, *The Genetical Theory of Natural Selection,* Oxford, Clarendon Press, 1930; New York, Dover Press, 1958 (revised ed.).

[11] W. M. Fitch, "Rate of change of concomitantly variable codons," *J. Molec. Evol.,* Vol. 1 (1971), pp. 84–96.

[12] ———, "Does the fixation of neutral mutations form a significant part of observed evolution in proteins," *Brookhaven Symp. Biol.,* in press.

[13] W. M. Fitch and E. Margoliash, "The usefulness of amino acid and nucleotide sequences in evolutionary studies," *Evol. Biol.,* Vol. 4 (1970), pp. 67–109.

[14] W. M. Fitch and E. Markowitz, "An improved method for determining codon variability in a gene and its application to the rate of fixation of mutations in evolution," *Biochem. Genet.,* Vol. 54 (1970), pp. 579–593.

[15] E. Freese, "On the evolution of base composition of DNA," *J. Theor. Biol.,* Vol. 3 (1962), pp. 82–101.

[16] T. C. Gibson, M. L. Schleppe, and E. C. Cox, "On fitness of an *E. coli* mutation gene," *Science,* Vol. 169 (1970), pp. 686–690.

[17] J. B. S. HALDANE, "The cost of natural selection," *J. Genet.*, Vol. 55 (1957), pp. 511–524.

[18] J. KERKIS, "Study of the frequency of lethal and detrimental mutations in *Drosophila*," *Bull. Acad. Sci. U.S.S.R.*, Vol. 1 (1938), pp. 75–96.

[19] M. KIMURA, "Evolutionary rate at the molecular level," *Nature*, Vol. 217 (1968), pp. 624–626.

[20] ———, "Genetic variability maintained in a finite population due to mutational production of neutral and nearly neutral isoalleles," *Genet. Res.*, Vol. 11 (1968), pp. 247–269.

[21] ———, "The rate of molecular evolution considered from the standpoint of population genetics," *Proc. Nat. Acad. Sci. U.S.A.*, Vol. 63 (1969), pp. 1181–1188.

[22] ———, "The number of heterozygous nucleotide sites maintained in a finite population due to steady flux of mutations," *Genetics*, Vol. 61 (1969), pp. 893–903.

[23] ———, "Theoretical foundation of population genetics at the molecular level," *Theor. Pop. Biol.*, Vol. 2 (1971), pp. 174–208.

[24] M. KIMURA and J. F. CROW, "The number of alleles that can be maintained in a finite population," *Genetics*, Vol. 49 (1964), pp. 725–738.

[25] M. KIMURA and T. MARUYAMA, "Pattern of neutral polymorphism in a geographically structured population," *Genet. Res.*, Vol. 18 (1971), pp. 125–132.

[26] M. KIMURA and T. OHTA, "The average number of generations until fixation of a mutant gene in a finite population," *Genetics*, Vol. 61 (1969), pp. 763–771.

[27] ———, "On the rate of molecular evolution," *J. Molec. Evol.*, Vol. 1 (1970), pp. 1–17.

[28] J. L. KING and T. H. JUKES, "Non-Darwinian evolution," *Science*, Vol. 164 (1969), pp. 788–798.

[29] D. E. KOHNE, "Evolution of higher-organism DNA," *Quart. Rev. Biophys.*, Vol. 3 (1970), pp. 327–375.

[30] C. D. LAIRD, B. L. MCCONAUGHY, and B. J. MCCARTHY, "Rate of fixation of nucleotide substitutions in evolution," *Nature*, Vol. 224 (1969), pp. 149–154.

[31] T. MARUYAMA, "On the rate of decrease of heterozygosity in circular stepping stone models of populations," *Theor. Pop. Biol.*, Vol. 1 (1970), pp. 101–119.

[32] J. MAYNARD SMITH, "Haldane's dilemma, and the rate of evolution," *Nature*, Vol. 29 (1968), pp. 1114–1116.

[33] ———, "Population size, polymorphism, and the rate of non-Darwinian evolution," *Amer. Natur.*, Vol. 104 (1970), pp. 231–237.

[34] N. E. MORTON, "The future of human population genetics," *Prog. Med. Genet.*, Vol. 8 (1972), pp. 103–124.

[35] N. E. MORTON, J. F. CROW, and H. J. MULLER, "An estimate of the mutational damage in man from data on consanguineous marriages," *Proc. Nat. Acad. Sci. U.S.A.*, Vol. 42 (1956), pp. 855–863.

[36] T. MUKAI, "The genetic structure of natural populations. I. Spontaneous mutation rate of polygenes controlling viability," *Genetics*, Vol. 50 (1964), pp. 1–19.

[37] T. MUKAI, S. I. CHIGUSA, L. E. METTLER, and J. F. CROW, "Mutation rate and dominance of genes affecting viability in *Drosophila melanogaster*," *Genetics*, 1972, in press.

[38] H. J. MULLER, "Our load of mutations," *Amer. J. Hum. Genet.*, Vol. 2 (1950), pp. 111–176.

[39] T. OHTA and M. KIMURA, "Functional organization of genetic material as a product of molecular evolution," *Nature*, Vol. 223 (1971), pp. 118–119.

[40] F. J. RYAN, "Natural selection in bacterial populations," *Atti del VI Cong Int. Microbiol. (Italy)*, Vol. 1 (1953), pp. 1–9.

[41] M. P. SHANNON, T. C. KAUFMAN, and B. H. JUDD, "Lethality patterns of mutations in the *zeste-white* region of *Drosophila melanogaster*," *Genetics*, Vol. 64 (1970), p. 58.

[42] N. SUEOKA, "On the genetic basis of variation and heterogeneity of DNA base composition," *Proc. Nat. Acad. Sci. U.S.A.*, Vol. 48 (1962), pp. 582–592.

[43] J. A. SVED, "Possible rates of gene substitution in evolution," *Amer. Natur.*, Vol. 102 (1968), pp. 283–292.

[44] R. G. Temin, H. U. Meyer, P. S. Dawson, and J. F. Crow, "The influence of epistasis on homozygous viability depression in *Drosophila melanogaster*," *Genetics*, Vol. 61 (1969), pp. 497–519.

[45] N. W. Timofeeff-Ressovsky, "Auslösung von Vitalitätsmutationen durch Röntgenbestrahlung bei *Drosophila melanogaster*," *Nachr. Ges. Wiss. Göttingen, Biol. N. F.*, Vol. 1 (1935), pp. 163–180.

[46] J. Watson, *Molecular Biology of the Gene*, New York, Benjamin, 1970.

[47] S. Wright, "Evolution in Mendelian populations," *Genetics*, Vol. 16 (1931), pp. 97–159.

[48] ———, "Adaptation and selection," *Genetics, Paleontology and Evolution* (edited by G. L. Jepson, G. G. Simpson, and E. Mayr), Princeton, Princeton University Press, 1949, pp. 365–389.

[49] ———, "The theoretical course of directional selection," *Amer. Natur.*, Vol. 103 (1969), pp. 561–574.

[50] ———, "Random drift and the shifting balance theory of evolution," *Mathematical Topics in Population Biology* (edited by K. Kojima), Berlin-Heidelberg-New York, Springer-Verlag, 1970, pp. 1–31.

COMPARATIVE EVOLUTION AT THE LEVELS OF MOLECULES, ORGANISMS, AND POPULATIONS

G. L. STEBBINS
UNIVERSITY OF CALIFORNIA, DAVIS
R. C. LEWONTIN
UNIVERSITY OF CHICAGO

1. Introduction

Nearly all modern biologists who study evolution agree in believing that the rate and direction of evolution is determined by interactions between different processes, all of which are significant and none of which can be neglected. A synthetic theory of evolution, in some form or another, is here to stay. Five processes are recognized to be of basic significance. Two of these, mutation and gene recombination, contribute to the genetic variability that exists in all populations of higher organisms. The other three, selection, chance variations or gene fixations, and reproductive isolation, tend to reduce genetic variability, but are most significant as determiners of the rates and directions of evolution.

In spite of general agreement that all of these processes exist and may be significant at least under some conditions, great disagreement exists as to their relative importance. These differences may be due in part to the fact that the relative importance of different processes differs considerably from one group of organisms to another. Furthermore, even in the same evolutionary line great differences exist with respect to the relative importance of mutation, recombination, selection and the effects of chance when different molecules or even parts of molecules are compared with each other. Consequently, agreement and understanding are most likely to be reached by intensive studies of comparative evolution. Comparisons must be made both with respect to the evolution of entire organisms that belong to different evolutionary lines, as well as with respect to the evolution of different individual molecules or molecular systems in organisms belonging to the same evolutionary line. In accordance with the subject of the present conference, the comparisons to be made in this paper will be largely at the molecular level. Nevertheless, analogies to the evolution of molecular systems and of organs will be made when relevant.

2. Neutral mutations at the molecular level

Knowledge of the genetic code makes unavoidable the conclusion that a certain proportion of mutations, consisting of nucleotide substitutions from one codon to another that is synonymous with it, have little or no effect on the phenotype, since they do not alter the amino acid sequence of the polypeptide chain that is the primary product of the gene [17], [23]. Some of these may affect the phenotype because the synonymous codons are complementary to different molecules of transfer RNA, and so may be subject to different mechanisms that regulate transcription [8], [34]. That all synonymous codons differ with respect to these regulatory mechanisms is, however, rather unlikely. Nevertheless, even mutations to codons that are completely synonymous with each other may in many instances alter the future mutational possibilities of the genotype [8], [34]. This is because the most synonymous nucleotide of triplet codons, that in the third position, may sometimes be synonymous for all possible substitutions, while in other instances substitution of a purine for a pyrimidine or vice versa causes the codon to code for an amino acid having very different properties. If, for instance a DNA codon complementary to the messenger codon GGC mutates to one that is complementary to GGG, the coding for glycine at that position will be unchanged. Nevertheless, the unmutated codon can mutate by a single change in its first nucleotide to the complement of either UGC (cysteine) or AGC (serine), which requires a two step change for the mutated codon; while the latter can mutate in a single step to the complement of UGG (tryptophan), which requires the two steps for the unmutated codon.

In addition, mutations to codons for amino acids having properties that are very similar to those specified by the unmutated codon, for example, GUU (valine) to GCU (alanine) are very unlikely to affect the phenotypic properties of many proteins, particularly large enzyme molecules, and so would be neutral in their effects on the phenotype. Since the relationships between properties of individual amino acids and the effects that they produce upon proteins into which they have become incorporated is highly complex and imperfectly understood [37], the limits of this class of neutral mutations are even harder to define than are the limits of those that are due to synonymy of codons. Nevertheless, although we may for a long time have difficulty in classifying individual mutations as neutral or significant, we can hardly deny the existence of phenotypically neutral mutations. The frequency of their occurrence may be fairly high. Although, as has been so clearly shown by Ayala, Powell, Tracey, Mourao, and Perez-Salas [2], natural populations contain a very large amount of genetic variability that can be detected at the molecular level, the total amount that they contain is probably much higher. One way in which this can be shown is by selecting artificially for increased frequency of certain rare abnormalities, as has been done by Huether [14] with respect to corolla structure in *Linanthus*. As will be discussed later, the relationship of this variability to neutrality and random fixation of genes depends to a large extent upon the relative positions of adaptive and neutral genes upon chromosomes.

3. The importance of adaptive shifts and of accessory adaptive characters

Most of this paper will be devoted to comparing the relative importance of natural selection as compared to random deviations, both in different evolutionary lines and in different molecules. At the outset, therefore, we need to have clearly in mind the relationship between adaptation and evolutionary change. To the extent that selection guides the course of evolution, it does so not by accepting or rejecting individual mutations, but through the effects of an altered environment upon the adaptive values of constellations of genes. The adaptive value of a particular allele or mutant gene, therefore, cannot be treated mathematically or statistically as a constant. Adaptive values of genes vary widely in relation to both the external environment to which the population is exposed, and the internal environment of the cell, in which the primary product of each gene must perforce interact with the products of many other genes.

The issue at hand, therefore, is not whether frequencies of individual genes are altered by the action of natural selection or by chance variations, but whether alterations of entire genotypes represent adaptive shifts of interacting genes that form integrated systems, or whether genotypes are altered by random changes in the frequencies of genes that act independently of each other. The case for believing that most observable changes in phenotypes and their determinant genotypes represent adaptive shifts that are due to natural selection in the presence of altered interactions between populations and their environment has been well presented and richly documented by many evolutionists [9], [30], [36]. Further discussion of it here would be superfluous. The question that must be asked here is: to what extent do the observed changes in molecules, particularly changes in the amino acid sequences of proteins, represent adaptive shifts in response to changing environments, and to what extent have they resulted from neutral mutations and random drift?

Before discussing this question, we must explore the consequences of recognizing that adaptive shifts based upon simultaneous alterations in the frequency of many genes, are far more significant than are alterations in the frequency of individual genes acting independently. For this purpose, we should like to extend the concept of homeostasis from animal physiology, where it was originally developed by Cannon [6] and from genetic variation in populations, for which it was developed by Lerner [27], to evolutionary variation in related groups of organisms. The basic principle of homeostasis, that can be applied at all three of these levels, is that certain essential functions, or processes can be kept constant over a wide variety of conditions by adaptive alterations of secondary or accessory characteristics.

A good example of evolutionary homeostasis is presented by the evolution of molecules and structures associated with photosynthesis in plants. This is shown in Table I. The basic molecules for the process are those of chlorophyll. The principal molecule, chlorophyll *a* has been constant and unchanged throughout the evolution of plants, so that the only significant difference at the level of these molecules is that between the photosynthetic bacteria and other photosynthetic organisms. On the other hand, secondary chlorophyll molecules (*b* and *c*), as well as the molecules that constitute the accessory pigments of plastids, such as carotene and xanthophyll, and the special pigments found in various algae, vary in chemical structure from one phylum to another, but are usually constant within phyla. At a higher level of organization, that of the fine structure of chloroplasts, most variation is at the level of phyla or classes, but an outstanding exception is the specialized kind of chloroplast that exists in certain tropical grasses and other plants, and is associated with a different pathway of photosynthesis [26]. In one example, this difference with respect to fine structure and metabolic pathway exists between species of the same genus, *Atriplex* [4]. Finally, variation with respect to gross characteristics that are accessory to photosynthesis, such as leaf anatomy and leaf shape, exists mostly between different genera of the same family, between species of a genus, between races of the same species, or even between different leaves on the same plant.

TABLE I

Adaptive Differences Associated with Photosynthesis and Levels of the Evolutionary Hierarchy at Which They Occur

Structure affected	Level at which differences occur
Chlorophyll molecule	Kingdom
Plastid pigments	Phylum
Fine structure of chloroplast	Phylum, class, order (family, genus, species)
Cellular organization	Order, family (genus, species)
Leaf anatomy	Order, family, genus, species
Leaf size and shape	Genus, species, local population

Consequently, neither the great constancy of a structure or molecule in an evolutionary line nor its excessive variability can by itself be regarded as evidence for its adaptiveness or nonadaptiveness. On the other hand, adaptive variability at the level of populations and species is more likely to occur with respect to accessory functions rather than basic metabolic processes.

4. Comparisons of variations in different molecules

The principle of evolutionary homeostasis may explain the differences in rate of evolution of different molecules, that have already been discussed in this conference. Cytochrome *c*, which now may be regarded as a classic example of evolutionary stability, catalyzes one of the most basic processes of cellular metabolism, that is equally important for all cellular organisms, and takes place at a site that is well protected from variations in either the external or the internal, cellular environment. Ferredoxin, which is somewhat more variable [5] catalyzes a reaction that is accessory to the main reaction of photosynthesis, but is closely associated with it. Hemoglobin, which is comparable in variability to ferredoxin, but may show significantly more variation, is also an accessory molecule, since its function is to transport oxygen through the vertebrate body. Finally, a much greater amount of variation at the level of family and genus is found in the fibrinopeptides [10], [31]. These molecules, since they are functionally associated with blood clotting, are definitely accessory, and in addition are particularly likely to interact with the external environment of the animal.

The question of adaptive shifts *versus* neutral mutations and random drift requires particular attention with respect to fibrinopeptides, for two reasons. In the first place, both King and Jukes [23] and Ohno [32] have maintained that their variations are neutral. Secondly, among the molecules in which variation of amino acid sequences has been recorded at the phyletic level, they are the only ones that are neither enzymes nor molecules with enzymelike functions, such as hemoglobin. The latter fact is important to those who wish to correlate phylogenies based upon phenotypes with those that are based upon changes in enzymes and similar molecules. The phenotype of an animal or plant is based chiefly upon the nature of two kinds of protein molecules. The first kind are those associated with regulation of growth: enzymes responsible for the biosynthesis of hormones and other regulators of gene action, growth and differentiation, as well as the proteins to which active hormones are frequently bound [15]. The second kind are the various structural proteins, such as collagen, actin and myosin of muscles, the proteins of nerve cells, keratin, chitin, cell wall proteins of plants, and many others. Many of these proteins differ widely in amino acid composition from the metabolic proteins that have been studied phylogenetically, and we cannot assume *a priori* that rates and degrees of adaptiveness in these different classes of molecules are necessarily the same. Possibly, the situation presented by the fibrinopeptides may foreshadow what will be found when other nonenzyme proteins are analyzed comparatively.

For this reason, the data on fibrinopeptides of artiodactyls presented by Mross and Doolittle [31] have been analyzed further by one of us (Stebbins). Two questions were asked: (1) Are amino acid substitutions randomly distributed over the peptide chains or are they localized in particular regions? (2) If the substitutions are divided into two groups, either (a) conservative, with little effect on the properties of the peptide, or (b) radical, with potentially large effects, are the relative distributions of these two kinds of substitutions random or nonrandom? The classification into conservative and radical was made according to the criteria set forth by Smith [37] and followed his arrangement of residues into groups (his Table I). For example, the substitution of arginyl for lysyl, both of them positively charged and hydrophilic, is conservative, as is also aspartamyl for glutamyl and alanyl for valyl. On the other hand, substitutions of glutamyl for lysyl, of aspartyl for glycyl, or of seryl for valyl, were all classified as radical, and would be expected to affect greatly the properties of the peptide, unless concealed by other residues when the peptide assumed its tertiary structure.

The classification of Sneath [38] is similar to that of Smith but more quantitative. Using Sneath's correlation coefficients for pairs of amino acids, substitutions between pairs of residues having coefficients higher than 0.600 are conservative, and between pairs having coefficients lower than 0.600 are radical.

The results of this analysis were unequivocal (Figure 1). The statement of Mross and Doolittle, that substitutions are unequally distributed, was amply confirmed. Moreover, the high concentration of substitutions is at the amino terminal end of the peptide, at which end are also concentrated the various deletions that have taken place during phylogeny. In addition, the radical substitutions were concentrated in those sites having the largest number of substitutions, while the conservative ones are mainly scattered over regions in which few or no substitutions have become established.

These results must be considered in the context of the extensive studies that various workers have made on the physiological action of fibrinopeptides. For example, Bayley, Clements, and Osbahr [3] found that bovine fibrinopeptide B and human fibrinopeptide A given in minute concentrations to rabbits, dogs and lambs caused pulmonary hypertension, decreased effective pulmonary blood flow, and had several other effects on the circulatory system.

Even more significant is the evidence reviewed by Chandrasekhar and Laki [7], indicating that the residues present near the amino terminal end of fibrinopeptide A, particularly at position 13, influence the rate at which thrombin splits the molecule. Moreover, they conclude (p. 127): "that thrombin can strictly discriminate the amino acid residue at position 13 only if a serine residue occurs at position 16 or if an aspartic acid is located at position 18." On the basis of these conclusions, we postulate that the radical substitutions recorded in Figure 1 for positions 12 to 19 in fibrinopeptide A and for positions 18 to 21 in fibrinopeptide B took place during artiodactyl evolution in association with adaptive shifts. We believe that they are associated with alterations in the accessory adaptive properties of the fibrinogen molecule and its derivatives, which have aided in the acquisition by these very different animals of their various physiological and ecological properties.

5. Adaptiveness *versus* chance variation with respect to enzymes and similar molecules

The next question to ask is: Can the concept of evolutionary homeostasis, as outlined at the beginning of this paper, be applied to the evolution of amino acid sequences in enzymes and other molecules having similar properties? Based upon available evidence, this question can tentatively be answered in the

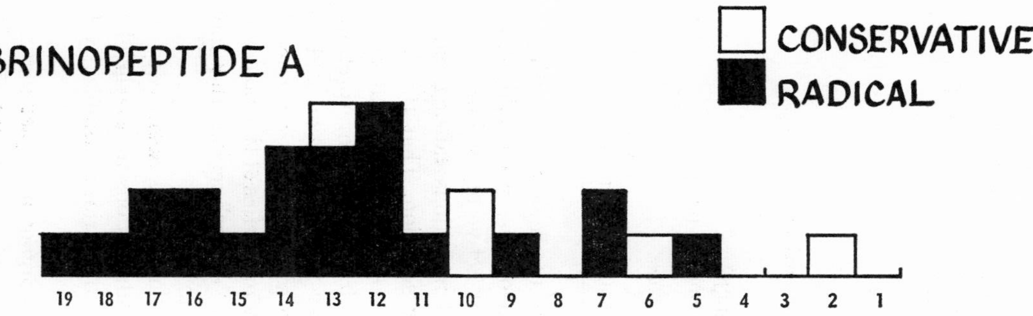

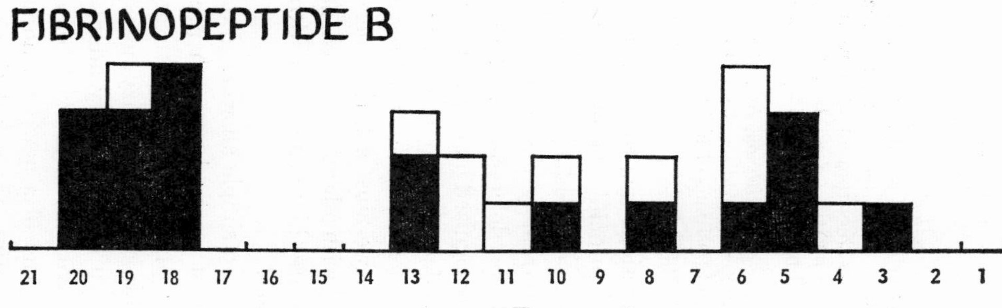

FIGURE 1

Diagram showing the numbers and positions of amino acid substitutions that occurred in fibrinopeptides A and B during the evolution of artiodactyls, as recorded by Mross and Doolittle ([31], Tables III and IV and Figure 3). White, open squares represent conservative substitutions; black squares represent radical substitutions. The numbers represent positions of residues from the position of the split by thrombin (1 = arginine in all peptides) and the amino terminal end of the peptide. Further explanation in the text.

affirmative. Moreover, basic functions as well as accessory adaptations are normally carried out by different parts of the same molecule. The different kinds of adaptations that an enzyme molecule can possess are listed in Table II. The

TABLE II

Adaptive Characteristics of Enzymes and Proteins with Similar Metabolic Functions

Basic to function
Configuration of active site
Configuration of binding sites
Accessory to function
Position of principal residues (cys) that determine tertiary structure
Position of residues that can form alpha helices
Position of residues that affect subunit association
Number and position of residues that affect solubility
Number and position of residues that affect hydrophobic properties
Related to enzyme activation, inhibition, and interactions
Position and configuration of allosteric site or sites
Number and positions of residues capable of forming hydrogen bonds

list of accessory functions was compiled largely from the review by Smith [37].

Based upon the principle of evolutionary homeostasis, we would expect residues or sequences of residues that are present at active or binding sites of an enzyme to be extremely conservative, since all mutations causing radical substitutions would be deleterious and rejected by selection. That this is the case is generally recognized and has already been stated by several of the speakers in this conference. On the other hand, differences of interpretation exist with respect to substitutions at regions other than the active and the binding sites. Hence, we must ask the question: To what extent are these substitutions associated with adaptive shifts in accessory properties of the molecules, and to what extent are they neutral substitutions in "spacer" regions that have no adaptive function?

This question has no simple answer. In order to illustrate its complexity, we would like to suggest an analogy with accessory adaptations at the level of macroscopic organs, and to present an example of adaptational complexity at the molecular level in a bacterial enzyme.

A widespread feature of accessory adaptations in organs of higher plants is that many alternative pathways exist for achieving the same adaptation. If, for instance, the impact of a new environment increases the adaptive value of increased seed production, the population can respond by either (1) an increase in the number of flowers per plant, (2) an increase in the number of gynoecial units or "carpels" per flower, or (3) an increase in the number of seeds per capsule. If the latter kind of adaptation takes place, it can be accomplished either by increasing the length of the capsule, its width or its thickness, or by various combinations of these changes. Hence, the same kind of selective pressure, acting simultaneously upon different populations that are isolated from each other, can cause them to diverge, if they respond at the outset in different ways. Which of the possible responses will be made may be determined by chance, depending upon the kinds of mutations that appear first, or that happen to be present in the gene pool. On the other hand, the adaptive structures that already exist may play the decisive role, since the most acceptable mutational changes will be those that cause the least disturbance of the developmental pattern [39].

There is good reason to believe that accessory regions of enzyme molecules, that are responsible for such characteristics as tertiary structure, temperature optima, solubility and interaction with other enzymatic processes would respond in a similar fashion to changes in either the external or the internal cellular environment. This is strongly suggested by the evidence that Yanofsky and his associates have obtained on the tryptophan synthetase enzyme in *Escherichia coli*. Figure 2 is a summary of some of their results [41]. They were obtained by analyzing a succession of mutations from full function to the auxotrophic condition that requires tryptophan in the medium, followed by reverse mutations to partial or full function.

The diagram shows that four different combinations of residues, involving substitutions at three different sites in two different peptides, can give full function. Two combinations at these same sites give partial function, and two of those obtained are completely nonfunctional. One could reason from these results that adaptive shifts to full function under a different set of environmental conditions could likewise be accomplished by several different combinations of mutational changes. The data obtained by Hardman and Yanofsky [13] suggest that the adaptive properties at this site affect substrate binding.

The probable existence of these complex alternative pathways of adaptation warns us that we must be very careful in assuming that a particular amino acid substitution is neutral unless we have positive evidence that this is so. In particular, we must avoid committing what may be called "the fallacy of omniscience." This fallacy, which has often been committed in the past by evolutionists and taxonomists who are comparing macroscopic characters of organisms, runs about as follows: "I can't see what adaptive value this character difference could have, therefore, it is inadaptive and was not influenced by natural selection." The fallacy here lies in the implication that the author of the statement knows everything that can be known about the adaptiveness of the organisms concerned. With respect to macroscopic organisms this is usually not true. With respect to enzyme action *in vivo*, about which our knowledge is still very rudimentary, it probably is never true. Evolutionists are, therefore, unable to prove, by the process of elimination, the null hypothesis that amino acid substitutions in certain portions of enzyme molecules are neutral and do not affect the function of the whole molecule.

At the same time, interpretations of enzyme structure and function must guard against the reverse fallacy, which would necessarily ascribe an adaptive significance to amino acid substitutions, without definite evidence in favor of it.

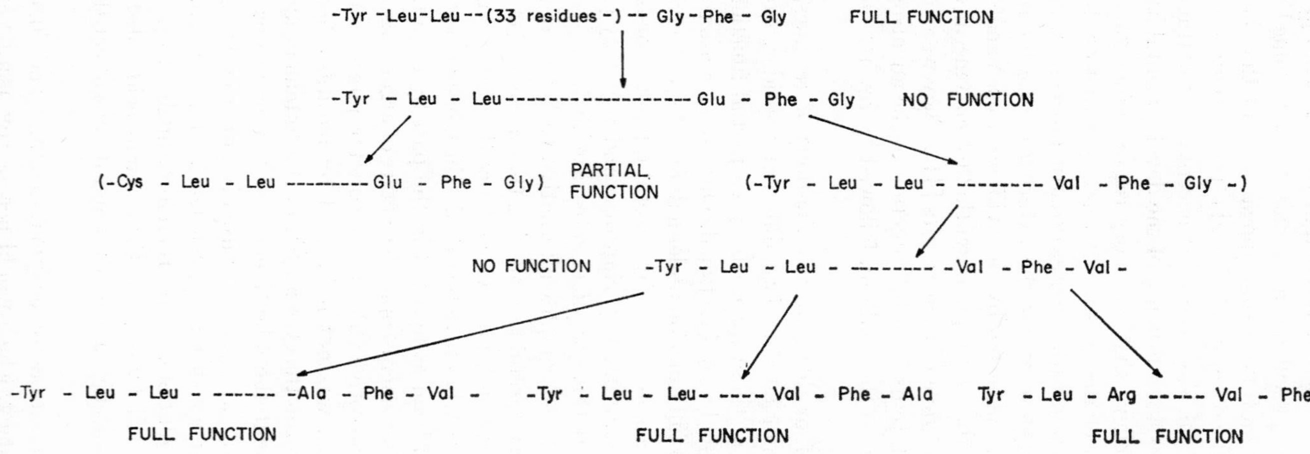

FIGURE 2

Functional combinations of residues at a single site of the tryptophan synthetase molecules. Data from Yanofsky, Horn, and Thorpe [41].

In the present state of our knowledge, molecular evolutionists must resign themselves to saying more often those three very difficult words: "I don't know."

Any distinction between hypotheses of evolution by random differentiation as opposed to natural selection must then take a more complex form. In particular, advocates of the neutrality hypothesis offer evidence that the observed facts of protein variation both within and between species, are consonant with an assumption of random variation, while at the same time being irreconcilable with a selective hypothesis. We will review separately the quality of these *positive* arguments for neutrality and the *negative* ones against selection.

6. Proposed positive arguments in favor of neutral amino acid substitutions

The first of these positive arguments, advanced by King and Jukes [23], is that the proportion of amino acid sites that have 0, 1, 2, $\cdots$, n substitutions in variants of globins, cytochromes *c*, and the variable ($S-$) regions of immunoglobins follow the Poisson distribution. This argument, however, is based upon an erroneous assumption, which includes the "fallacy of omniscience," as has been pointed out by Clarke [8]. Since they recognize that some sites of enzyme molecules have great adaptive significance, they arbitrarily divide the polypeptide chains into "adaptive" or "invariant" regions and "nonadaptive" or "variant" regions. Clarke shows that the distribution of residues and substitutions in the latter regions is not random. Moreover, the assumption of King and Jukes, that only the invariant sites are concerned with adaptation includes a second, hidden assumption that is surely unwarranted. This is that the action of enzymes is an all-or-none affair under any set of environmental conditions, and that it cannot be gradually modified in adaptation to shifting conditions of the external or the internal environment.

The second argument, advanced by the same authors, is that when the average amino acid frequencies are tabulated among 53 vertebrate polypeptides, their frequencies agree remarkably well with those to be expected from random permutations of nucleic acid bases. From this they conclude that the average amino acid composition of proteins reflects, in a neutral fashion, the genetic code. Even if this interpretation is correct, however, one cannot conclude from it that amino acid substitutions in all or even in the majority of proteins have been largely neutral. The 53 proteins are in all probability a nonrepresentative sample, as has been pointed out by Clarke [8]. In structural proteins, the deviation from frequencies expected on the basis of the code is very great. More important, however, the interpretation represents a gross misunderstanding of the selective argument. No one suggests that on the average over a heterogeneous collection of proteins and species, particular amino acids should be in excess or deficiency. Indeed, if there were some general necessity for a particular amino acid to be very frequent in most proteins, the code probably would have evolved to take account of this physicochemical necessity. Even if the selective hypothesis were true in the most extreme form for every amino acid position in every protein, on the average for many unrelated proteins and distantly related species, the gross frequency of amino acids would represent the codon frequencies. Stone walls in Vermont reflect, on the average, the naturally occurring frequencies of stones of different sizes (Vermonters being sensible and frugal), but different walls have different size distributions depending upon whether they are rough boundary markers, pasture fences, farm plot fences, or garden fences. Moreover, when a stone falls out, only some stones will fit into the hole. This is not the only case when the advocates of the neutrality hypotheses have been confounded by the law of large numbers.

The third argument, that has been advanced not only by Jukes and King, but also by several of the speakers at this conference, is that when the numbers of amino acid substitutions in a group of related organisms are compared with the times when the ancestors of these modern forms are believed to have diverged from each other, a high correlation is found between numbers of substitutions and presumable ages of divergence. Based upon these correlations, uniform rates of amino acid substitutions are inferred. Admittedly, however, these "rates" are only approximately uniform, and even with the relatively small sample that is now available, striking exceptions can be found, such as the rapid rate of substitution in guinea pig insulin [8]. Moreover, some data suggest that these inequalities of rate are greater when we compare organisms that have diverged relatively recently than when comparisons are made between divergences that occurred long ago [29]. An example is the chart of differences with respect to hemoglobin prepared by Kimura and Ohta [22] for this Symposium. The figures for mammals belonging to different orders differ from each other by a factor of up to 65 per cent (17 in mouse-human *versus* 28 in mouse-rabbit) but the differences between each of these mammals and the fish (carp) are much more similar. If this situation exists generally, we can conclude that the so-called "constancy" of rates over millions of years is nothing but the law of large numbers. In the intervening times, there have been many speedings-up and slowings-down of the rate of amino acid substitution differentially in different lines as within the mammals. On the average, over vast stretches of time, however, we expect that different phylogenetic lines will have similar *average* numbers of substitutions in molecules having similar basic functions. The entire argument is based upon a confusion between an average and a constant. This point is clearly illustrated by fossil evidence. Simpson [36] shows that many phyletic lines have both fast and slow periods of evolution such that the average rate gives no index to the detailed history of the line. There are even periods when many independent lines simultaneously show very high rates of extinction and speciation, as in the Pleistocene, yet no one would suggest that the similarity in taxonomic rates of such lines indicates randomness!

A fourth "permissive" argument for neutrality concerns the variation *within* species as revealed by recent studies of "allozyme" variation at various structural gene loci [33], [35]. Typically, within any population of *Drosophila*, mice, or man, among other species, between one quarter and one half of a random sample of loci is found to be polymorphic, and 10 to 15 per cent of the loci are heterozygous in a typical individual. The question is whether such variation is a result of the accumulation of mutations subject to random variation in frequency in finite populations, or whether one needs to invoke some selective balance to explain the maintenance of the variation. Kimura [18] has asked how much variation is to be expected in a population at equilibrium under recurrent mutation and loss of alleles from genetic drift. If one assumes that each mutation is unique, then in a population of size N and a mutation rate to new alleles of u, the effective number of alleles (the reciprocal of the probability that two alleles in the population are identical) is, to a close order of approximation $n_e = 4Nu + 1$. Kimura, in several publications, makes use of "reasonable" values of N and u to show that the observed allelic frequencies in populations are consonant with this prediction. The difficulty with this formula is its remarkable behavior with respect to the two parameters N and u, whose values we can only guess to an order of magnitude. Provided that Nu is of smaller order than unity, the effective number of alleles is 1, irrespective of N and u. On the other hand, if Nu is of larger order than unity, the number of effective alleles is very much larger than anything ever observed. Only when Nu is of order 1 is there any interesting sensitivity of n_e to the actual values. Thus, all that can be said from evidence on allozyme polymorphism is that if these polymorphisms are neutral, Nu is smaller than unity for most loci and about order 1 for a few (the esterase locus in *D. pseudoobscura,* for example). This theory is then *too* permissive.

The great power, and therefore weakness as a testable hypothesis, of the random theory can also be seen in another aspect of the observations on genic polymorphism. The simplest form of the random theory predicts that although each population of a species will have about the same value of n_e for each locus, the particular alleles that are in high or low frequency in a population will be random, so that allele frequencies will vary considerably from population to population. The contrary is observed, in *D. pseudoobscura* [33] and in *D. willistoni* [2]. These species display a remarkable similarity in allele frequencies in widely separated populations. Such an observation would seem to rule out the random theory, but a small change saves it. If there is a small amount of migration from one population to another, there will be virtually no differentiation between populations under a pure drift model. If m is the migration rate between neighboring populations, then if Nm is of order 1 or greater, differentiation will be effectively prevented. But if $Nm = 1$, that is the same as saying that one or less migrant individual is exchanged per generation! Such tiny migration rates are in practice unmeasurable, or rather, could never be ruled out by any observation. Thus, the random theory, augmented by a small but unmeasurable migration rate, is so powerful a prediction that no observation can be in contradiction to it. That is what we mean by a theory so powerful that it is weak as a testable hypothesis. In the terms used by Popper, the neutral theory is "empirically void" because it has no set of potential falsifiers. All observations can be made consonant with it. For that reason, a different quality of evidence is necessary for distinguishing a selective from a neutral theory. In recognition of this necessity, the advocates of the neutral theory have tried to show that a selective theory is *contradicted* by the evidence, so that the neutral theory is confirmed by elimination.

7. Positive and negative evidence on selection

The history of evolutionary science during the last half century has demonstrated clearly the fact that an understanding of the processes of evolution can best be acquired by examining first actual evolutionary events at the level of populations and gene frequencies, and synthesizing from this firm base theories and concepts about the broader aspects of macroevolution, rather than by beginning with attempts to construct all encompassing phylogenies. We believe that this basic principle of studying evolution upward from the population rather than downward from a postulated phylogeny must be applied to molecules as well as to organisms. Because of this fact, data on the distribution of allozymes and other variants of protein molecules has the greatest possible relevance to biochemical evolution.

As we have said, a large amount of data have been accumulated on the variability of homologous proteins, particularly allozymes, in both natural and artificial populations. Good samples have been presented in this conference by Drs. Ayala [2] and Allard [1]. The opinion of these authors, that most of this variation is adaptive or at least associated with adaptation and maintained by natural selection, is shared by the majority of the workers in this field.

Nevertheless, direct evidence for this point of view is still scanty. The selection experiments with populations of barley, described by Dr. Allard, constitute an example. Another is the distribution of allozymes of serum esterase in populations of catastomid fishes, in which an allozyme having a more southerly distribution, in a warmer climate, has a higher temperature optimum *in vitro* than the predominant allozyme of populations adapted to cooler climates [24], [25].

Indirect evidence that points toward the adaptiveness of molecular polymorphism in populations is, however, much more extensive. Dr. George Johnson [16], in a manuscript that he has kindly let us read before publication, has discovered some very interesting correlations and parallelisms between molecular and phenetic evolution in a group that is now becoming a classic example of rapid evolution in response to environmental diversity: the Hawaiian species of *Drosophila*. In the first place, after comparing the number of allozyme types per locus in species inhabiting four different islands (Hawaii, Maui, Oahu, Kauai), he has found an indirect correlation between this number and the number of different species found on each island. Interestingly enough, these numbers are not correlated with the sizes of the islands, as would be predicted on the basis of some models that ecologists have constructed, but with ecological diversity.

By far the largest amount of indirect evidence for the selective basis of molecular polymorphism, however, comes from the consistent observation that in nearly every species studied certain polymorphisms occur in a similar fashion over wide stretches of territory. As previous speakers have already pointed out, the only possible explanations for this constancy are (1) that it is maintained by selection, or (2) that it is due to extensive and continuous migration between widely separated populations. In at least some of the examples, this latter explanation can be ruled out, as Drs. Ayala and Allard have explained.

The crux of the evidence on selection, however, is of the *negative* variety. That is, it is contended by Kimura, Crow, King, and others that any selective hypothesis that is not trivially different from a neutral theory contains certain irresolvable contradictions. In particular, it is maintained that the differential production of offspring necessary to account either for rates of gene replacement in evolution or for the maintenance of observed polymorphism in a balanced state, is vastly greater than any real populations can actually afford. This is the problem of "genetic load," but it might equally well be regarded as the problem of "fitness variance." If selection is operating with respect to a particular locus, such that different genotypes have, on the average, different numbers of offspring, two results follow. First, the mean rate of offspring production of a segregating population will be less than the rate in a population consisting of only the most fit genotype. The difference between this mean rate and the rate for the best genotype is the *genetic load* associated with selection at that locus. Second, there is a certain variation in reproductive rate among genotypes which is reflected in the *variance in fitness* in the population. Both of these quantities are necessarily limited in any real population by the biological facts of the reproductive cycle of a species. While it is not certain how many fertilized eggs a human female could produce in a lifetime, it is obviously less than, say, 1,000, and this puts an absolute ceiling on the variance in fitness possible in a human population. What is at issue is whether a selective hypothesis applied to known rates of gene substitution in evolution and the standing genic polymorphism, would require more fitness variance than is biologically possible.

We will not repeat here the evidence, so well presented by Kimura in this Symposium, and by Haldane [12], Kimura [18], [19], [22], and Kimura and Crow [20], [21] to the effect that evolutionary rates and degrees of polymorphism are too high to be consonant with a selective hypothesis. Rather, we will examine briefly some of the suppositions inherent in these models that make them rather less convincing than is usually assumed.

First, the argument about genetic load is based upon a model that is clearly incorrect. It assumes that genes segregate independently from each other, and so neglects the very powerful effects of linkage over short distances [11]. If it is really true that at least one third of the genome is segregating in *Drosophila,* then the average recombination between segregating loci is on the order of one in a thousand and the result is that strongly linked interacting blocks will be built up. Before we can discard selection on the basis of genetic load, we must examine more thoroughly the whole theory of linked loci. When our present understanding of linkage is coupled with more sophisticated understanding of gene interaction, as for instance the possibility that many genes are maintained by selection for an intermediate optimum, the argument against the adaptive nature of a large genetic load fails to be convincing.

For example, Lewontin [28] found that selection for an intermediate optimum caused profound linkage disequilibrium, even for genes that are quite far apart on the chromosome, yet the variance of fitness in the equilibrium population was quite small. The conclusions of Franklin and Lewontin are clearly on the conservative side since they were based on a very weak model of gene interaction. Bringing in optimum model selection will magnify these linkage effects many times and thus reduce the genetic load associated with gene substitution and polymorphism. The meaning of these findings is that genes are not substituted in evolution independently of each other, but in correlated blocks.

A second and important objection to genetic load arguments is that they are based on a special and unrealistic model of population growth and regulation. While it is part of the model that the mean reproductive rate of a segregating population is lower than that of a population made up only of the best genotype, this assumption is, in general, false. Because of nongenetic causes of mortality and the existence of an elastic "ecological load," there is usually no effect on population size or growth rate from the segregation of selected genes [40], until there is a complete saturation of the available reproductive excess. With the exception of a few species on the verge of extinction, there is no evidence of such a saturation. Such an argument does not deny that a saturation of the "load space" is possible if genetic load were high enough.

There is a third perspective that does not seem to have been appreciated by the supporters of evolution by random drift. When gene frequencies change over the course of time, there is a variance in fitness of genotypes *a fortiori*. This can best be appreciated in a haploid population. Suppose we observe a gene replacement in such a population over the entire course of its substitution, beginning at a very low frequency and finally reaching the frequency of 1. The gene will have increased and decreased during the evolution of the population, but since it eventually went to fixation, there was an average positive rate of increase of the allele. Now the question is: "How, from this sample path can one distinguish selection from random drift?" The answer is: "In no way." The total variance in "fitness" and the total integrated "genetic load" can be *tautologically* calculated from the sample path, since at every generation the change in gene frequency *tautologically* defines the net selection at that generation. Now it might be objected that an erratic path is evidence of drift, while a monotone path to fixation indicates selection. But this misses the point. First, of course, there is no requirement that the environment be uniform over the history of the population. More important, the total integrated load calculated for an erratic path is *greater* than for a monotone one. The longer the gene frequency bounces around in an unfixed state, the greater the eventual total load, as tautologically defined. This is because, *a posteriori*, it is impossible to know the difference between a "selective" and a "neutral" sample path. This is because a selection coefficient, if it has any biological meaning, must be defined from the expectation of gene frequency change over replicated populations simultaneous in time. It cannot be defined from the expectation of gene frequency change over a *time* ensemble for one population, since then all processes are selective *a fortiori*.

In a diploid population, one might imagine a change in genotype composition, even though family size, that is, number of individuals per generation, remained constant. In each successive generation, a different sample of segregates from heterozygotes might survive and reproduce. This, however, is never the case. Even on a pure random model with a Poisson distribution of offspring number, half of the change in gene frequency arises from variance in family size between genotypes and half from variance in composition among families of heterozygotes. Thus, half of the genetic variance in a diploid population changing its gene frequency will be ascribable *a posteriori* to selection, without any evidence about the numbers of offspring expected in replicate populations that are subject to selection. Thus, it is hard to see how, given a certain rate of gene substitution, one avoids the problem of genetic load simply by postulating a "neutral model." Selection, in its tautological sense, does not thereby disappear unless it is also proposed that there is no variance in offspring number between genotypes and all the variance arises from segregation in heterozygotes. But that would require a device for producing negative correlations between families, so that even the sampling variance between genotypes can be suppressed. No one has yet suggested this as a possibility.

8. Conclusion

In our opinion, the relative position of population geneticists and molecular evolutionists who are dealing with the comparative molecular structure of proteins and nucleic acids can be stated about as follows. Both groups agree that some evolutionary changes have been due to adaptive shifts mediated by natural selection, while other changes have been the result of chance events. The principal differences between the two groups are the three following ones.

(1) Most geneticists who have studied actual populations believe that the effects of chance are minor relative to those of selection, and can be regarded as a kind of "evolutionary noise." Many comparative molecular evolutionists, on the other hand, believe that most changes have been brought about by neutral mutations randomly fixed.

(2) Most evolutionists who have studied macroevolutionary changes from the point of view of comparative morphology, physiology, and ecology believe that these changes represent chiefly adaptive shifts mediated by natural selection, while many comparative molecular evolutionists believe that at the molecular level most macroevolutionary changes are the result of chance events, and that natural selection has acted chiefly to stabilize highly adaptive molecules,

certain amino acid sequences that are parts of molecules, and, presumably, highly adaptive phenotypes.

(3) Finally, many evolutionists believe that the greatest progress will be made by working upward from an understanding of actual populations of organisms living in natural environments or known and controlled conditions in the laboratory, while others prefer to make inferences and extrapolations from comparisons between distantly related organisms, using various mathematical devices to make these extrapolations plausible.

From this summary, we believe that the nature of evolutionary processes is not yet resolved. However, the pathways toward a resolution are available. We believe that many exciting new results will be obtained by both groups in the near future. As long as pathways of communication remain open, as they have been at this conference, a resolution of the differences is bound to come. Most likely it will be accomplished by younger scientists, whose opinions have not yet become congealed.

REFERENCES

[1] R. W. Allard and A. L. Kahler, "Patterns of molecular variation in plant populations," *Proceedings of the Sixth Berkeley Symposium on Mathematical Statistics and Probability*, Berkeley and Los Angeles, University of California Press, 1972, Vol. 5, pp. 237–254.

[2] F. J. Ayala, J. R. Powell, M. L. Tracey, C. A. Mourão, and S. Pérez-Salas, "Enzyme variability in the *Drosophila willistoni* group. III. Genetic variation in natural populations of *Drosophila willistoni*," *Genetics*, Vol. 70 (1971), pp. 113–139.

[3] T. Bayley, J. A. Clements, and A. J. Osbahr, "Pulmonary and circulatory effects of fibrinopeptides," *Circ. Res.*, Vol. 21 (1967), pp. 469–485.

[4] O. Björkman, J. E. Boynton, M. A. Nobs, and R. W. Pearcy, "Physiological Ecology Investigations," *Annual Rept. Dept. Plant Biol., Carnegie Inst. Year Book*, No. 69 (1970), pp. 624–648.

[5] D. Boulter, E. W. Thompson, J. A. M. Ramshaw, and M. Richardson, "Higher plant cytochrome c," *Nature*, Vol. 228 (1967), pp. 552–554.

[6] W. B. Cannon, *The Wisdom of the Body*, New York, Norton, 1932.

[7] N. Chandrasekhar and K. Laki, "Evolution and the fibrinogenthrombin interaction," *Fibrinogen* (edited by K. Laki), New York, M. Dekker, Inc., 1968, pp. 117–130.

[8] B. Clarke, "Darwinian evolution of proteins," *Science*, Vol. 168 (1969), pp. 1009–1011.

[9] T. Dobzhansky, *Genetics of the Evolutionary Process*, New York, Columbia University Press, 1970.

[10] R. F. Doolittle and B. Blombäck, "Amino-acid sequence investigations of fibrinopeptides from various mammals: evolutionary implications," *Nature*, Vol. 202 (1964), pp. 147–152.

[11] I. Franklin and R. C. Lewontin, "Is the gene the unit of selection?" *Genetics*, Vol. 65 (1970), pp. 707–734.

[12] J. B. S. Haldane, "The cost of natural selection," *J. Genet.*, Vol. 55 (1957), pp. 511–524.

[13] J. Hardman and C. Yanofsky, "Substrate binding properties of mutant and wild-type A proteins of *Escherichia coli* tryptophan synthetase," *Science*, Vol. 156 (1967), pp. 1369–1371.

[14] C. A. Huether, Jr., "Exposure of natural genetic variability underlying the pentamerous corolla constancy in *Linanthus androsaceus* ssp. *androsaceus*," *Genetics*, Vol. 60 (1968), pp. 123–146.

[15] E. V. Jensen, M. Numata, S. Smith, T. Suzuki, P. I. Brechner, and E. R. DeSombre, "Estrogen-receptor interaction in target tissue," *Developmental Biol. Suppl. 3, Communication in Development* (edited by Anton Lang), 1969, pp. 151–171.

[16] G. B. Johnson, "The relationship between enzyme polymorphism and species diversity," *Evolution,* Vol. 26 (1972), in press.

[17] T. H. Jukes, *Molecules and Evolution,* Columbia University Press, New York, 1966.

[18] M. Kimura, "Genetic variability maintained in a finite population due to mutational production of neutral and nearly neutral isoalleles," *Genet. Res.*, Vol. 11 (1968), pp. 247–269.

[19] ———, "The rate of molecular evolution considered from the standpoint of population genetics," *Proc. Nat. Acad. Sci. U.S.A.*, Vol. 63 (1969), pp. 1181-1188.

[20] M. Kimura and J. F. Crow, "The number of alleles that can be maintained in a finite population," *Genetics,* Vol. 49 (1964), pp. 725–738.

[21] ———, "Natural selection and gene substitution," *Genet. Res.*, Vol. 13 (1969), pp. 127–141.

[22] M. Kimura and T. Ohta, "Population genetics, molecular biometry and evolution," *Proceedings of the Sixth Berkeley Symposium on Mathematical Statistics and Probability,* Berkeley and Los Angeles, University of California Press, 1972, Vol. 5, pp. 43–68.

[23] J. L. King and T. H. Jukes, "Non-Darwinian evolution," *Science,* Vol. 164 (1969), pp. 788–798.

[24] R. K. Koehn, "Esterase heterogeneity: dynamics of a polymorphism," *Science,* Vol. 163 (1969), pp. 943–944.

[25] R. K. Koehn and D. I. Rasmussen, "Polymorphic and monomorphic serum esterase heterogeneity in Catostomid fish populations," *Biochem. Genet.*, Vol. 1 (1967), pp. 131–144.

[26] W. M. Laetsch, "Chloroplast specialization in dicotyledons possessing the C_4-dicarboxylic acid pathway of photosynthetic CO_2 fixation," *Amer. J. Bot.*, Vol. 55 (1968), pp. 875–883.

[27] I. M. Lerner, *Genetic Homeostasis,* New York, Wiley, 1954.

[28] R. C. Lewontin, "Selection in and of populations," *Proceedings of the Twelfth Congress of Zoology, Ideas in Biology,* Vol. 6, 1965.

[29] E. Margoliash and W. M. Fitch, "Evolutionary variability of cytochrome *c* primary structures," *Ann. N.Y. Acad. Sci.*, Vol. 151 (1968), pp. 359–381.

[30] E. Mayr, *Animal Species and Evolution,* Cambridge, Harvard University Press, 1963.

[31] G. A. Mross and R. F. Doolittle, "Amino acid sequence studies on artiodactyl fibrinopeptides II. Vicuna, elk, pronghorn, antelope and water buffalo," *Arch. Biochem. Biophys.*, Vol. 122 (1967), pp. 674–684.

[32] S. Ohno, *Evolution by Gene Duplication,* New York and Berlin, Springer, 1970.

[33] S. Prakash, R. C. Lewontin, and J. L. Hubby, "A molecular approach to the study of genetic heterozygosity in natural populations of *Drosophila pseudoobscura,*" *Genetics,* Vol. 61 (1969), pp. 841–858.

[34] R. C. Richmond, "Non-Darwinian evolution: A critique," *Nature,* Vol. 225 (1970), pp. 1025–1028.

[35] Robert K. Selander and Suh Y. Yang, "Protein polymorphism and genic heterozygosity in a wild population of the house mouse," *Genetics,* Vol. 63 (1969), pp. 653–667.

[36] G. G. Simpson, *The Major Features of Evolution,* New York, Columbia University Press, 1953.

[37] E. L. Smith, "Evolution of enzymes," *The Enzymes,* New York, Academic Press, Vol. 1 (1970), pp. 267–339.

[38] P. H. A. Sneath, "Relations between chemical structure and biological activity in peptides," *J. Theor. Biol.*, Vol. 12 (1966), pp. 157–195.

[39] G. L. Stebbins, "Adaptive radiation and trends of evolution in higher plants," *Evolu-*

tionary Biology (edited by Th. Dobzhansky, M. K. Hecht, and Wm. C. Steere), 1967, pp. 101–142.

[40] B. WALLACE, "Polymorphism, population size and genetic load," *Population Biology and Evolution* (edited by R. C. Lewontin), Syracuse, Syracuse University Press, 1968, pp. 87–108.

[41] C. YANOFSKY, V. HORN, and D. THORPE, "Protein structure relationships revealed by mutational analysis," *Science*, Vol. 146 (1964), pp. 1593–1594.

POPULATION GENETICS, MOLECULAR BIOMETRY, AND EVOLUTION

MOTOO KIMURA and TOMOKO OHTA
NATIONAL INSTITUTE OF GENETICS, MISHIMA, JAPAN

1. Introduction

It has been said that Darwin's theory of evolution by natural selection is one of the greatest intellectual triumphs of our civilization (Crick [7]). Equally important is the recent discovery that the instruction to form an organism is encoded in DNA (or sometimes RNA) with four kinds of nucleotide bases. It is natural, therefore, that attempts be made to understand evolution in molecular terms.

Studies of evolution always contain two aspects. One is historical and is concerned with the reconstruction of past processes. The other is causal in that the underlying mechanism is pursued. Although these two are intimately connected, we are mainly concerned in this paper with the latter aspect of molecular evolution and we shall discuss several problems from the standpoint of population genetics.

As a branch of genetics, population genetics investigates the laws which govern the genetic composition of Mendelian populations (reproductive communities), and through such study, we intend to clarify the mechanism of evolution. The fundamental quantity which is used here is the gene frequency or the proportion of a given allelic gene in the population.

Because of the particulate nature of Mendelian inheritance, gene frequencies change only gradually with time under the influence of mutation, migration, selection, and random sampling of gametes in reproduction in any reasonably large population. The mathematical theory which treats such processes of change as stochastic processes was founded by the great works of R. A. Fisher [15] and Sewall Wright [66], and since then has been considerably extended under the name of diffusion models (Kimura [25]; see also Crow and Kimura [10], Chapters 8 and 9).

Although population genetics theories in general, and especially their deterministic aspects such as those initiated by J. B. S. Haldane [18], have promoted greatly the development of neo-Darwinian theory of evolution (see Haldane [21]), the real impact of the mathematical theory of population genetics has not been felt in the study of evolution. The main reason for this is that popula-

Contribution No. 820 from the National Institute of Genetics, Mishima, Shizuoka-ken, 411 Japan.

tion genetics theory is built on the concept of gene frequencies and the actual studies of evolution are conducted at the phenotypic level, and there is no direct way of unambiguously connecting the two. This has often made the study of microevolution a victim of loose jargon and facile generalizations, to the discouragement of the time consuming efforts to build mathematical models and check them with observable quantities.

It is fortunate, therefore, that the study of molecular evolution has opened a new field where the mathematical theory of population genetics can be introduced (Kimura [27], [31]). We now know, thanks to the pioneering work of Zuckerkandl and Pauling [67], that mutant substitutions have proceeded within the gene locus (cistron) coding for the alpha chain of hemoglobins at an average interval of roughly ten million years in the course of vertebrate evolution. Similar estimates of evolutionary rate are now available for several cistrons [38], [45]. In addition, the estimation of the rate of nucleotide substitution in evolution has begun using DNA hybridization techniques [39], [40].

For many years, attempts have been made in vain to estimate the number of gene substitutions that actually occurred in the course of evolution, transforming one species into another, one genus into another, and so forth. But now, by the methods that can measure gene differences in molecular terms, this has become feasible.

An exciting possibility confronting us is, by synthesizing comparative studies of informational macromolecules and modern studies of paleontology by the methods of population genetics and biometry, to go far back into the history of life and to penetrate deep into the mechanism of evolution at the molecular level. Certainly, there is much to be done by statisticians and applied mathematicians in this new venture.

2. Population genetics of gene substitution

From the standpoint of population genetics, the process of molecular evolution consists of a sequence of events in which a rare molecular mutant increases its frequency and spreads into the species, finally reaching the state of fixation. They represent a lucky minority among a tremendous number of mutants that actually appear in the species in the course of evolution.

Before we present the dynamics of mutant substitution in the population, let us summarize briefly the nature of genes and mutations at the molecular (DNA) level. A gene, or more precisely a cistron, may be thought of as a linear message written with four kinds of nucleotide bases (A, T, G, C) from which a polypeptide chain is transcribed (A = adenine, T = thymine, G = guanine, C = cytocine). The message is so composed that a set of three consecutive letters (triplet) form a code word or codon for an amino acid. With four possible letters at each position of a triplet, there are 4^3 or 64 codons. Of these, 61 are used to code for 20 amino acids, while the remaining three codons serve as punctuation marks ("chain termination"). The entire 64 code words have been deciphered (see Table I).

TABLE I

STANDARD RNA CODE TABLE

Ala = Alanine; Arg = Arginine; Asn = Asparagine; Asp = Aspartic acid; Cys = Cysteine; Gln = Glutamine; Glu = Glutamic acid; Gly = Glycine; His = Histidine; Ile = Isoleucine; Leu = Leucine; Lys = Lysine; Met = Methionine; Phe = Phenylalanine; Pro = Proline; Ser = Serine; Thr = Threonine; Try = Tryptophan; Tyr = Tyrosine; Val = Valine; Term. = Chain terminating codon.

1 \ 2	U	C	A	G	2 / 3
U	Phe	Ser	Tyr	Cys	U
	Phe	Ser	Tyr	Cys	C
	Leu	Ser	Term.	Term.	A
	Leu	Ser	Term.	Try	G
C	Leu	Pro	His	Arg	U
	Leu	Pro	His	Arg	C
	Leu	Pro	Gln	Arg	A
	Leu	Pro	Gln	Arg	G
A	Ile	Thr	Asn	Ser	U
	Ile	Thr	Asn	Ser	C
	Ile	Thr	Lys	Arg	A
	Met	Thr	Lys	Arg	G
G	Val	Ala	Asp	Gly	U
	Val	Ala	Asp	Gly	C
	Val	Ala	Glu	Gly	A
	Val	Ala	Glu	Gly	G

A typical cistron (gene) may consist of some 500 nucleotide bases. Mutations, then, are changes in the DNA message and they can be classified into two groups. One is base replacement and the other is structural change. The latter consists of deletion and insertion of one or more nucleotide bases as well as transposition and inversion of larger DNA segments. These tend to produce drastic effects on fitness.

In what follows, we shall be concerned mainly with the former type of change, that is, base replacement within a cistron. In terms of occurrence, base replacement seems to be the most common kind of mutation. A replacement of nucleotide base will lead to one of the following changes in the corresponding polypeptide chain: (1) no change occurs; this is known as synonymous mutation; (2) one of the amino acids is replaced; this has been called missense mutation; (3) polypeptide becomes incomplete in length due to one of the codons within the cistron changing into a terminating codon; this is known as chain terminating mutation.

Among the three types of mutations, synonymous mutations amount to roughly 25 per cent of cases and must be the least damaging type to the organism. It is possible that most of them are selectively neutral. Missense (amino acid substitution) mutations may also affect the biological activity of the polypeptide very little unless amino acid substitutions occur at the active sites that are crucial for the function of the molecule. This class of mutations is particularly important in the study of molecular evolution since they lead to changes that are found by comparative studies of amino acid sequences. Also, roughly one third of these mutations can be detected by electrophoresis. Chain terminating mutations amount to roughly five per cent of the cases and they must usually be very damaging to the function of the protein, so that they are readily eliminated by natural selection.

In considering population consequences of mutations at the molecular level, two very important points that we must keep in mind are: (1) the number of possible allelic states at any locus (cistron) is so large as to be practically infinite, and (2) the back mutation in the strict sense is so rare as to be negligible for any short interval of time. As an example, let us take the cistron coding for the α chain of the mammalian hemoglobins. This polypeptide consists of 141 amino acids, and so its cistron is made up of 423 nucleotide sites. This allows 4^{423} or some 10^{254} allelic states through base replacements alone, because each nucleotide site may be occupied by one of the four kinds of nucleotide bases. Thus, for any one of these alleles, there are 3×423 or 1269 other alleles that can be reached by a single step base replacement. The probability of returning to the original allele from any one of the latter alleles by further single base replacement is only one in 1269, assuming that all base replacements occur with equal probability.

This example brings to light the inadequacy of the conventional model in which a pair of alleles (usually denoted by A and a) are assumed with reversible mutations at comparable rates at each gene locus. Also, we must note that the mutation rate per nucleotide site must be several hundred times lower than the conventional figure of 10^{-5} usually assumed for a gene. Clearly, we need more realistic models to treat problems of population genetics at the molecular level. So far, two models have been devised to meet such a need. One is the model used by Kimura and Crow [32] who assumed that the number of possible allelic states at a locus is so large that each new mutant represents an allelic state not pre-existing in the population. Another is the model used by Kimura [28] who assumed that the number of nucleotide sites making up the genome is so large, while the mutation rate per site is so low, that whenever a mutant appears (within a limited evolutionary time period), it represents a mutation at a new site. These two models may be called "the model of infinite alleles" and "the model of infinite sites," respectively, [31]. The latter is especially useful when we consider the rate of mutant substitution in evolution.

Let us denote by k the rate of mutant substitution (incorporation) and define this as the long term average of a number of mutants that become fixed, per unit time (year, generation, and so forth) in the course of evolution. To avoid confusion, we must emphasize here that this rate is different from the rate at which an individual mutant increases its frequency in the population. We wait long enough so that the length of time taken for each substitution does not influence the result. Thus, as long as the average interval between occurrences of consecutive mutants (considering only those that are destined to reach fixation) is the same, two populations have the same k value.

Consider a panmictic population consisting of N diploid individuals and having the effective number N_e (for the meaning of N_e, see [10], p. 345). Let v be the mutation rate of a cistron per gamete per unit time and let u be the probability of ultimate fixation of an individual mutant. Then, the rate of mutant substitution at this locus is given by

$$k = 2Nvu, \tag{1}$$

because $2Nv$ new mutants appear per unit time in the population and the fraction u reach ultimate fixation. Here, the model of infinite sites is appropriate and we assume that mutants at different sites behave independently.

Using the formula for the probability of gene fixation by Kimura [24], and assuming that the mutant has selective advantage s in heterozygote and $2s$ in homozygote, we have

$$u = \frac{1 - \exp\{-2N_e s/N\}}{1 - \exp\{-4N_e s\}}. \tag{2}$$

Also, in this and in the subsequent formulae, we assume that each mutant is represented only once at the moment of appearance.

If the mutant has a definite selective advantage so that $4N_e s \gg 1$ but $s \ll 1$, this reduces approximately to

$$u = \frac{2N_e s}{N}. \tag{3}$$

On the other hand, if the mutant is selectively neutral such that $|4N_e s| \ll 1$, then, taking the limit $s \to 0$, we have

$$u = \frac{1}{2N}. \tag{4}$$

First, consider the neutral case since this leads to a very simple result. Substituting (4) in (1), we have

$$k = v, \tag{5}$$

namely, the rate of mutant substitution in evolution is equal to the mutation rate per gamete [27], [38], [8]. Note that this is independent of population size. On the other hand, if the mutant has a definite selective advantage, substituting (3) in (1), we have

$$k = 4N_e sv. \tag{6}$$

In this case k depends on N_e and s, as well as on v.

In addition to the probability of fixation and the rate of mutant substitution, we need to know the average length of time involved for each substitution. A general theory on this subject has been worked out by Kimura and Ohta [35] based on the diffusion models. The theory gives the average number of generations until fixation (excluding the cases of eventual loss), assuming the initial frequency of the mutant is p. In the special case of selectively neutral mutant, taking $p = 1/2N$, the average number of generations until fixation is approximately

$$\bar{t}_1 = 4N_e. \tag{7}$$

Namely, it takes, on the average, four times the effective population number for a selectively neutral mutant to reach fixation by random frequency drift. Actually, in this particular case of neutral mutants, the probability distribution of the length of time until fixation has been obtained [30]. For selected mutants, if they have selective advantage both in homozygotes and heterozygotes, the average length of time until fixation is shorter, while if they are overdominant, the time is prolonged.

When mutant substitutions are carried out by natural selection rather than by random drift, the population must stand the load of gene substitution. This was first pointed out by Haldane [19] in his paper entitled "The cost of natural selection." He showed that the sum of the fraction of selective deaths over all generations for one gene substitution is given by $D = -2 \log_e p$ if the mutant is semidominant in fitness and has initial frequency p. A remarkable point is that the cost D is independent of the selection coefficient s (>0). For example, if $p = 10^{-6}$, we have $D = 27.6$. If mutant substitutions proceed independently at the rate k per generation, the fraction of selective elimination per generation or selection intensity is $I = kD$. Haldane conjectured that the selection intensity involved in the standard rate of evolution is of the order of 0.1, so that $k = 1/300$ is a typical figure for the rate of gene substitution. He believed this explains the observed slowness of evolution (at the phenotypic level). To what extent species can stand the load of substitution depends on the reproductive excess that the species can afford. Haldane's result is based on the deterministic model that disregards the effect of random sampling of gametes in finite populations. The problem of obtaining the cost or the substitutional load in a finite population was solved by Kimura and Maruyama [34] using the diffusion models. For semidominant mutants having definite selective advantage, the load for one gene substitution is approximately

$$L(p) = -2 \log_e p + 2, \tag{8}$$

where we can put $p = 1/2N$ for molecular mutants. This approximation formula is valid under the same condition for which formula (6) on the rate of mutant substitution is valid. By comparing this with Haldane's formula, we note that in a finite population the cost is larger by 2, although this difference is usually relatively small.

Since Haldane's original formulation on this subject, a number of papers have been published criticizing it. However, Haldane based his principle of cost of natural selection on his deep consideration of the ecology of the living species, as well as on their genetics and evolution. In our opinion, nothing biologically significant has been added to Haldane's original papers [19], [20] by these criticisms. Meanwhile, further developments of Haldane's principle of the cost have been made by Kimura and Crow [33], Crow [9], Felsenstein [14], and Nei [50].

3. Neutral mutation-random drift theory as the first approximation

It is customary in the literature of molecular evolution to ascribe amino acid differences of homologous proteins simply to "accepted point mutations." From our standpoint, however, one amino acid difference is the result of at least one mutant substitution in which a rare molecular mutant increases its frequency and finally spreads to the whole species. Not only a large number of generations are involved for such a substitution, but also a significant amount of substitutional load is imposed if it is carried out by natural selection. Also, such a mutant represents a lucky minority among a large number of mutants that actually occur in the population. The majority of mutants are lost from the population within a small number of generations [15], [36]. It is often not realized that this applies not only to deleterious and selectively neutral mutants, but also to advantageous mutants unless the advantage is very large.

When the rate of molecular evolution is analyzed from such standpoint, we find two salient features in it. One is a remarkable uniformity for each molecule and the other is a very high rate for the total DNA.

The remarkable uniformity of the evolutionary rate is particularly evident when we analyze amino acid substitutions in hemoglobins among diverse lines of vertebrate evolution [29].

Figure 1 illustrates the amino acid differences of hemoglobin α between carp and four mammalian species together with their phylogeny. It may be seen that, with respect to this molecule, the mammals have diverged among themselves less than the group has diverged from carp. Taking into account the estimated time since divergence, we obtain a rate of amino acid substitution k_{aa} of approximately 10^{-9} per amino acid site per year. That the rate of substitution is proportional to chronological time rather than the number of generations becomes apparent when we compare the number of amino acid substitutions in the two lines, one leading to the mouse and the other leading to man from their common ancestor B. The former is estimated to be only about 50 per cent larger than the latter. If the rate of substitution is proportional to the number of generations, the number of substitutions in the line leading to the mouse should be larger by a factor of some 40 or so. Extensive calculations based on various comparisons involving β hemoglobin and lamprey globin as well as α hemoglobin reveal the remarkable uniformity of the rate of amino acid substi-

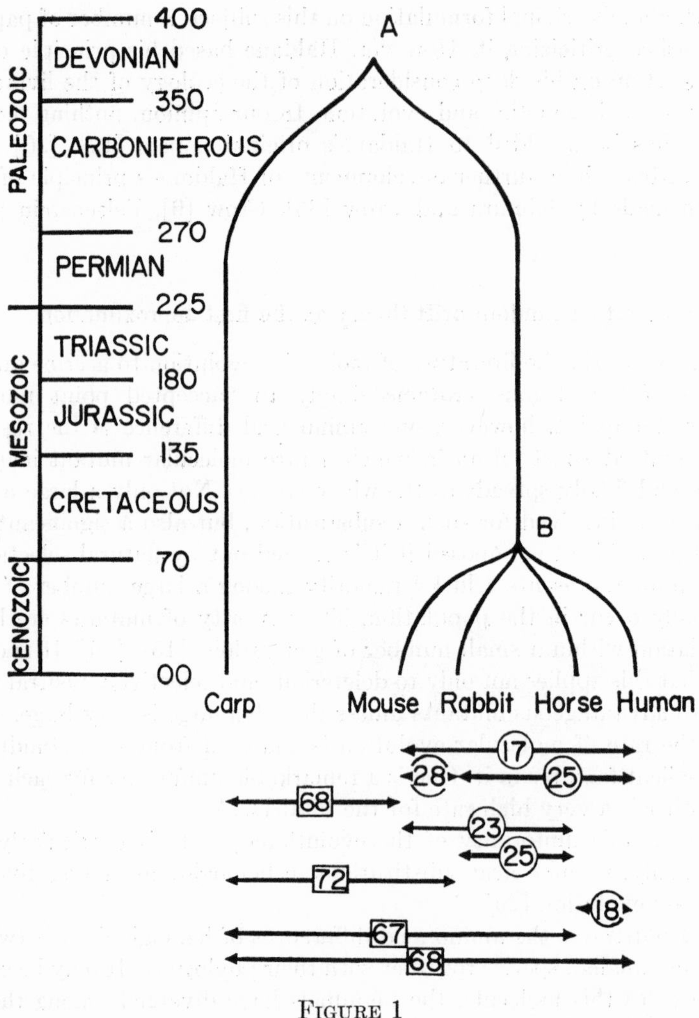

FIGURE 1

A phylogenetic tree of carp and four mammalian species together with geologic time scale. Numbers of different amino acid sites with respect to the α chain of hemoglobin are also given for various comparisons.

tution in vertebrate evolution, always giving approximately $k_{aa} = 10^{-9}$ per amino acid site per year. Particularly noteworthy, in this context, are the results obtained when the human β chain is compared with the human, mouse, rabbit, horse, bovine, and carp α chains. As shown in Table II, relative to the β chain these α chains have differentiated almost equally. It is remarkable that the two structural genes coding for the α and β chains, after their origin by duplication, have diverged from each other independently and to the same

TABLE II

FRACTIONS OF DIFFERENT AMINO ACID SITES WITH RESPECT TO COMPARISONS BETWEEN α AND β HEMOGLOBIN CHAINS

Comparison	Fraction of different sites
Human β-Human α	75/139
Human β-Mouse α	75/139
Human β-Rabbit α	79/139
Human β-Horse α	77/139
Human β-Bovine α	76/139
Human β-Carp α	77/139

extent, whether we compare α and β chains taken from the same organism (man) or from two different organisms (man and carp) which have evolved independently for some 400 million years.

The uniformity of the rate as well as the fortuitous nature of amino acid substitution in evolution are also evident in cytochrome *c*. In this case, however, the rate per year is about one third that of the hemoglobins (Figure 2). In Figure 2 note that, compared to the wheat, the various animals have differentiated to about the same extent. The estimates of the rate of amino acid substitutions are now available for several proteins [38], [11]. According to King and Jukes, the average rate is 1.6×10^{-9} per amino acid site per year, or using the terminology proposed by Kimura [29], it is 1.6 paulings (one pauling standing for the substitution rate of 10^{-9} per site per year).

Next, let us examine the second characteristic, namely, the very high overall rate of mutant substitution. In mammals (including man), the total number of nucleotide sites making up the haploid DNA is roughly 3×10^9. Not all of them may code for proteins but as long as they are self reproducing entities, they are members of the genome in a broad sense. There are also "repeating sequences" whose function is not understood at the moment. In some species such as in the mouse they amount to as much as 40 per cent of the total DNA [2]. In the following discussion we will disregard such sequences, for this will not alter our conclusion.

If we take 1.6×10^{-9} as the average substitution rate per amino acid site (per year) in cistrons, this corresponds roughly to the substitution rate of 6.3×10^{-10} per nucleotide site in which synonymous mutations are taken into account. Extrapolating this to the total nucleotide sites, the rate of mutant substitution amounts to roughly two per haploid DNA per year. Then, for mammals which take three years for one generation, the rate of mutant substitution amounts to some half dozen per generation.

Comparing this figure with Haldane's earlier estimate of 1/300, we note that it is unbelievably high. In fact, if the majority of such substitutions are carried out by natural selection and if each substitution entails a load of about 30, the total load per generation is 180.

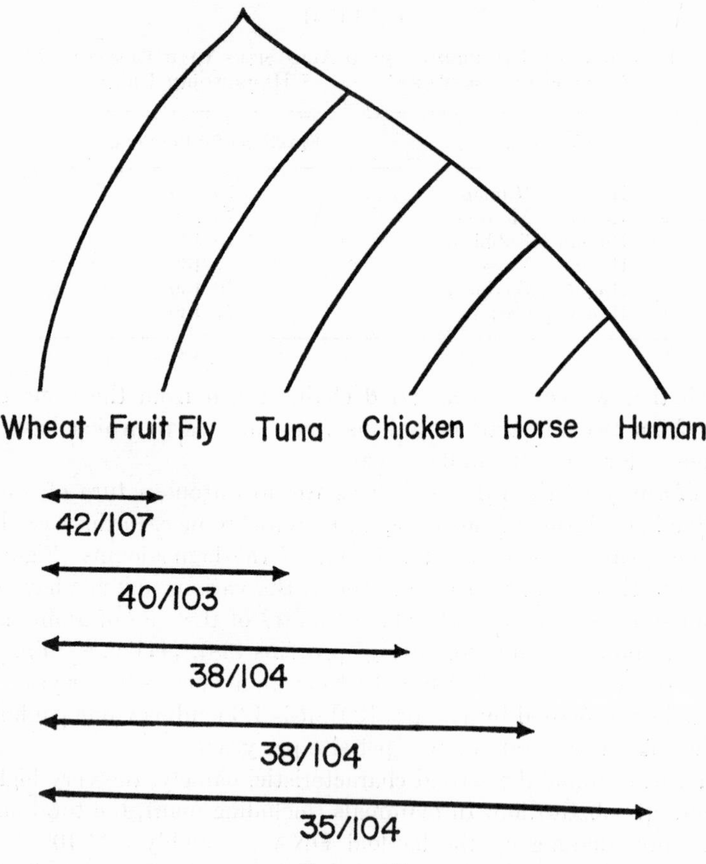

FIGURE 2

A phylogenetic tree and fraction of different amino acid sites with respect to cytochrome *c* of wheat and various animals.

This means that in order to carry out independent gene substitutions at this rate and still maintain the same population number, each parent must leave e^{180} or about 10^{78} offspring for only one of them to survive. It is obvious that no mammalian species can stand such a heavy load of substitution. This reasoning led one of us [27] to put forward for the first time the neutral mutation-random drift theory as the main cause of molecular evolution. For neutral mutants, there is no selection, and hence, no genetic load.

Since then, models assuming truncation selection were proposed to avoid the heavy load of substitution [63], [42]. With very small (yet effective) selective advantage, mutant substitution can be carried out at a high rate in such models without excessive substitutional load. However, these models encounter the same kind of difficulty as the ordinary selection model when we try to explain the constancy of the evolutionary rate of cistrons. In addition, the assumption of a very small selective advantage requires a very high production of such mutants per generation in order to carry out substitutions at a high rate. The main reason for this is that the probability of fixation of such mutants is very low. To see this point more quantitatively, let us consider a mammalian species having an effective population number of 5×10^4 with actual population number possibly much larger. This is a realistic value for effective population size of mammals having a large body size and a generation time of three years.

Let us suppose that the selective advantage of the mutants is $s = 0.001$. Then, putting $k = 6$, $N_e = 5 \times 10^4$, and $s = 10^{-3}$ in formula (6) and solving for v, we get $v = k/4N_e s = 3/100$. This means that in order to carry out mutant substitution at the rate of six per generation, the mutation rate for such advantageous mutations must be three per cent per gamete. This is comparable to the total rate for lethal and semilethal mutations per gamete. If k is larger, or if either N_e or s is smaller, a higher value for v is required. We believe that such a high rate of production of advantageous mutations is unlikely. These models even seem to contradict the principle of adaptive evolution, since the models require that advantageous mutations occur under rapidly changing environments and also under constant environments at an equal rate (provided that N_e is the same).

The contrast between the hypothesis of neutral substitution and that of adaptive substitution becomes quite pronounced when we try to interpret a large difference in evolutionary rate between fibrinopeptides and histones. It is estimated that fibrinopeptides evolved some 1500 times as fast as histones [45]. According to the neutral theory, a majority of amino acid substitutions in fibrinopeptides is selectively neutral, while in histones virtually all mutations are deleterious. On the other hand, under the adaptive substitution theory, as pointed out by Dr. Sewall Wright (personal communication), one might have to make the following interpretation: one particular amino acid sequence in histone is so perfect that any mutation is deleterious, irrespective of changes in the rest of the organism, while in fibrinopeptides there is so much functional dependence on other evolving molecules that mutations have been 1500 times as likely to be favorable compared to histone during the course of evolution. This view appears to encounter difficulty in explaining the uniformity of substitution rate. Also, it appears to contradict the fact that the function of fibrinopeptides is nonspecific [53].

An additional example that is instructive in this context is the rapid evolutionary change observed in the middle portion of the proinsulin molecule. This molecule is a precursor of insulin and consists of three parts, A chain, B chain, and a middle segment connecting the two. When active insulin is formed, the middle segment (amounting to roughly one third of the total in length) is removed. According to Nolan, Margoliash, and Steiner [52], bovine proinsulin differs from porcine proinsulin with respect to the middle segment by about 50 per cent in structure, but only by two residues with respect to the remaining portion. Assuming that bovine and pig lines were separated about 8×10^7

years ago (note that bovine, pig, and human have differentiated from each other to the same extent with respect to hemoglobin α and β chains), we get about $k_{aa} = 4.4 \times 10^{-9}$ as the rate of amino acid substitution per year in this middle segment. This is not very different from the rate in fibrinopeptides. On the other hand, the corresponding evolutionary rate is estimated to be about $k_{aa} = 0.4 \times 10^{-9}$ for insulin A and B [45]. It is interesting that the rate of substitution is very high in this middle segment which appears to be functionally unimportant.

This example shows clearly that the rate of amino acid substitution in evolution can be very different in different parts of a molecule. Recently, Fitch and Markowitz [16] carried out a detailed statistical analysis of evolutionary change of cytochrome *c* and arrived at the important conclusion that in this molecule only about 10 per cent of the amino acid sites (codons) can accept mutations at any moment in the course of evolution. They called such codons the concomitantly variable codons. This method of analysis was applied to the hemoglobins by Fitch [17], who found that the number of concomitantly variable codons ("covarions" in short) is about 50 in the mammalian alpha hemoglobin genes. Furthermore, he has noted the remarkable fact that if the rates of mutant substitution in evolution are calculated only on the basis of covarions, hemoglobin, cytochrome *c*, and fibrinopeptide A are all evolving at roughly the same rate.

Therefore, we believe that the neutral mutation-random drift theory is much more plausible, as a scientific hypothesis, than the conventional positive selection model in explaining the great majority of amino acid substitutions in evolution.

The basic idea of our neutral mutation-random drift theory as succinctly reviewed by Maynard Smith [43] is as follows. At each cistron, a large fraction of mutations are harmful and they will be eliminated by natural selection. A small but significant fraction is selectively neutral and their fate is controlled by random frequency drift. The main cause of molecular evolution is thus random fixation of neutral mutants. The fraction of such neutral mutations differs from cistron to cistron depending on the functional requirement of the protein molecule. Favorable mutations may occur, and although they are extremely important in adaptive evolution, they are so rare that they influence very little the estimates of the rate of amino acid substitution.

4. Nearly neutral mutations and constancy of evolutionary rate

In the foregoing sections we regarded mutants as neutral if their selection coefficients s are so small that $|4N_e s|$ is much smaller than unity. It is likely that in reality the borderline between neutral and deleterious mutations is not distinct but, rather, continuous. Clearly, many amino acid substitutions in proteins are deleterious, and we need to clarify the relationship between neutral and deleterious amino acid substitutions in molecular evolution. But since the overall fitness of a mutant consists of a great many components, it is hard to believe that neutral and deleterious mutations are distinctly separated.

For nearly neutral mutations having a slight advantage or disadvantage (but not necessarily $|4N_e s| \ll 1$), the rate of evolutionary substitution is determined not only by the mutation rate but also by such factors as effective population number N_e and selection coefficient of the mutant s. It is convenient, for the following treatment, to express the fixation probability given by formula (2) as $u(s)$ indicating its dependence on s. Figure 3 illustrates the fixation probability as a function of $4N_e s$. For example, as compared with a completely neutral mutant, a disadvantageous mutant having $4N_e s = -2$ has about one third probability of eventual fixation, while a mutant with $4N_e s = -1$ has about three fifths probability of fixation.

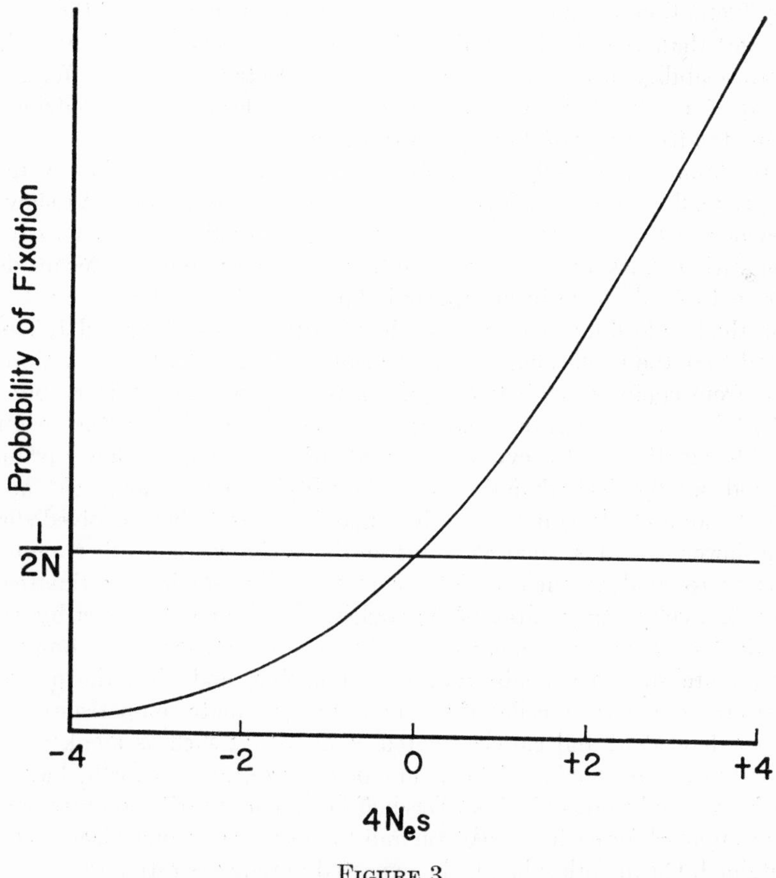

FIGURE 3

Probability of ultimate fixation of a mutant as a function of $4N_e s$, where $N_e s$ = product of effective population size and selective advantage.

When we consider the rate of nucleotide substitution in evolution, we must take into account all mutations that have finite chance of eventual fixation. Thus, the evolutionary rate is the sum of the product of the mutation rate and the corresponding fixation probability of the mutant over all possible kinds of mutations, that is,

$$k = 2N \int_{-4/N_e}^{4/N_e} u(s)v(s)\, ds, \tag{9}$$

where $u(s)$ and $v(s)$ are the mutation rate and fixation probability as functions of selection coefficient s, and we take into account all the mutants whose $N_e s$ ranges from -4 to $+4$. This integral has been called the "effective neutral mutation rate" by Ohta and Kimura [57].

Actually, however, the selection coefficient may not remain constant over a very long period, but may fluctuate from generation to generation due to random fluctuation in environmental conditions. In such a case, if the variance V_s is larger than the absolute value of the mean selection coefficient $|\bar{s}|$, the fixation probability of the mutant, even if it is selected, does not differ greatly from that of a neutral mutant [54]. This must, at least partly, contribute to increasing the frequency of "nearly neutral mutations."

Let us examine more fully the problem of whether the evolutionary rate of cistrons is really proportional to the simple chronological time. As shown in the previous section, the evolutionary rate of individual cistron is mostly uniform for various lines over vast geologic time. However, some significant deviations from this rule have been reported. For example, insulins evolved much faster in the line leading to guinea pig than in other lines. King and Jukes [38] estimated that the evolutionary rate of insulins is 5.31×10^{-9} per amino acid per year from comparisons between guinea pig and other organisms, but only 0.33×10^{-9} from comparisons among the other mammals. Another example, although less distinct, is the evolutionary rate of hemoglobins in lower primates. As pointed out by Zuckerkandl and Pauling [67], Buettner-Janusch and Hill [3], and Nolan and Margoliash [51], hemoglobins seem to have evolved slightly faster in lower primates, such as lemur and tree shrew, than other mammals.

In order to analyze such variations statistically, we have estimated the variance in evolutionary rates of hemoglobins and cytochrome c by taking independent comparisons among relatively close organisms. The comparisons among remote species available are not numerous, and also, the deviations appear to be somehow cancelled if one makes very remote comparisons.

Seven independent and closely related comparisons such as monkey-mouse, human-rabbit, horse-bovine (fetal), human δ sheep, and so forth, have been chosen for β type hemoglobin from Dayhoff [11]. The substitution rate per year k_{aa} was computed for each comparison and the variance among seven k_{aa} values was obtained. On the other hand, the expected variance is estimated by

$$\sigma^2_{k_{aa}} = \frac{\bar{p}_d}{4\bar{T}^2(1 - \bar{p}_d)\bar{n}_{aa}}, \tag{10}$$

where $\bar{p}_d$ and $\bar{n}_{aa}$ are, respectively, the averages of p_d (fraction of different amino acids) and n_{aa} (total number of amino acid sites per protein compared), and $\tilde{T}$ is the harmonic mean of the time since divergence T. Similarly, five independent comparisons were chosen for α type hemoglobin and seven comparisons for cytochrome c to calculate observed and expected variances in the rates of substitution.

Whether the observed variance is significantly larger than expected was tested by the F test. It turned out that the F value is highly significant for β type hemoglobins and cytochrome c, but not for the hemoglobin α. The estimated time since divergence may not be accurate, and this will inflate the observed variance somewhat, but this effect should not be very large. For details see [57]. We conclude, then, that the variations in evolutionary rates among highly evolved animals are sometimes larger than expected from chance. However, the uniformity of the evolutionary rate is still valid as a first approximation.

In the remainder of this section, we shall examine theoretically the problem of constancy of the rate of substitution per year with respect to nearly neutral mutations. Let us denote by g the generation time in years and by v the mutation rate per gamete per generation. Then, for completely neutral mutations, the evolutionary rate per year is,

$$k_1 = v/g. \tag{11}$$

If most of the gene substitutions are selectively advantageous, then

$$k_1 = 4N_e sv/g. \tag{12}$$

On the other hand, if most of the amino acid substitutions are slightly deleterious ($s < 0$) such that $|N_e s| < 1$, then

$$k_1 \propto v/N_e g, \tag{13}$$

since the fixation probability is negatively correlated with $|N_e s|$ as in Figure 3.

If v/g (mutation rate per year) is really constant among various lines of higher organisms, the simple neutral theory of the previous section is appropriate to explain the observed uniformity of the evolutionary rate. On the other hand, if v, the mutation rate per generation, rather than v/g, is generally constant, slightly deleterious mutations are more likely to be the main source of gene substitution as suggested by Ohta and Kimura [57]. They pointed out that organisms having a larger body size tend to have a smaller population number and longer generation time and vice versa. Hence, the value of $N_e g$ in formula (13) could be nearly constant among various lines of organism.

Also, from formula (12), we can easily see that the adaptive gene substitution will create significant variations in the evolutionary rate, if N_e and g are inversely correlated. Thus, we conclude again that mutant substitutions at the molecular level are mostly selectively neutral or nearly so (very slightly disadvantageous) and that Darwinian (definitely adaptive) mutant substitutions should represent only a minor fraction.

5. Molecular biometry of amino acid composition

The amino acid composition of proteins and the base composition of DNA are products of molecular evolution. Since the nucleotide sequence of a gene specifies the amino acid sequence of a protein; in other words, because of "colinearity" between the two sequences, it is clear that the amino acid composition of proteins reflects the base composition of DNA and vice versa.

As early as 1961, when the three letter coding system was not yet exactly known, Sueoka [62] noticed a correlation between the frequency of a particular amino acid and base frequency, such as between alanine and G-C content. Also Jukes [22] estimated the base composition of hemoglobin genes from their amino acid composition. Since then, the DNA code words have been completely deciphered, and we are able to carry out such analyses on a firmer basis.

Kimura [26] showed that average amino acid composition of proteins can be predicted fairly well from the knowledge of the genetic code and by assuming random arrangements of the four kinds of bases within a cistron. This approach was improved by King and Jukes [38] who estimated frequencies of four bases (A, U, G, C) directly from the amino acid composition and then used these frequencies for their calculation of the expected amino acid composition. With this improvement, the overall agreement between the observed and the expected compositions becomes much better. The only exception is arginine whose observed frequency is only half as high as expected under random arrangement of bases.

The method was further refined by Ohta and Kimura [55], but they arrived at the same conclusion as King and Jukes with respect to the deficiency of arginine. In addition, they developed a method of estimating base frequencies of individual cistron using data on amino acid composition of its protein. At present it is not feasible by means of biochemical method to measure directly the base composition of individual genes. Thus, the refined method of estimating base frequencies of individual cistron as developed by Ohta and Kimura has some use in the study of molecular evolution.

Now, by inspection of the code table (Table I), one finds that the base composition at the first and second positions of the codon could be estimated from the amino acid composition. However, with respect to the third position this appears to be impossible because of high "degeneracy." (For example, four codons having G in both the first and second positions always code for Gly independent of the third position.)

Let us examine these points in more detail. Consider the second position, since the estimation is simplest in this position. The relative frequency of A (adenine) can be estimated without error by adding the relative frequencies of tyrosine, histidine, glutamine, asparagine, lysine, aspartic acid, and glutamic acid; in other words, by adding amino acids occurring in the column under letter A in the table (but disregarding two terminating codons). In symbols,

$$(14)\quad A2 = \frac{1}{n}([\mathrm{Tyr}] + [\mathrm{His}] + [\mathrm{Gln}] + [\mathrm{Glu}] + [\mathrm{Asp}] + [\mathrm{Asn}] + [\mathrm{Lys}]),$$

where n is the total number of amino acids composing a protein, and [Tyr], and so forth, denote the number of tyrosine, and so forth, within the protein. Similarly, the sum of the frequencies of G and C (usually called "G-C content") can be estimated without error by

$$(15)\quad G2 + C2 = \frac{1}{n}([\mathrm{Ser}] + [\mathrm{Pro}] + [\mathrm{Thr}] + [\mathrm{Ala}] + [\mathrm{Cys}] + [\mathrm{Trp}] + [\mathrm{Arg}] + [\mathrm{Gly}]).$$

The frequency of U (uracil) is then given by $1 - A2 - G2 - C2$.

Separation of G2 from C2, however, can only be achieved indirectly, since serine contributes both to C and G. One way of achieving this is through iteration. Of the six codons of serine, four contribute to C and the remaining to G. By assuming that base frequencies at the first and second positions of codons are statistically independent, we estimate the frequency of the group of four serine codons having C in their second position by

$$(16)\quad [\mathrm{Ser}]_1 = \frac{2(U1 \times C2)}{2(U1 \times C2) + (A1 \times G2)}[\mathrm{Ser}],$$

where U1 is the frequency of U in the first position. The frequency of a group of two codons having G in their second position is then

$$(17)\quad [\mathrm{Ser}]_2 = [\mathrm{Ser}] - [\mathrm{Ser}]_1.$$

This type of separation is required to estimate all the base frequencies in the first position. For example, leucine enters both in U and C and we must separate six codons coding for leucine into two groups. The frequency of the group of two codons having U in the first position may be estimated by

$$(18)\quad [\mathrm{Leu}]_1 = \frac{U1}{U1 + 2C1}[\mathrm{Leu}].$$

Using this type of separation method and iteration by a computer, we have estimated the base frequencies in the first position, as well as G and C in the second position. This is one place where professional statisticians can develop a much better method of estimation.

At any rate, from the analysis of 17 vertebrate proteins, we obtain results that indicate that the base composition is generally different in the first and second position of the codon. Also, the results on the second position showed clearly that the base composition of the informational strand of DNA (that is, the strand actually used for transcription) is different from that of its complementary strands. How such differences evolved, we believe, is an interesting and puzzling problem that needs further investigation. Previously, King and Jukes [38] reported that G + A is not generally equal to C + U, indicating some nonrandomness in base arrangement. Our analyses confirmed their results, and in particular, we found the relation A > U (in terms of RNA code). Also, the frequency distribution of G-C content at the second position shows that its mean is approximately 42 per cent, agreeing well with the results of chemical analysis of vertebrate nuclear DNA. However, its variance is about two to three times larger than that expected under random arrangement of bases, suggesting some nonrandomness.

Such nonrandomness in base composition and arrangement must be, at least partly, due to the functional requirement of each cistron (gene). In other words, cistrons must keep their own characteristic base composition for their function. Molecular evolution by nucleotide substitution must proceed without impairing such functional requirements, otherwise mutation cannot be tolerated by natural selection (that is, cannot be selectively neutral). It is natural to think, then, that there should be some restriction or nonrandomness in the pattern of amino acid substitutions in evolution.

Actually, Clarke [5] and Epstein [13] reported some nonrandomness in amino acid substitutions. They pointed out that the substitutions between similar amino acids are more likely to occur than those between dissimilar ones. Clarke considered this fact to indicate Darwinian (that is, positively selected) substitutions. However, a much more plausible interpretation is that it implies non-Darwinian (that is, neutral) substitutions, as claimed by Jukes and King [23]. They pointed out that important adaptive changes should contain substitutions between dissimilar amino acids.

Let us investigate the pattern of amino acid substitution in the evolutionary change of proteins. As we have mentioned already, the average amino acid composition of proteins can be predicted fairly well by the knowledge of the genetic code and by assuming random arrangement of nucleotide bases within the genes. However, there are some significant deviations. In particular, the arginine content is much smaller than expected. In order to determine the cause of arginine deficiency, Ohta and Kimura [59] used the transition probability matrix method. It consists of a 20×20 matrix giving transition probabilities from any one of the 20 amino acids to any other during a unit length of time in evolution. It is desirable to construct such a matrix based on a large body of data. However, as a preliminary attempt, we used the "mutation probability matrix" of Dayhoff, Eck, and Park [12]. This matrix was constructed by counting the number of "accepted point mutations" among closely related sequences, amounting to 814 mutations, taken from cytochrome *c*, globins, virus coat proteins, chymotrypsinogen, glyceraldehyde 3-phosphate dehydrogenase, clupeine, insulin, and ferredoxin (see [12], p. 75 for details).

By comparing this matrix M with the corresponding random matrix R made by allowing only single random base substitutions (using the code table), a clear difference was recognized. The difference reflects the differential survival of amino acid substitutions. Most noteworthy is the deficiency of evolutionary input of arginine, and this indicates that mutation to arginine is largely selected against in evolution of proteins. It is interesting to note, in this context, that arginine is substituted not infrequently for other amino acid among hemoglobin variants found in human populations. The eigenvector of M reflects the approximate amino acid composition used for the construction of this matrix [12]. It also represents the equilibrium composition. In order to compare the two equilibrium compositions corresponding to M and R, we also computed the eigenvector of R. Figure 4 illustrates the relationship between the two eigenvectors.

We also examined the hypothesis that the actual amino acid composition of proteins represents a quasi-equilibrium of neutral mutations. We compared the

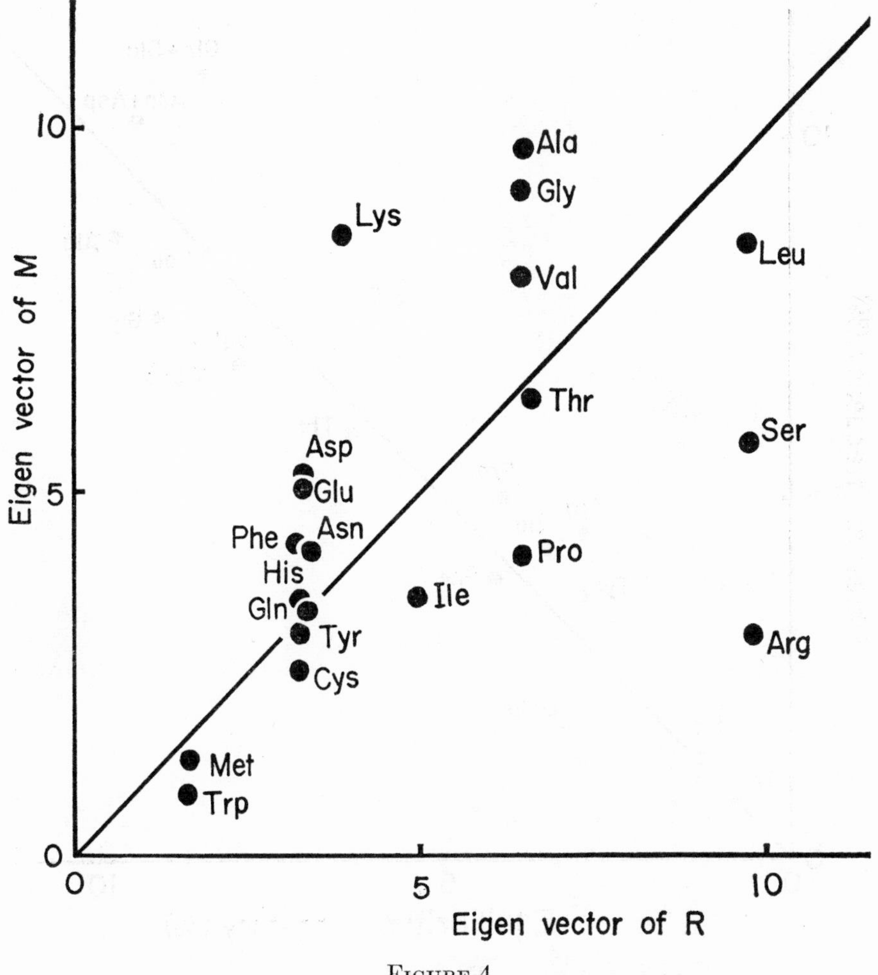

FIGURE 4

Graph showing relationship between eigenvector of M and that of R with respect to amino acid composition of an average protein.

eigenvector of M with the observed amino acid composition. For the observed values we used the average values obtained from Smith's data [61]. He compiled amino acid compositions of 80 proteins taken from various organisms including vertebrates, bacteria, and viruses. Figure 5 illustrates the relationship between the observed and the equilibrium compositions. Agreement between the two seems to be satisfactory, and there is no marked discrepancy with respect to Arg.

Through these analyses we have been led to the view that the amino acid composition of proteins is determined largely by the existing genetic code and

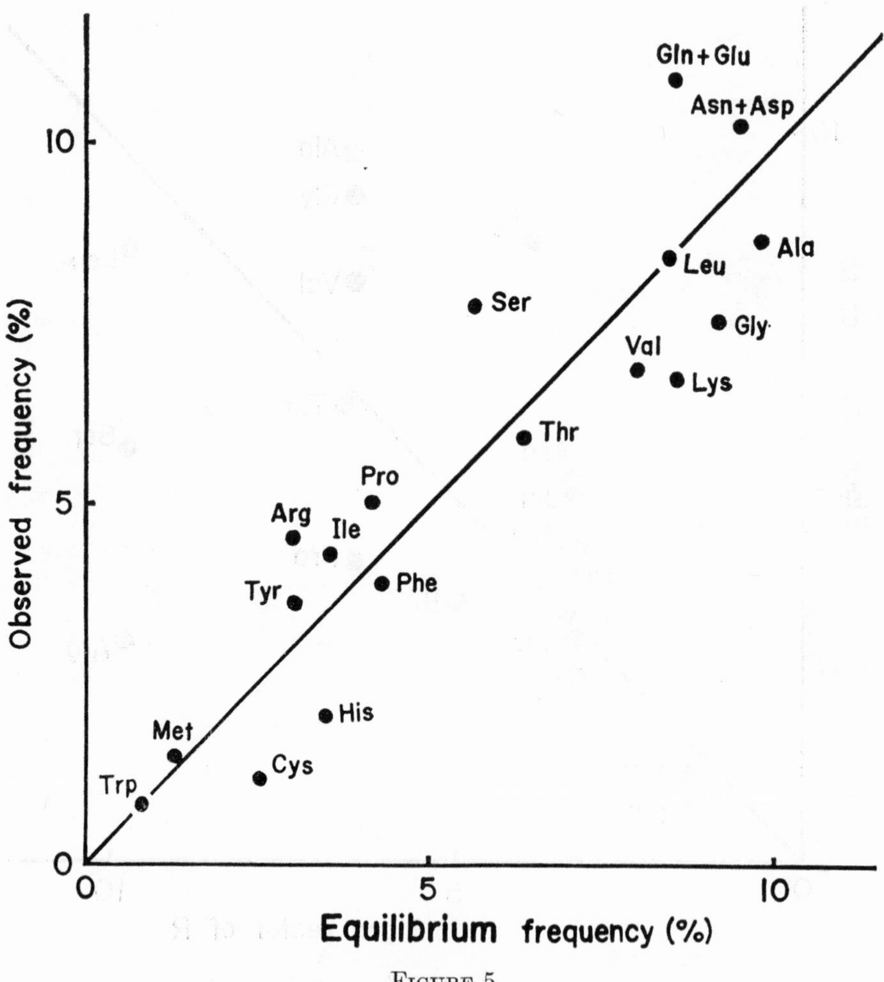

FIGURE 5

Graph showing relationship between observed and equilibrium amino acid compositions.

the random nature of base changes in evolution. Small but significant deviations from such expectation can be accounted for satisfactorily by assuming selective constraint of amino acid substitutions in evolution.

6. Evolutionary change of genes and phenotypes

Since Darwin, a great deal has been written about evolution at the phenotypic level. However, it is only in the past few years that we have started to understand evolution at the molecular level.

In the field of evolutionary genetics, consensus appears to have been reached among leading evolutionists of the world (except possibly Sewall Wright) that natural selection is omnipotent and is the most prevailing factor for evolutionary change. This orthodox view (formed under the dominating influence of R. A. Fisher) also asserts that neutral mutant genes are very rare if they ever exist, and random genetic drift is negligible in determining the genetic structure of biological populations, except possibly for the case of the colonization of a new habitat by a small number of individuals—the founder effect (see Mayr [44]).

The neutral mutation-random drift theory, therefore, would be an open challenge to this view if it were concerned with the same subject. However, we must realize that we are concerned here with changes in DNA base (and, therefore, amino acid) sequences that may have no clear cut and straightforward correspondence with phenotypic change. It is with respect to the level of information macromolecules that random drift plays a dominant role in large population as well as small.

In the present paper, we have mainly concerned ourselves with amino acid substitutions, but in addition to these, duplication of DNA segments must be very important in evolution, although they occur much less frequently. The main evolutionary significance of duplication lies in the fact that it allows one of the duplicated segments to accumulate mutations and acquire a new function, while another segment retains the old function necessary to survive through the transitional period. This idea, which goes as far back as the great *Drosophila* workers of the Morgan school (see Bridges [1]), has recently been much extended by Ohno [53] in relation to vertebrate evolution. It is likely that many evolutionary innovations owe their origin to gene duplications.

In addition to such positive change, duplication must have caused a great deal of degeneration of genetic material due to accumulation of mutants by random drift that would have been harmful before duplication but have become neutral after duplication [49], [58]. This should be taken into account when we consider functional organization of genetic material in higher organisms. It is a well-known fact that if we estimate the total number of genes in man by simply dividing the total number of nucleotide pairs per haploid DNA (some 3×10^9) by the average number of nucleotide pairs per cistron (assuming around 500), a very large number, amounting to several million is obtained (see [47], [65]). On the other hand, considerations based on classical genetic analyses and mutational load lead us to the estimated gene number of some 3×10^4 [48]. Ohta and Kimura [58] claim that such discrepancy can be understood by noting that, as a whole, degeneration after duplication is more frequent than progressive organization so that a large fraction of DNA is noninformational in the sense that the base arrangement therein is irrelevant to the organism's life.

Thus, the correspondence between phenotype and nucleotide sequence must, in general, be an extremely complicated one. It is possible that both loss of function and acquisition of a new function must occur alternately at the same site, and the adjustments involved must be intricate beyond our comprehension. For example, some mutants which were originally neutral and fixed by random drift might later become essential for the organism, after a series of gene substitutions by natural selection, whose very advantage presupposes the existence of originally neutral mutants.

The neutral mutation-random drift theory allows us to make a number of predictions, so we shall present some of them. First, through random fixation of selectively neutral mutants, genes of "living fossils" must have undergone as many nucleotide (and amino acid) substitutions as the corresponding genes in more rapidly evolving species. Thus, underneath the constant morphology that has been kept unchanged by incessant action of natural selection for hundreds of millions of years, a great flow of neutral or nearly neutral mutants transforms the base sequence of genes tremendously in any organism. By studying a suitable molecule and using the observed changes as an "evolutionary clock," and analyzing its information by sensitive biometrical methods with the aid of computers, we hope to understand more thoroughly the early histories of life and living organisms. Also, the method of "minimum evolution" by Cavalli-Sforza and Edwards (see [4]) should have more relevance at the molecular than phenotypic level, as exemplified by the MBDC (minimum base difference per codon) method of Jukes [22].

Second, we should find in every species (with sufficiently large population size), ample evidence for molecular evolution in progress in the form of protein polymorphism. Although accompanied by a spurious effect of balancing selection due to associative overdominance, we believe that polymorphic alleles themselves are selectively neutral [56]. In this sense, protein polymorphisms are transient rather than permanent. However, for our ephemeral existence, they are almost permanent, persisting millions of years before disappearing. This view, first put forward by one of us [27], has been revised and extended by Kimura and Ohta [37]. One remark that we would like to make here is that the alternative model assuming overdominance plus truncation selection such as the one proposed by Sved, Reed, and Bodmer [64], although widely accepted at the moment, contains several difficulties. First, there is no assurance that natural selection mimics artificial selection in such a way that the number of heterozygous loci is counted and population is sharply divided into two groups based on such loci ([10], p. 307). Secondly, this model predicts that the rate of inbreeding depression decreases as the inbreeding coefficient increases, but this is contrary to most observational results. Thirdly, according to recent work of Mukai and Schaffer [46], the broad sense heritability H^2 with respect to fitness is very low. They extracted chromosomes from a natural population of *Drosophila melanogaster*, and by making random heterozygotes, obtained $H^2 = 0.002$. Then, they showed by simulation experiments on a computer that with such low heritability, truncation selection at the phenotypic level is not effective enough to explain a large occurrence of isozyme polymorphisms without creating a considerable magnitude of genetic load.

Despite several criticisms (for example, [60], [6]), we believe that evidence is growing in our favor in support of the neutral mutation-random drift theory of molecular evolution and polymorphism. Mather [41], commenting on the neutral theory, says that its acceptability depends on the "credibility of selective neutrality." However, the history of the development of quantum mechanics amply indicates that predictability and consistency are very much more important than credibility for a scientific theory to be valid. In fact, it shows that an apparently incredible theory can still be successful in science. We believe that our theory has now reached the stage where it should be put to thorough, critical test to determine its validity.

We would like to thank Doctors K. Mayeda, J. L. King, T. Jukes, and S. Wright for valuable comments and criticisms.

REFERENCES

[1] C. B. Bridges, "Genes and chromosomes," *The Teaching Biologist*, Nov. (1936), pp. 17–23.

[2] R. J. Britten and D. E. Kohne, "Repeated segments of DNA," *Sci. Amer.*, Vol. 222 (1970), pp. 24–31.

[3] J. Buettner-Janusch and R. L. Hill, "Evolution of hemoglobin in primates," *Evolving Genes and Proteins* (edited by V. Bryson and H. J. Vogel), New York, Academic Press, 1965, pp. 167–181.

[4] L. L. Cavalli-Sforza and A. W. F. Edwards, "Phylogenetics analysis: Models and estimation procedures," *Amer. J. Hum. Genet.*, Vol. 19 (1967), pp. 233–257.

[5] B. Clarke, "Selective constraints on amino acid substitutions during the evolution of proteins," *Nature*, Vol. 228 (1970), pp. 159–160.

[6] ———, "Darwinian evolution of proteins," *Science*, Vol. 168 (1970), pp. 1009–1011.

[7] F. Crick, *Of Molecules and Men*, Seattle and London, University of Washington Press, 1967.

[8] J. F. Crow, "Molecular genetics and population genetics," *Proceedings of the Twelfth International Congress on Genetics*, Idengaku, Fukyukai, Mishima, Shizuoka-ken, Japan, 1969, Vol. 3, pp. 105–113.

[9] ———, "Genetic loads and the cost of natural selection," *Mathematical Topics in Population Genetics* (edited by K. Kojima), Berlin-Heidelberg, Springer-Verlag, 1970, pp. 128–177.

[10] J. F. CROW and M. KIMURA, *An Introduction to Population Genetics Theory,* New York, Harper and Row, 1970.
[11] M. O. DAYHOFF, *Atlas of Protein Sequence and Structure,* Vol. 4, Silver Spring, Md., National Biomedical Research Foundation, 1969.
[12] M. O. DAYHOFF, R. V. ECK, and C. M. PARK, "A model of evolutionary change in proteins," *Atlas of Protein Sequence and Structure,* Vol. 4, Silver Spring, Md., National Biomedical Research Foundation, 1969, pp. 75–83.
[13] C. J. EPSTEIN, "Non-randomness of amino acid changes in the evolution of homologous proteins," *Nature,* Vol. 215 (1967), pp. 355–359.
[14] J. FELSENSTEIN, "On the biological significance of the cost of gene substitution," *Amer. Nat.,* Vol. 105 (1971), pp. 1–11.
[15] R. A. FISHER, *The Genetical Theory of Natural Selection,* Oxford, Clarendon Press, 1930.
[16] W. F. FITCH and E. MARKOWITZ, "An improved method for determining codon variability in a gene and its application to the rate of fixation of mutations in evolution," *Biochem. Genet.,* Vol. 4 (1970), pp. 579–593.
[17] W. F. FITCH, "Evolutionary variability in hemoglobins," *Haematologie und Bluttransfusion* (edited by H. Martin), Munich, J. F. Lehmanns Verlag, in press.
[18] J. B. S. HALDANE, "A mathematical theory of natural and artificial selection, Part 1," *Trans. Camb. Phil. Soc.,* Vol. 23 (1924), pp. 19–41.
[19] ———, "The cost of natural selection," *J. Genet.,* Vol. 55 (1957), pp. 511–524.
[20] ———, "More precise expressions for the cost of natural selection," *J. Genet.,* Vol. 57 (1960), pp. 351–360.
[21] ———, "A defense of beanbag genetics," *Perspect. Biol. Med.,* Vol. 7 (1964), pp. 343–359.
[22] T. H. JUKES, "Some recent advances in studies of the transcription of the genetic message," *Advan. Biol. Med. Phys.,* Vol. 9 (1963), pp. 1–41.
[23] T. H. JUKES and J. L. KING, "Deleterious mutations and neutral substitutions," *Nature,* Vol. 231 (1971), pp. 114–115.
[24] M. KIMURA, "Some problems of stochastic processes in genetics," *Ann. Math. Statist.,* Vol. 28 (1957), pp. 882–901.
[25] "Diffusion models in population genetics," *J. Appl. Probability,* Vol. 1 (1964), pp. 177–232.
[26] ———, "Genetic variability maintained in a finite population due to mutational production of neutral and nearly neutral isoalleles," *Genet. Res.,* Vol. 11 (1968), pp. 247–269.
[27] ———, "Evolutionary rate at the molecular level," *Nature,* Vol. 217 (1968), pp. 624–626.
[28] ———, "The number of heterozygous nucleotide sites maintained in a finite population due to steady flux of mutations," *Genetics,* Vol. 61 (1969), pp. 893–903.
[29] ———, "The rate of molecular evolution considered from the standpoint of population genetics," *Proc. Nat. Acad. Sci. U.S.A.,* Vol. 63 (1969), pp. 1181–1188.
[30] ———, "The length of time required for a selectively neutral mutant to reach fixation through random frequency drift in a finite population," *Genet. Res.,* Vol. 15 (1970), pp. 131–133.
[31] ———, "Theoretical foundation of population genetics at the molecular level," *Theor. Pop. Biol.,* Vol. 2 (1972), pp. 174–208.
[32] M. KIMURA and J. F. CROW, "The number of alleles that can be maintained in a finite population," *Genetics,* Vol. 49 (1964), pp. 725–738.
[33] ———, "Natural selection and gene substitution," *Genet. Res.,* Vol. 13 (1969), pp. 127–141.
[34] M. KIMURA and T. MARUYAMA, "The substitutional load in a finite population," *Heredity,* Vol. 24 (1969), pp. 101–114.
[35] M. KIMURA and T. OHTA, "The average number of generations until fixation of a mutant gene in a finite population," *Genetics,* Vol. 61 (1969), pp. 763–771.

[36] ———, "The average number of generations until extinction of an individual mutant gene in a finite population," *Genetics*, Vol. 63 (1969), pp. 701–709.
[37] ———, "Protein polymorphism as a phase of molecular evolution," *Nature*, Vol. 229 (1971), pp. 467–469.
[38] J. L. KING and T. H. JUKES, "Non-Darwinian evolution: random fixation of selectively neutral mutations," *Science*, Vol. 164 (1969), pp. 788–798.
[39] D. KOHNE, "Evolution of higher-organism DNA," *Quart. Rev. Biophys.*, Vol. 33 (1970), pp. 327–375.
[40] C. D. LAIRD, B. L. MCCONAUGHY, and B. J. MCCARTHY, "Rate of fixation of nucleotide substitutions in evolution," *Nature*, Vol. 224 (1969), pp. 149–154.
[41] K. MATHER, "The nature and significance of variation in wild populations," *Variation in Mammalian Populations* (edited by R. J. Berry and H. N. Southern), New York, Academic Press, 1970, pp. 27–39.
[42] J. MAYNARD SMITH, " 'Haldane's dilemma' and the rate of evolution," *Nature*, Vol. 219 (1968), pp. 1114–1116.
[43] ———, "The causes of polymorphism," *Variation in Mammalian Populations* (edited by R. J. Berry and H. N. Southern), New York, Academic Press, 1970, pp. 371–383.
[44] E. MAYR, *Animal Species and Evolution*, Cambridge, The Belknap Press of Harvard University Press, 1963.
[45] P. J. MCLAUGHLIN and M. O. DAYHOFF, "Evolution of species and proteins: A time scale," *Atlas of Protein Sequence and Structure* (edited by M. O. Dayhoff), Vol. 3, Silver Spring, Md., National Biomedical Research Foundation, 1969, pp. 39–46.
[46] T. MUKAI and H. E. SCHAFFER, "Genetic consequences of truncation selection at the phenotypic level in *Drosophila melanogaster*," in preparation.
[47] H. J. MULLER, "Evolution by mutation," *Bull. Amer. Math. Soc.*, Vol. 64 (1958), pp. 137–160.
[48] ———, "The gene material as the initiator and the organizing basis of life," *Heritage from Mendel* (edited by R. A. Brink), Madison, University of Wisconsin Press, 1967, pp. 419–447.
[49] M. NEI, "Gene duplication and nucleotide substitution in evolution," *Nature*, Vol. 221 (1969), pp. 40–42.
[50] ———, "Fertility excess necessary for gene substitution in regulated populations," *Genetics*, Vol. 68 (1971), pp. 169–184.
[51] C. NOLAN and E. MARGOLIASH, "Comparative aspects of primary structures of proteins," *Ann. Rev. Biochem.*, Vol. 37 (1968), pp. 727–790.
[52] C. NOLAN, E. MARGOLIASH, and D. F. STEINER, "Bovine proinsulin," *Fed. Proc.*, Vol. 28 (1969), p. 343.
[53] S. OHNO, *Evolution by Gene Duplication*, Berlin-Heidelberg, Springer-Verlag, 1970.
[54] T. OHTA, "Fixation probability of a mutant influenced by random fluctuation of selection intensity," *Genet. Res.*, in press.
[55] T. OHTA and M. KIMURA, "Statistical analysis of the base composition of genes using data on the amino acid composition of proteins," *Genetics*, Vol. 64 (1970), pp. 387–395.
[56] ———, "Development of associative overdominance through linkage disequilibrium in finite populations," *Genet. Res.*, Vol. 16 (1970), pp. 165–177.
[57] ———, "On the constancy of the evolutionary rate of cistrons," *J. Mol. Evol.*, Vol. 1 (1971), pp. 18–25.
[58] ———, "Functional organization of genetic material as a product of molecular evolution," *Nature*, Vol. 233 (1971), pp. 118–119.
[59] ———, "Amino acid composition of proteins as a product of molecular evolution," *Science*, Vol. 174 (1971), pp. 150–153.
[60] R. C. RICHMOND, "Non-Darwinian evolution: A critique," *Nature*, Vol. 225 (1970), pp. 1025–1028.

[61] M. H. SMITH, "The amino acid composition of proteins," *J. Theor. Biol.*, Vol. 13 (1966), pp. 261–282.

[62] N. SUEOKA, "Compositional correlation between deoxyribonucleic acid and protein," *Cold Spring Harbor Symp. Quart. Biol.*, Vol. 26 (1961), pp. 35–43.

[63] J. A. SVED, "Possible rates of gene substitution in evolution," *Amer. Nat.*, Vol. 102 (1968), pp. 283–292.

[64] J. A. SVED, T. E. REED, and W. F. BODMER, "The number of balanced polymorphisms that can be maintained in a natural population," *Genetics*, Vol. 55 (1967), pp. 469–481.

[65] F. VOGEL, "A preliminary estimate of the number of human genes," *Nature*, Vol. 201 (1964), p. 847.

[66] S. WRIGHT, "Evolution in Mendelian populations," *Genetics*, Vol. 16 (1931), pp. 97–159.

[67] E. ZUCKERKANDL and L. PAULING, "Evolutionary divergence and convergence in proteins," *Evolving Genes and Proteins* (edited by V. Bryson and H. J. Vogel), New York, Academic Press, 1965, pp. 97–166.

THE ROLE OF MUTATION IN EVOLUTION

JACK LESTER KING
UNIVERSITY OF CALIFORNIA, SANTA BARBARA

This paper is dedicated to retiring University of California Professors Curt Stern and Everett R. Dempster.

1. Introduction

Eleven decades of thought and work by Darwinian and neo-Darwinian scientists have produced a sophisticated and detailed structure of evolutionary theory and observations. In recent years, new techniques in molecular biology have led to new observations that appear to challenge some of the basic theorems of classical evolutionary theory, precipitating the current crisis in evolutionary thought. Building on morphological and paleontological observations, genetic experimentation, logical arguments, and upon mathematical models requiring simplifying assumptions, neo-Darwinian theorists have been able to make some remarkable predictions, some of which, unfortunately, have proven to be inaccurate. Well-known examples are the prediction that most genes in natural populations must be monomorphic [34], and the calculation that a species could evolve at a maximum rate of the order of one allele substitution per 300 generations [13]. It is now known that a large proportion of gene loci are polymorphic in most species [28], and that evolutionary genetic substitutions occur in the human line, for instance, at a rate of about 50 nucleotide changes per generation [20], [24], [25], [26]. The puzzling observation [21], [40], [46], that homologous proteins in different species evolve at nearly constant rates is very difficult to account for with classical evolutionary theory, and at the very least gives a solid indication that there are qualitative differences between the ways molecules evolve and the ways morphological structures evolve. Finally, there is the amazing complexity of each gene and every protein, and the superastronomical numbers of combinatorial possibilities in theoretically possible genes and proteins, which together appear to make the evolution of specific macromolecules utterly impossible with undirected mutation and natural selection [33], [45].

At present there appear to be two approaches to a resolution of these differences. One is to conclude that nearly all molecular polymorphism and molecular evolution is due to origin by mutation, and fixation by random drift, of molecular variants (alleles) that are completely neutral with regard to the processes of natural selection [20], [21], [24], [6]. Then one is left with an unspecified minority of adaptive molecular changes which, one is free to hypothesize, behave exactly as dictated by established theory. As it happens, neo-Darwinian theory itself has included the prediction that no genetic change can be selectively neutral, but upon close examination this idea, attributable largely to R. A. Fisher, is only an unsupported opinion.

The second approach to attempting a resolution is to modify neo-Darwinian theory to accommodate the new observations. An example of this approach is a rejection of the mathematically simplifying assumption that minor deleterious effects are independent in action, and the substitute proposal that selection on genes with such effects acts by simultaneously eliminating many deleterious genes with the genetic death of a few individuals that by chance exceed some threshold level of deleterious effects (so-called truncation selection, see [22], [30], [53]). Such threshold effects are commonly encountered in developmental genetics studies, but at present the idea of a generalized threshold in fitness has rather little observational support, either classical or molecular.

There is no inherent contradiction in these two approaches, and both may be valid. A synthesis of approaches may be most constructive. I think it somewhat unlikely that molecular polymorphisms and molecular amino acid substitutions are each of two discrete classes, one due entirely to natural selection and the other due entirely to neutral mutation and random drift; rather, it is likely that mutation, random drift, and natural selection are often (each to a greater or lesser degree in individual instances) important in molecular evolution and in natural variation on the molecular level.

2. The role of mutation in evolution: the classical view

A few decades ago there was a lively controversy over the role of mutation in evolution. Did mutation have any directive influence on evolution? The controversy was resolved in favor of R. A. Fisher and other population genetics theorists, and until now evolutionary biologists have held the following opinions virtually unanimously.

(a) There is always sufficient genetic diversity present in any natural population to respond to any selection pressure. Therefore actual mutation rates always are in excess of the evolutionary needs of the species.

(b) There is no relationship between the mutation rate and the rate of evolutionary change.

(c) Because mutations tend to recur at reasonably high rates, any clearly adaptive mutation is certain to already have been fixed. Therefore, natural populations are at, or very near, either the best of all possible genetic constitutions, or an adaptive peak of genotype frequencies [56].

(d) Since all possible adaptive mutations are fixed, and since neutral mutations are unknown, virtually all new mutations are deleterious, unless the environment has changed very recently. Even a recent change in the environment does not make new mutations necessary, because of (a).

(e) Evolution is directed entirely by natural selection, acting on genetic variability that is produced by recombination, from "raw materials" produced by recurrent mutation a long time ago.

(f) Mutation is random with respect to function.

The remarkable thing about this consensus of opinion on the role of mutation in evolution is that it is generally true on the level at which it was formulated, namely, morphological and physiological evolution; at the same time, every statement is untrue at the level of molecular change in evolution. Thus, as we hope to document in the remainder of this paper:

(a) most specific allelic states achievable by even the simplest and most common form of mutation—single nucleotide substitution—are highly unlikely to be present within a species at any point in time;

(b) there is a simple and direct relationship between the mutation rate and the rate of evolutionary change on the molecular level;

(c) specific mutations do not recur at reasonably high rates; a species may have to wait millions of years before a specific adaptive mutation occurs and begins to increase toward fixation;

(d) an evolutionarily significant proportion of new mutations are either neutral or very slightly advantageous;

(e) an increased mutation rate may be beneficial to a population or a species;

(f) mutation is not random with respect to function on the molecular level.

Let us look first into the question of recurrent mutation. To do this adequately, we must first determine the fundamental mutation rate.

3. The fundamental nucleotide substitution mutation rate

There are many classes of mutation that are evolutionarily significant. On the molecular level, these include deletions, insertions, gene duplications, and nucleotide substitutions, the last sometimes but not always resulting in amino acid substitutions. Nucleotide substitutions are the most common kind of mutational event and the most common kind of evolutionary change. What is the mutation rate, in humans, of nucleotide substitutions? I shall present four independent estimates of this rate, all quite consistent.

3.1. *The rate of DNA divergence.* Kohne [25], [26] finds that the difference in mean melting point temperature between native DNA and human-green monkey hybrid DNA (after the removal of redundant sequences) is 7.0°C, corresponding to approximately 10.5 per cent nonhomology of nucleic acid sites. If the last common ancestor of the old world monkeys and the hominid line lived 30 million years ago [46], [47], the mean rate of evolutionary substitution in primate unique sequence DNA is approximately 18.4×10^{-10} substitutions per nucleotide per year in each line of descent. DNA hybridization studies between man and new world monkeys, and between new world monkeys and old world monkeys, indicate an average of 18.0 per cent nonhomology; using Sarich and Wilson's estimate of 50 million years since the last common ancestor of new world monkeys, old world monkeys, and man, this gives a very similar nucleotide substitution rate of 19.8×10^{-10} per nucleotide per year (allowing for repeated changes at some sites).

This *evolutionary* rate of nucleotide substitution per species is almost certainly very close to the nucleotide substitution *mutation* rate per gamete. This is because most nucleotide substitutions in total DNA are probably selectively neutral, and the evolutionary rate of selectively neutral substitutions per species is the same as the mutation rate for neutral mutations per gamete, as has been shown elsewhere [20], [24]. Reasons have also been given elsewhere [24], [21] for the conclusion that more than 99 per cent of all mammalian DNA is nongenetic in the sense that *all* point mutations occurring within that portion are selectively neutral. This DNA does not code for protein and appears to have no known function. Kohne [25], [26] provides a further convincing argument by presenting evidence that most "unique sequence" DNA appears in fact to be the degenerate remains of past "repeated sequence" families, the members of which have lost detectable homology because of random divergence. If most random divergence in DNA is not subject to natural selection, the fundamental nucleotide substitution mutation rate is approximately 19×10^{-10} per nucleotide per genome per year.

3.2. *Measured mutation rates for human genetic pathologies.* For decades geneticists have measured "recurrent" mutation rates in many species, including man. Generally these mutations have been recurrent only in that they occur in the same gene and tend to inactivate it; on the molecular level, many different mutations may cause gene inactivation, while other mutations occurring within the cistron may go undetected. A reasonable estimate of the average mutation rate of lethal or visible mutations per cistron for human pathologies is about 10^{-6} per generation [51]. This is a lower figure than is usually reported. Since I wish to show that the mutation rate for specific changes is very low, let us conservatively take a higher (but still reasonable) mutation rate of 5×10^{-6} per gamete per generation. For the basic nucleotide substitution rate, however, we need somewhat different information. The only experimental basis for translating lethal mutation rates into estimated nucleotide substitution rates is provided by the work of Ames [1] and Whitfield, Martin, and Ames in Salmonella [55]. They found that only ten per cent of mutations known to occur in the histidine loci investigated were recoverable as lethals; the remainder, all base substitutions, either had no effect or only partially inactivated the gene loci and hence were unrecoverable in their test situation. The general conclusion is that a detectable mutation rate of 5×10^{-6} implies a nucleotide substitution rate of about 5×10^{-5} per cistron per generation. Taking the mean generation time to be 25 years and a representative cistron size of 1000 nucleotides, this gives an estimated nucleotide substitution rate of 20×10^{-10} per year.

3.3. *Concomitantly variable codons in cytochrome c.* Fitch and Markowitz [11] observed that calculations of the probable number of potentially variable sites in the evolution of cytochrome *c* depended on the evolutionary time span covered. As they consider groups of animals more closely related in evolutionary time, the estimated number of potentially variable codons decreases in a regular way (Figure 1). The regression line, extrapolated to zero, indicated to Fitch and Markowitz that only ten per cent of cytochrome c's 104 codons were free to vary in one species at one point in time. King and Jukes [24] estimated the evolutionary rate of cytochrome c in mammals to be about 4.3×10^{-10} amino acid substitutions per codon per year; if Fitch and Markowitz are correct, this is the equivalent of 43×10^{-10} amino acid substitutions per *variable* codon per year. Allowing for three nucleotides per codon, and for the fact that about one fourth of nucleotide substitutions within codons do not change amino acids, the calculated rate of change for nucleotides free to vary in the cytochrome c cistron is $43 \times 4/3 \times 1/3 \times 10^{-10} = 19 \times 10^{-10}$ per nucleotide per year. This is nearly identical with previously calculated rates, indicating that variable codons in cytochrome c vary at the mutation rate; this may be taken as circumstantial evidence favoring the possibility that mutations in the variable codons are neutral and that nearly all cytochrome c evolution in mammals may have been due to mutation and random drift.

King and Jukes [24] suggested earlier that 9/10 of cytochrome c mutations are selected against. They calculated that fibrinopeptide A in mammals evolved at the rate of 43×10^{-10} per codon per year, the equivalent of 19×10^{-10} per nucleotide per year and ten times the rate of change of cytochrome c. This would appear to indicate that nearly all of the fibrinopeptide A codons are free to vary, a conclusion also reached by Fitch on different grounds. Estimates of the evolutionary rate of the fibrinopeptides are probably not too reliable, however, because of the numerous evolutionary gaps and insertions and the consequent ambiguity in homology, and as pointed out previously [24], portions of the fibrinopeptide molecule tend to be relatively conservative, presumably because of selective restraints [3], [4]. Even so, the consistency of these comparisons tends to support the estimated annual nucleotide mutation rate of 19×10^{-10}.

3.4. *Estimates based on human hemoglobin variants.* An upper limit on the fundamental base substitution rate can be derived from the frequencies of human hemoglobin electrophoretic variants. These are found among Northern Europeans at the frequency of about 5×10^{-4} per individual, or about 2.5×10^{-4} per haploid genome (both alpha and beta chains together). Not all these variants represent new mutations, of course. Motulsky [35] states that considerably fewer than ten per cent of all carriers have both parents unaffected—that is, considerably fewer than ten per cent carry new mutations. Motulsky assumes that the proportion of carriers with new mutations is between 0.4 per cent and 4 per cent; if so, the mutation rate per gamete for electrophoretic variants is then between 10^{-6} and 10^{-5} for the combined alpha and beta chain cistrons. Only about one fourth of all amino acid substitutions, and only about 3/16 of all nucleotide substitutions, result in electrophoretically detectable protein changes. There are 861 nucleotides in the two cistrons; the human generation span is about 25 years; so, for the annual nucleotide mutation rate the above range

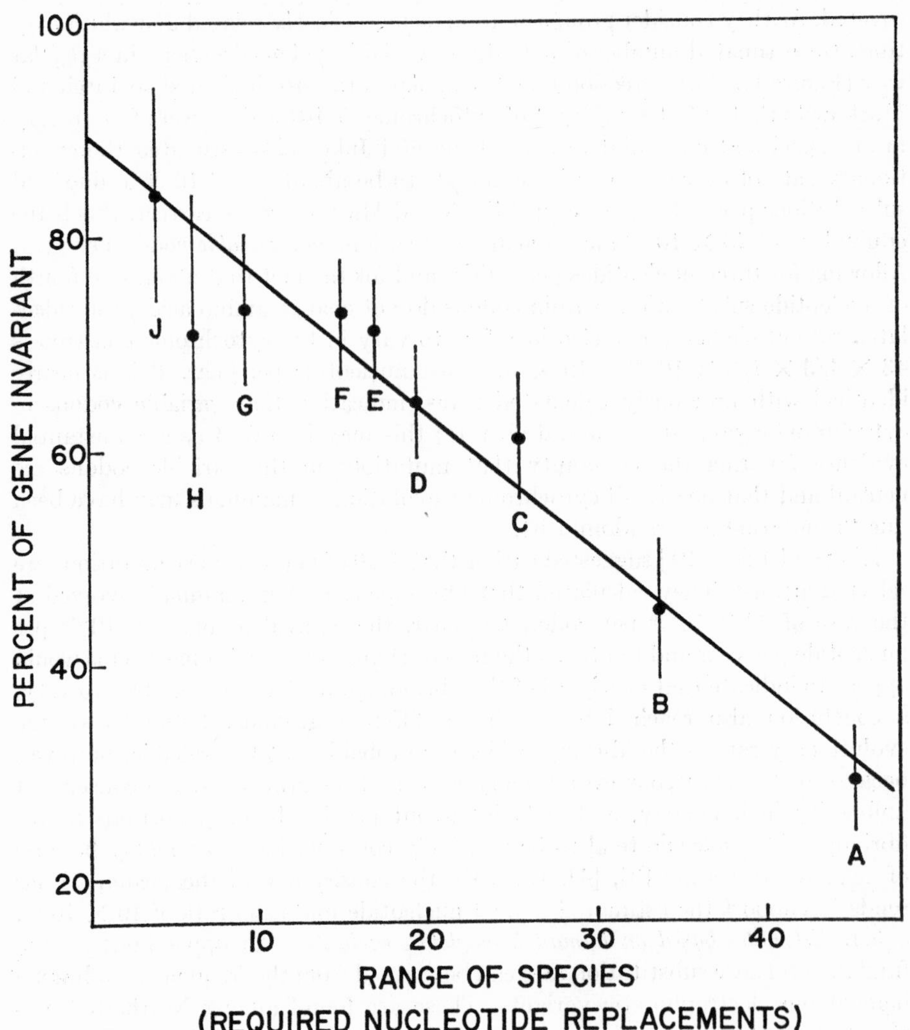

FIGURE 1

The number of invariant positions in cytochrome *c* can be estimated from the approximation of the distribution of evolutionary changes among variable positions to modified Poisson distributions. The calculated number of invariant positions is a function of the evolutionary distance between species compared; extrapolating back to zero, Fitch and Markowitz [11] concluded that only ten per cent of the amino acid positions in cytochrome *c* are free to vary in one species at any one point in time. This corresponds well with the observation that mammalian cytochrome *c* evolves at a rate equal to ten per cent of the mutation rate.

should be divided by $3/16 \times 861 \times 25$. The range of likely mutation rates per nucleotide per year is thus between 2.5×10^{-10} and 25×10^{-10}, with an average estimated value of 13.75×10^{-10}. This is not very different from the other estimates. The principal uncertainty is in the actual proportion of variant hemoglobin carriers lacking a carrier parent. A prediction can be made: since the basic nucleotide substitution rate appears to be approximately 19×10^{-10} per year, the proportion of variant hemoglobin carriers with both parents normal will be approximately three per cent.

The approximate agreement of these four estimates indicates that the annual nucleotide mutation rate in humans is probably close to 19×10^{-10}. Furthermore, this internal consistency tends to confirm the original contention that most mammalian DNA is not functioning as genetic material: the total evolutionary rate of DNA is approximately equal to what has been found, by other methods, to be the fundamental nucleotide mutation rate. This indicates that most DNA is not subject to restraints of natural selection.

3.5. *The mutation rate as a function of astronomical time.* Kohne's DNA hybridization studies [25], [26] show that the DNA divergence rate is about the same in the two lines connecting humans and old world monkeys with the more distantly related capuchin: the DNA melting temperature depression for human-capuchin hybrid DNA is 11.6°C, while that for green monkey-capuchin hybrid DNA is 12.3°C. The difference between these two values is close to the limit of resolution of the experimental procedure. The melting point depression for man-green monkey hybrid DNA is 7.0°C; the three values together indicate that the human line has diverged 3.15°C and the old world monkey line has diverged 3.85°C from their common ancestor. Taken at face value, this means that the human line has evolved at an average rate about 18 per cent slower than the monkey line. However, similar comparisons between the chimpanzee and the rhesus indicates that, since the same common ancestor, the chimp has diverged 3.25°C and the rhesus has diverged 3.55°C, a rate difference of only eight per cent. Both chimpanzees and humans have long average generation spans relative to those of monkeys, so the indication is that the DNA evolutionary rate is not closely related to generation span. The further inference is that the DNA base substitution mutation rate is approximately constant with astronomical time.

Other papers presented at this conference [6], [21] document the general observation that the evolutionary rates for homologous proteins appear to be nearly constant with time in different vertebrate lines of descent, despite wide variations in rates of morphological evolution, in population size, and in mean generation spans. The simplest explanation is that the rate of molecular evolution is directly related to the mutation rate, and that the mutation rate is constant per unit time in any one species and very similar in different vertebrate species.

It has been suggested by various investigators that mutation might be a direct function of the number of generations (meioses) or of the number of cell generations (mitoses). There is little evidence to support either of these opinions. It is true that the mutation rate is nearly proportional to the number of cell generations in very rapidly dividing bacteria [29], but it quickly becomes a linear function of astronomical time in somewhat more slowly dividing bacteria [38] and continues to occur at a fairly high rate in bacteria that are not dividing at all [43], [44]. Cells in vertebrate germ line probably divide less than once a month on the average, and most of the mutations that occur are probably unrelated to DNA replication or cell division. The mutation rate in stored *Drosophila* sperm is exactly the same, per unit time, as the rate in rapidly developing larvae and pupae [16]. Mutations also occur in stored fungal spores at an approximately constant rate [54]. Until contrary evidence is presented, one must suspend any hypotheses of any direct relationships between mutation and either generation number or cell generation number.

Mutation rates per unit time do differ greatly in very distantly related life forms. Ryan [43], [44] calculated the mutation rate for autotrophy in nondividing histidine minus *E. coli* to be 1.2×10^{-9} per hour. This gives an annual mutation rate of 10^{-5} for histidine reversions. If one estimates (generously) that ten different specific mutations would lead to histidine reversion, the annual mutation rate for *specific* mutations in *E. coli* is about 10^{-6}, compared with an annual rate of 6×10^{-10} for specific mutations in humans. Thus, there appears to be a difference of at least a thousandfold in the mutation rates per unit time in these two life forms. Similarly, there is unquestionably a large discrepancy between the mutation rates per unit time in man and in *Drosophila*. The mean rate of lethal mutation per gene per generation is about 3×10^{-6} [37]; assuming that the rate of base substitutions per locus is ten times greater, that the average gene size is 1000 nucleotides, and that the average generation time in Muller's experiments was two weeks, the per-nucleotide annual mutation rate in *Drosophila* is estimated to be $3 \times 10^{-5} \times 10^{-3} \times 25 = 7.5 \times 10^{-7}$, more than 300 times greater than the annual mutation rate for humans *estimated by means of approximately the same procedure.* The mutation rate per locus per generation is about the same in both species, but there is about five hundredfold difference in generation spans.

3.6. *Evolutionary rates in insects and mammals.* This presents a conundrum, first pointed out to me by J. F. Crow. In the 103 positions of cytochrome *c* that are homologous in all three species, the number of amino acid differences between wheat and human is 34, while the number of differences between wheat and fruit fly is 42. Considering the large difference in annual mutation rates, why have the two animal species diverged so nearly the same distance from the plant species? The question is not easily answered; however, the difficulty may not be so great as it first appears. The corresponding number of differences between man and fly is 24. If one subtracts the (approximately) 30 invariable sites from 103, and makes a first order correction for multiple changes at individual sites by estimating the frequency of changes per variable site from the negative log of the proportion of variable sites unchanged [24], [40], the estimated number of evolutionary substitutions between man and fly is 29; between man and wheat,

46; between fly and wheat, 63. These corrections are known to be inadequate, since they assume that all variable sites are equally variable; the real numbers of evolutionary events must be greater and the proportional differences between them much greater. But if one accepts provisionally the evolutionary distances of 29, 46, and 63, the calculated number of evolutionary steps since the fly-human divergence would be only six in the human line and 23 in the fly line, indicating an evolutionary rate difference of 400 per cent. The complete inadequacy of these estimates is made evident by the fact that *more* than six evolutionary substitutions are almost certain to have occurred in the human line in just the last 100 million years or less, which is a small fraction of the total time since the human-fly divergence. Also, well more than 12 changes appear to have occurred in the human line since the fish-tetrapod divergence. With distant comparisons of homologous proteins, such as those between kingdoms and between major phyla, it is virtually impossible to translate amino acid differences into rates of evolutionary change. Too much information is lost and obscured by sequential changes at sites with very different intrinsic rates of change, as well as by the wholly coincidental identity of some sites in different species. Fitch and Markowitz [11] emphasize that a large proportion of the cytochrome *c* molecule is invariant among mammals, and a different but also large proportion is invariant among insects. Perhaps it is not impossible that cytochrome *c* has in fact been undergoing amino acid replacements at a 300-fold greater rate in insects than in mammals during the last 100 million years, if these changes occur only in restricted portions of the molecule in each group.

4. The specific mutation rate

Each nucleotide in DNA can mutate to each of three other nucleotides. Thus, if the fundamental base substitution rate is approximately 19×10^{-10} per nucleotide per year, the mutation rate of *specific* mutations is one third of this, or about 6×10^{-10} per year. This rate is so low that it is almost nonsensical to consider recurrent mutations to specific alleles to be evolutionarily significant. It is also nonsensical to consider back mutation rates to specific alleles [21], [40]. Each amino acid specifying codon, on the average, can mutate to codons specifying approximately seven other amino acids. A gene coding for 287 amino acids—the size of the hemoglobin alpha and beta chains together—can mutate to 2000 other states by single base replacement, each mutation occurring at the same rate of approximately six per ten billion per genome per year. New mutations rarely persist more than a very few generations [7]. It is therefore extremely unlikely, even in a very large population (of mammals, at least) that even one copy of a specific mutant form would be present at any given point in time. It is quite false to suppose that the requisite genetic variability is present to meet every evolutionary need, if one conceives of evolutionary needs in terms of specific molecular changes.

A species of one million individuals would have to wait almost a thousand years before a *specific* adaptive mutation would occur by mutation in just one member. But even then the mutant allele would probably be lost; the probability of fixation of a beneficial mutation is less than twice its selective advantage, or more specifically, $2s(N_e/N)$ where s is the relative selective advantage, N is the actual size of the population, and N_e is the effective size of the population after certain adjustments are made for sex ratio, temporal fluctuations, and variations in fecundity [17], [7]. The N_e/N for a species over evolutionary time may be a fairly small number, because of such likely events as the expansion of small population isolates and the displacement and extinction of competing isolates. A species with an evolutionarily effective population size of 10^5 would have to wait, on the average, 40 million years before a specific mutation with a selective advantage of 10^{-4} would occur and begin to increase toward fixation. A species with an effective size of one million—and an actual population size perhaps many times as large—would only have to wait an average of four million years! In either case, it is highly likely that the environment, the background genotype, and perhaps even other codons in the same cistron would have changed meanwhile, and with them the selection coefficient.

If this can be said to be an evolutionary mechanism at all, it is surely a very inefficient one. Some species are very much larger than one million in size—some insect species, for example—and for these it might be true that all possible single step mutant alleles are always present and that any possible advantageous mutations will be ready to compete in any new situation, the most fit eliminating all others. But other species have small populations and evolve just the same.

The basic observations that led to the conclusion that genetic variability is always present are that artificial selection on previously unselected metric traits is always successful, and that canalized traits can often be shown to conceal a large amount of potential genetic variability. That is to say, the genetic variability is there all right, when measured through its effect on morphology and physiology; but one cannot extrapolate from gross morphology to molecular structure and expect the same relationships to hold. Only a tiny fraction of the genetic variability potentially available by mutation is actually present in a species at any one time, and this fraction rapidly turns over, to be replaced by another array of genuinely rare mutant alleles. Yet this changing small sample of the genetic potential provides all the raw material for the adaptive (as well as neutral) evolution of protein molecules.

It is not known how much of the genetic variability that is uncovered by selection experiments is due to variation in protein specifying structural genes. It seems plausible that the genes that control rates of synthesis and developmental patterns are more important in response to artificial selection, and are also responsible for nearly all of the adaptive evolution of the kind usually considered [50]. Not much is known about controlling genes in higher organisms, but they are made of DNA, and the mutation rate considerations and the rarity of mutations to specific allelic forms must be the same for controlling genes as for protein specifying genetic material. On the other hand, adaptive evolution must occur in structural genes also, as proteins do appear to be highly adapted to their specific functions. But the mode and tempo of molecular adaptation may be very different from that of morphological evolution. While it is certainly wrong to conclude that the genetic variability required to make *specific* molecular changes is always present in the population, there are definite relationships between available genetic variability and the patterns of molecular evolution. It is *because* natural selection acts immediately and largely deterministically on the morphological and physiological level that it acts stochastically and quasi-randomly on the molecular and genetic level.

5. The beneficial role of mutation in evolution

When it was believed that there was always ample genetic variability to meet every selective need; that all possible beneficial mutations were certain to occur and certain to become fixed; and that there was no relationship between the rate of mutation and the rate of evolution, it was easy to believe that mutation was generally a harmful thing. It could be held that while some small amount of mutation was a prerequisite to evolution, this amount was in fact far below the actual mutation rate. It followed that any increase in mutation was always bad and any decrease was always good. The conclusion one could reach from these premises was that the mutation rate represented an irreducible minimum set by the physical inability of the organism to prevent all mutation. Others have proposed that, to the contrary, the mutation rate was itself subject to adjustment by natural selection, and represented an optimal balance between the beneficial effects of mutation on the population and the deleterious effects of mutation on the individual [32], [18]. For while most mutations may be slightly deleterious when they occur (as shown by the fact that the average rate of protein evolution is lower than the rate of DNA mutation), harmful mutant alleles are soon lost, while beneficial mutant alleles tend to increase in number. The *net* effect of mutation is clearly beneficial.

Several experiments have indicated that an increased mutation rate may be markedly beneficial for a population, at least for one that is adapting to an unaccustomed environment. Ayala [2] found that radiation induced mutation enabled populations of *Drosophila pseudoobscura* to adapt more rapidly to unusual conditions of crowding. Although radiation induced mutations tend to be rather more drastic than spontaneous mutations, in this case their net effect was clearly beneficial, as indicated by increases in both biomass and number. Gibson, Scheppe, and Cox found that a single mutator allele in *E. coli,* known to cause a thousandfold increase in DNA base substitution mutations, enables the mutant bearing strain to consistently outcompete otherwise coisogenic wild type strains under a variety of conditions [12]. These competition experiments were also conducted under "unnatural" conditions, that is, outside the human gut. One can still assume that natural populations are more closely adapted to their environments and would not benefit by greatly increased mutation rate, but the point is made that mutation rates can be too low for the good of the population and, as in the *E. coli* experiments, natural selection can successfully increase mutation favoring genotypes (but see [27]).

A rate that is optimal over relatively short periods of evolutionary time may be disastrous over longer periods of time. An organism that tracks its changing environment too closely can be led into an evolutionary blind end. It is conceivable, for instance, that the highly adaptive *E. coli* strains in Gibson, Scheppe, and Cox's competition experiment may have permanently lost many complex enzyme systems that were not needed in a chemostat, but which would be needed in another environment.

6. Nearly neutral mutations in evolution

The analogy between artificial selection and natural selection can be quite misleading in some aspects. Artificial selection works effectively only on allelic differences with quite large selection coefficients, only on alleles with intermediate gene frequencies, and only on pre-existing genetic variability. Sometimes natural selection may operate under similar restrictions, for instance, genetic "tracking" of a cyclic environment. But long term adaptive evolutionary change is able to involve newly arising variability, vanishingly low gene frequencies, and vanishingly small selection differences.

New mutations appear to be distributed around two means of selective effect: one mean close to lethality and another close to neutrality. Most near neutral mutations are slightly disadvantageous, but there is a continuum of selective effects. Of those mutations with positive selective advantages, probably very few have selective advantages greater than, say, one per cent; much smaller selective advantages may be much more frequent. Neutral and nearly neutral mutations appear to have a numerically large role in molecular evolution. Adaptive, Darwinian evolution may utilize nearly neutral mutations to a significant extent. In the region of near neutrality, drift effects will predominate over selection effects when the absolute value of the selection coefficient is less than the reciprocal of the population size (Figure 2). For completely neutral alleles, the probability of eventual fixation is equal to $1/2N$ for new mutations. When the selection coefficient is greater than about $1/N_e$, the effects of natural selection will predominate; with larger positive values, the probability of eventual fixation of new mutations asymptotes to $2s(N_e/N)$.

Figure 3 shows one possible distribution of the selection coefficients of new mutations around the region of selective neutrality; here there is a significant class of truly neutral mutations. This hypothetical distribution probably describes the true situation fairly accurately, if all synonymous changes and all base substitutions occurring in nonfunctional DNA are included. There may—or may not—also be a significant class of mutations that cause amino acid substitutions in functional proteins but are still truly selectively neutral.

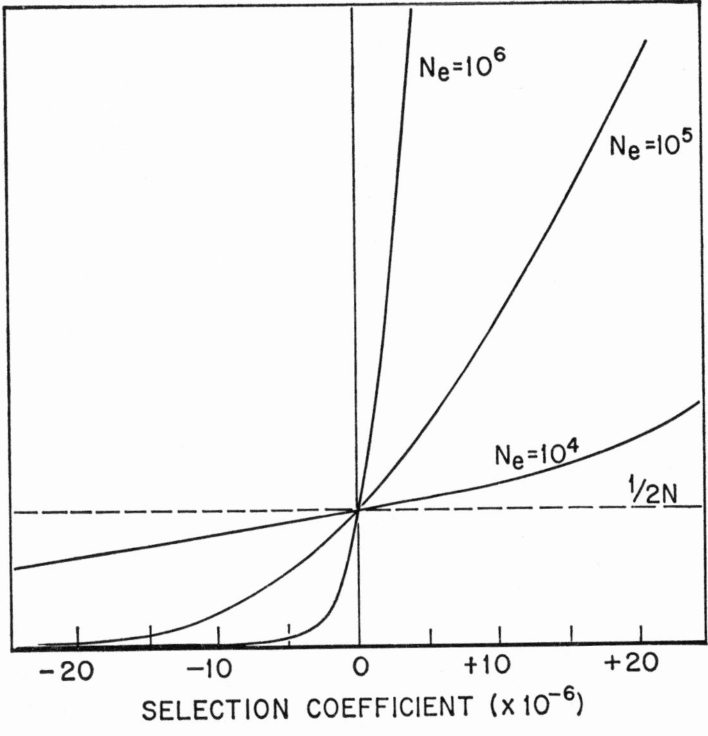

FIGURE 2

Distribution of the probability of fixation of new mutations with nearly neutral selection coefficients, relative to the probability of fixation for absolutely neutral mutations ($1/2N$). Distributions for three different effective population sizes are superimposed. After Ohta and Kimura [40].

Instead of there being a discrete class of selectively neutral amino acid substitution mutations, there might instead be a continuum of mutant selection coefficients in the vicinity of zero. Since we know that the majority of mutations are slightly deleterious, the frequency distribution of mutations with nearly neutral selection coefficients might have a mode at some negative selective value, and a steep negative slope in the vicinity of zero (Figure 4). The shape of the distribution of selective coefficients is relatively independent of the population size. The frequency distribution of evolutionarily successful mutant incorporations, however, is different than that of the unselected mutation distribution, and its shape definitely is a function of the population size. It is in fact the product of the probability of fixation and the mutation frequency for each value of the selection coefficient. Schematic distributions of evolutionarily successful mutations are shown in Figures 5 and 6.

Note that with increasing populations size there are fewer fixations of slightly

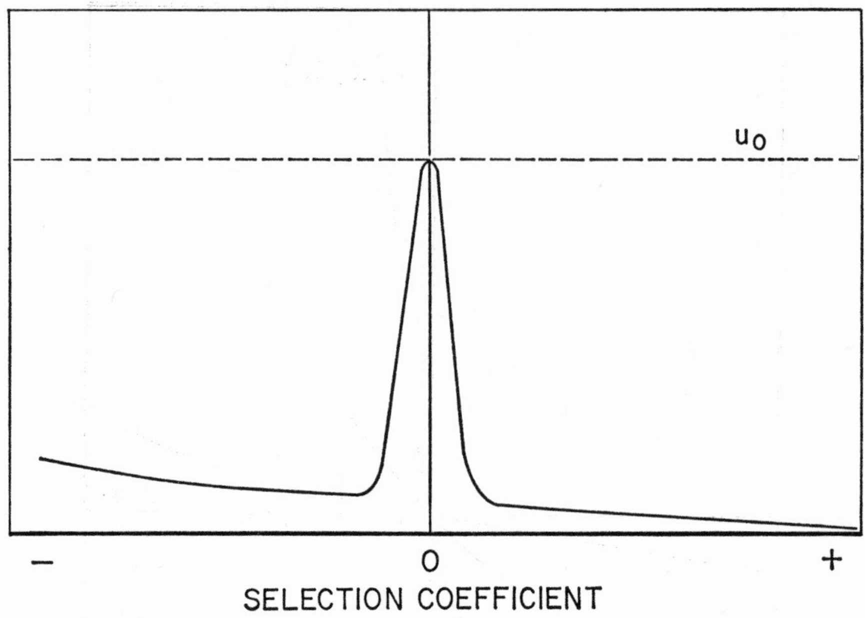

FIGURE 3

One hypothetical frequency distribution of new mutations according to selection coefficients (schematic). In this hypothesis there is a discrete class of absolutely neutral mutations.

deleterious mutants and a greater proportion of fixations of slightly beneficial mutants, while the rate of fixation of absolutely neutral mutants remains constant. In relatively small populations, the majority of near neutral mutants that are fixed may be slightly deleterious, simply because many more deleterious than beneficial mutations occur and the difference in the probability of fixation is not sufficiently great to reverse this distribution completely [40], [21] (see Figure 5). As the effective population size increases, the difference in the probability of fixation between the slightly deleterious and slightly beneficial mutants increases, and the shift is toward a preponderance of slightly beneficial fixations (Figure 6). Since the actual distribution of mutations in the vicinity of selective neutrality is not known, one cannot predict in detail the net effect of population size on the total rate of evolutionary change. It is quite possible that, over a considerable range of effective populations sizes, the negative effect of population size on the fixation of slightly unfavorable mutants just about balances the positive effect of size on the fixation of slightly favorable mutants, leaving relatively little net effect of populations size on the total rate of molecular evolution.

Slightly adaptive mutations might be numerically important in adaptive evolution without actually making very much of a contribution to the adaptive evolution of the species. For instance, suppose that the mutation rate for each

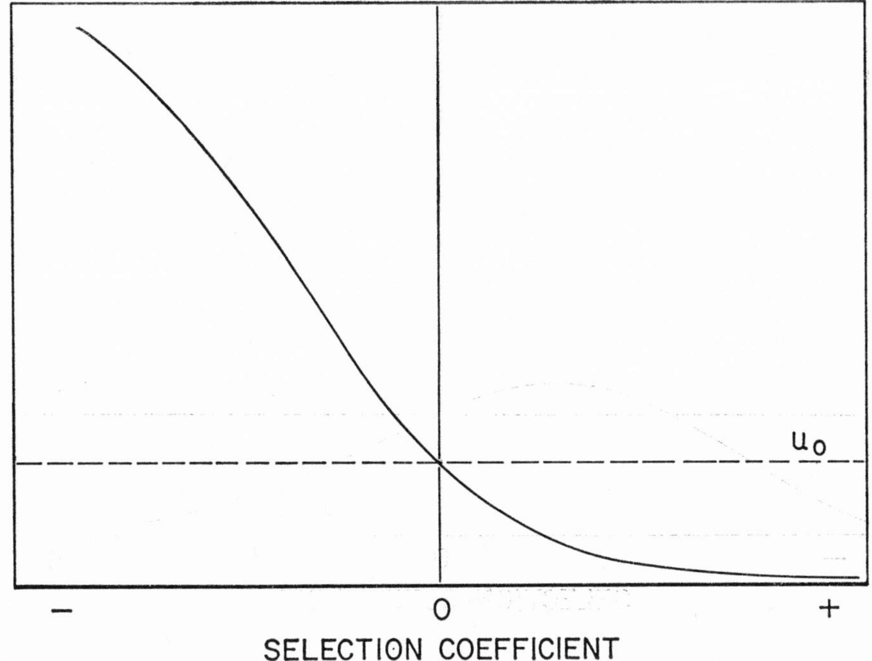

FIGURE 4

An alternate hypothesis of the frequency distribution of new mutations according to selection coefficients (including only amino acid change mutations). The mean selection coefficient is negative, but there is a continuum of selection effects with a steep slope through the region of selective neutrality (schematic).

class of advantageous mutation were roughly proportional to the reciprocal of the selection coefficient. Then mutations with $s = 0.001$ would be ten times more numerous than those with $s = 0.01$, while mutations with selection advantages of $s = 0.0001$ would be one hundred time as numerous. The probability of eventual fixation is proportional to the selective coefficient, so equal numbers of each of these three classes would eventually become fixed. The net advantage of one fixed allele with $s = 0.01$ is ten times that of one fixed allele with $s = 0.001$ and one hundred times that of the fixed allele with $s = 0.0001$ (assuming no dominance). The contribution of each class of adaptive mutations to the fitness of the species, in other words, is proportional to $u_{(s)}s^2$, where $u_{(s)}$ is the mutation rate to adaptive alleles with selection coefficients s. The total functional effect of adaptive mutation would be equal to $8N_e \int_{s=0}^{\infty} u_{(s)}s^2\,ds$ over all positive values of s.

These calculations assume a constant selection coefficient with no dominance and no epistasis. If $u(s) > k/s^2$ for some constant k and for values of s greater than $1/N_e$, then nearly neutral mutations may make a relatively large functional

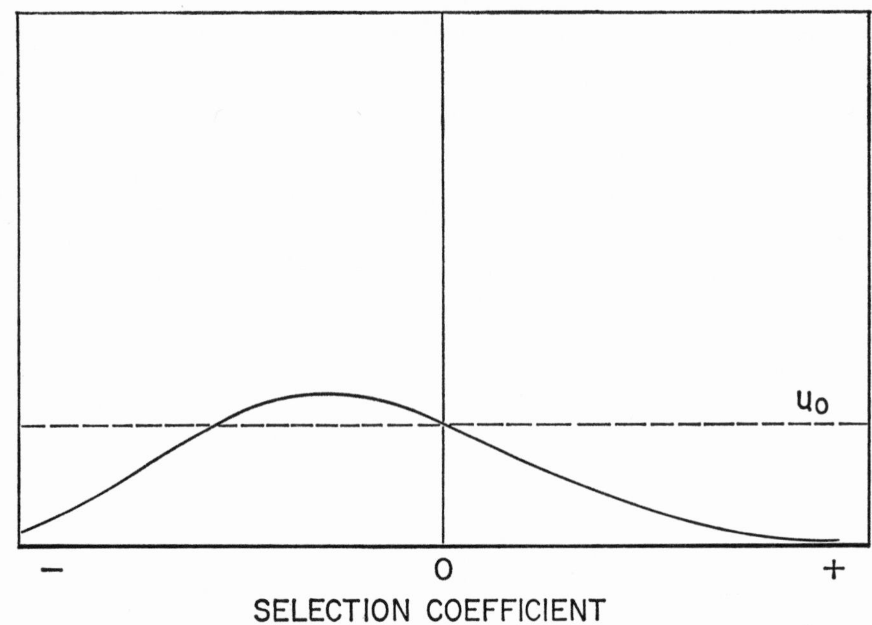

FIGURE 5

The distribution of selection coefficients of evolutionary fixations in the neighborhood of selective neutrality, for the distribution of mutations shown in Figure 4, with a small population: the mean is still somewhat negative. See references [21] and [40] (schematic).

contribution to the adaptive evolution of a species. This relationship does not seem likely, however. Even if the mutation rate for beneficial alleles were directly proportional to the reciprocal of the square of the selective advantage (that is, $u_{(s)} = k/s^2$) over the range of, say, $0.1 > s > (1/N_e)$, each interval of s would have the same net effect on adaptive evolution, and new mutations occurring in the very small intervals of the near neutral range would still have only correspondingly small effect on the net adaptedness of the species. Still, such alleles would constitute the overwhelming numerical majority of molecular changes in evolution.

7. Determinism and randomness in evolution

On the level of morphology and physiology, there is a strong element of determinism in an evolutionary response to the environment. Nocturnal habits impose successful selection for larger eyes and/or more sensitive hearing. Grazing on rough forage forces an evolutionary trend to teeth more resistant to abrasion. The organism adapts to its environment. The predictable evolutionary response of a population of insects to exposure to DDT is the development of DDT

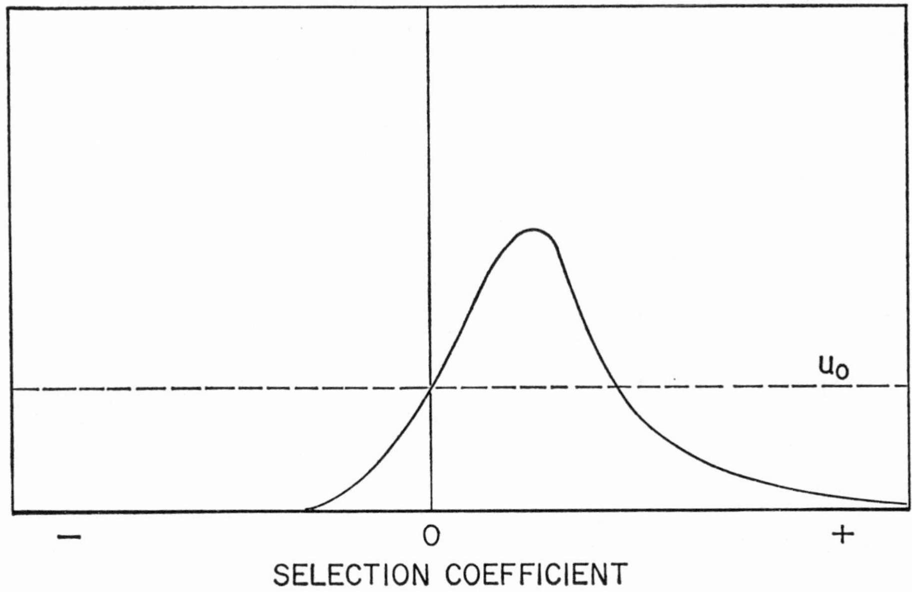

FIGURE 6

Same as Figure 5, but with a larger population. With increasing population size, the relative probability of fixation of slightly beneficial mutations increases; the mean shifts to positive. The net effect of population size on total rate of evolutionary fixation of nearly neutral mutations remains unknown (schematic).

resistance. Deterministic responses will always occur when there is the requisite genetic variability present in the population. For most continuous phenotypic variables, the genetic resources are usually present. The exact genes involved will differ from time to time and from population to population.

The determinism which is seen at the phenotypic level does not occur at the molecular level. A protein forming gene does not respond in specific ways to particular selective circumstances, because the requisite specific variability is rarely present. Rather, a selective requirement can usually be met by changes at any of a number of different gene loci, and which loci will actually change will depend upon which loci happen to come up with appropriate alleles during the time in which the selective requirement exists. Within a single gene, it is likely that any of a number of possible changes might be selected for—for instance, any of a number of amino acid substitutions would change the isoelectric point or modify the secondary or tertiary structure. A gene coding for a polypeptide of 150 amino acids contains 450 nucleotide pairs, each of which can mutate three ways: 1350 possible alleles that can be attained by means of single base mutations in a very small gene. Of these, most will be at least slightly harmful; some will not affect the fitness of the organism to any appreciable extent; a few may be slightly beneficial. Which if any of the beneficial changes that might occur will actually become evolutionary events may depend primarily upon which occurs first through mutation and happens also to survive the high probability of being lost while it is still rare. The mutation which becomes fixed as an evolutionary event might not be the best of the possible mutations; the best possible mutation may simply fail to occur. Once the gene has changed, the former spectrum of evolutionary possibilities is closed. An entirely different spectrum of 1350 possible alleles is available through single base change [31]. Perhaps among these there will be a much larger proportion of significantly beneficial alleles than formerly. One change in a gene, even a selectively neutral or slightly deleterious change fixed through accident and drift, can open up evolutionary possibilities previously unavailable.

J. Maynard Smith has envisioned a multidimensional "protein space" in which all possible proteins exist, connected by single base changes and other single evolutionary steps [31]. Evolutionary change is achievable only by going from functional proteins to nearby proteins that are either selectively superior or approximately equivalent. Natural selection speeds the passage from one protein to a superior form, but only in conjunction with stochastic processes and only after the pathway has been opened by mutation. The actual path traveled has a large element of randomness and is largely determined by the patterns of mutation and drift. As in Sewall Wright's visionary gene frequency adaptive

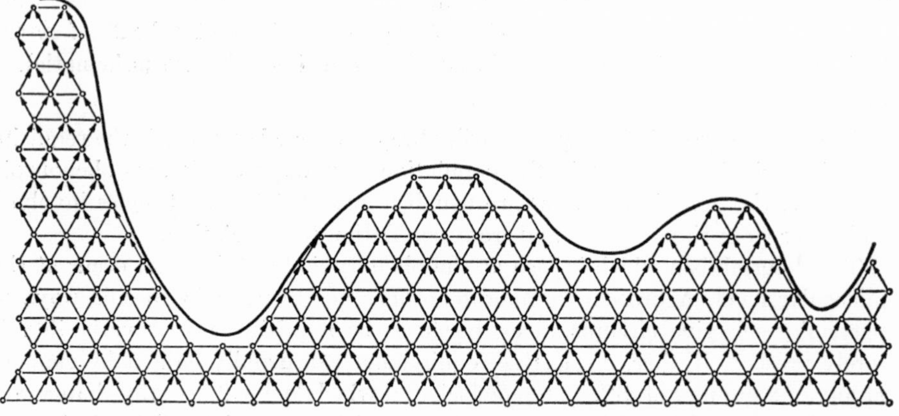

FIGURE 7a

Maynard Smith's "concept of the protein space" [30] in schematic diagram. All possible proteins are connected to one another by paths of possible mutation. In evolution, proteins change primarily either by mutation to superior forms (upward arrows) or to selectively equivalent forms (horizontal lines). The precise path, as well as the form achieved at any point in time is determined jointly by mutation, selection and drift. Few proteins reach "adaptive peaks" from which all change is deleterious. The vertical structure changes with time.

landscapes [56], there may be local adaptive peaks, at the pinnacles of which evolutionary change ceases at least for a while. Perhaps Histone IV has achieved such an adaptive peak, where all readily available changes are maladaptive. Maynard Smith's concept is shown schematically in Figures 7a and 7b; however, this static presentation is incomplete. The protein space diagram shows all possible mutational paths, whereas these occur only briefly and then disappear over long periods of evolutionary time. The relative selective values of different possible alleles make up the adaptive landscape, but these values are constantly changing with the environment and the residual genotype arrays. The protein space is dynamic, flexible, and multidimensional.

It has been pointed out on various occasions [33], [45] that the virtually infinite number of possible proteins or DNA molecules make the achievement of any specific one—say human hemoglobin—essentially impossible if the method

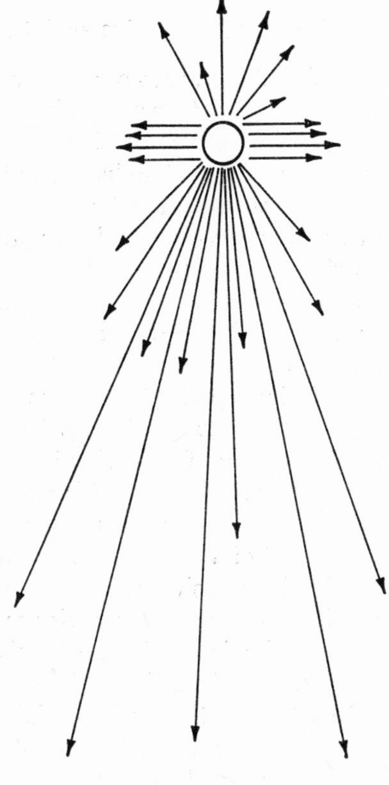

FIGURE 7b

A detail of 7a, this time showing deleterious as well as adaptive and neutral mutation possibilities (schematic). In actuality each protein is related to thousands of other proteins by single mutational steps.

of achievement is through the approved neo-Darwinian method of random mutation followed by natural selection. And this is perfectly true, before the fact. At the time of the origin of life, the probability of the eventual attainment of human hemoglobin was in fact infinitesimal, but it happened. The probability of any organism evolving exactly the same molecule again in the next billion years, starting from an unrelated protein, is likewise effectively zero. Biological molecules do not have any kind of predictable uniqueness, achieved by a deterministic evolution; no allele is the "best of all possible alleles." Presumably, a virtually infinite number of theoretically possible proteins could function as well or better than human hemoglobin; but the combination of past selection, random mutation, and drift has instead achieved the molecule as we know it, which by no accident happens to function quite well enough to allow its carrier to reproduce and compete with other imperfect forms in a changing and unpredictable world.

8. Randomness and nonrandomness in adaptive evolution

Although the first beneficial mutation that occurs in a cistron and is not lost by chance may not be the most advantageous of the changes possible for the cistron, the relative selective advantages of possible beneficial changes do have an overall effect. This is because the probability of not being lost is directly proportional to the selective advantage. Most mutations are lost by chance whether or not they occur first in time, so the relative probability of occurrence and fixation for a specific change is proportional to the product of its rate of occurrence and its probability of fixation for each occurrence. Thus, if two possible beneficial mutations are equally likely to occur, but one has twice the selective advantage of the other, the corresponding probability of one possible change actually being realized in evolution is twice that of the other.

8.1. *The evolutionary origin of dominance and of overdominance.* The operative selective advantage in the probability of not being lost is that associated with the mutation when it first occurs, namely, when it is rare and heterozygous. The potential selective advantage of the homozygous mutant is not relevant to the probability of initial establishment of a new allele. A mutant allele whose potential beneficial effect is completely recessive will not be likely to become established, relative to the probability of success for one with some degree of dominant advantage. The "evolution of dominance" quite likely lies in the relative probabilities for dominant and recessive alleles of becoming fixed in the first place [41]. Similarly, a mutant allele that is beneficial in the heterozygote is already well on its way to becoming established before it first occurs in a homozygous genotype. Just as most mutant *heterozygous* genotypes are likely to be deleterious when they first occur, this first mutant *homozygous* genotype is likely to turn out to be deleterious. One then has a case of balanced heterosis that might survive indefinitely, or perhaps only until modifying genes favorably alter the fitness of one of the homozygous genotypes, or until yet more favorable alleles arise [41].

9. Random mutation specifies amino acid composition of proteins

Since adaptive mutations which are fixed are those which arise first and are not lost before they are able to increase to sufficient numbers to assure evolutionary retention, the overall pattern of evolutionary change will strongly reflect the patterns of mutation. In an earlier article [24], we wrote that, in the case of adaptive evolution, "one particular amino acid will be optimal at a given site in a given organism, and it matters little whether there are six possible codons (as there are for serine) or only one (as there is for methionine)." On further reflection, we must reject this deterministic view of adaptive molecular evolution. King and Jukes [24] and Kimura [19], [39] showed that the frequencies of 19 of the 20 amino acids could be predicted with surprising accuracy by random permutations and combinations of nucleotide bases, as read by the genetic code (Figures 8 and 9). We originally interpreted this evidence of randomness in molecular evolution as indirect support for the idea of non-Darwinian evolution [24], but have come to realize that it is evidence only against a deterministic view of molecular evolution. It is quite consistent with alternative views of adaptive evolution [23]. Suppose that at a given time, there are several possible amino acid substitutions that might improve a protein, and among these are changes to serine and to methionine; serine, with its six codons, has roughly six times the probability of becoming fixed in evolution. Once this has occurred, the mutation to methionine may no longer be advantageous. Ultimately, the amino acid frequency composition of proteins will reflect the frequencies at which the various amino acids arise by mutation, which in turn depends on the genetic code and DNA nucleotide frequency composition.

These studies have turned up four major and a few minor systematic discrepancies from complete randomness of structural gene DNA and amino acid composition. The minor discrepancies can be seen in Figures 8 and 9; although the fit between expected and observed frequencies is close, it is not exact. Note that all the amino acids that are above the line (more common than predicted) in the graph of microorganismal protein composition are also on the line or above in the graph of mammalian protein composition; those that are below the line in one graph are on or below the line in the other. Glutamine is the one exception out of twenty amino acids. This simply indicates that some amino acids are more likely to be deleterious when they arise by mutation than are other amino acids.

The major systematic discrepancies from randomness are all worth investigating in some detail. They are: (1) arginine is present at about one third its expected frequency in both groups [24]; (2) as noted by Ohta and Kimura [39], the calculated base composition of mRNA is rather different for the first nucleo-

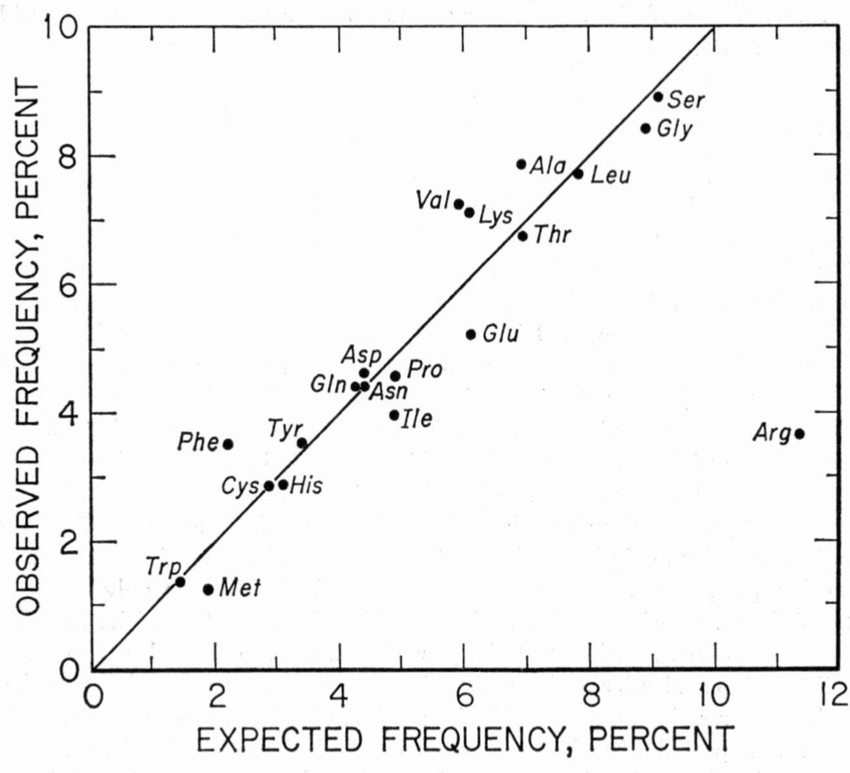

FIGURE 8

29 unrelated mammalian proteins, 3780 residues. The amino acid composition of 29 mammalian proteins was analyzed to determine the base composition of the relevant mRNA. Random permutations and combinations of these base frequencies predicted amino acid frequencies rather well except for arginine ($r = 0.96$ for the other 19 amino acids), indicating a large element of randomness in protein composition. Base frequency estimations were corrected for selection against arginine and chain terminating codons.

tide position of the implied codons than it is for the second nucleotide position; (3) as previously noted by King and Jukes [24], structural DNA is not symmetrical with regard to base composition; from the codons implied by the observed amino acid frequencies, it appears that the purine content of the transcribed strand is 43.6 per cent and that of the nontranscribed strand is 56.4 per cent; (4) the work of Josse [15] and Subak-Sharpe [52] on nearest neighbor frequencies have shown that some sequential pairs of bases are unexpectedly rare in vertebrate DNA, specifically CpG and TpA dinucleotides.

9.1. *The marked deficiency of arginine.* The frequency of arginine departs from random expectation for essentially the same reason that other amino acid

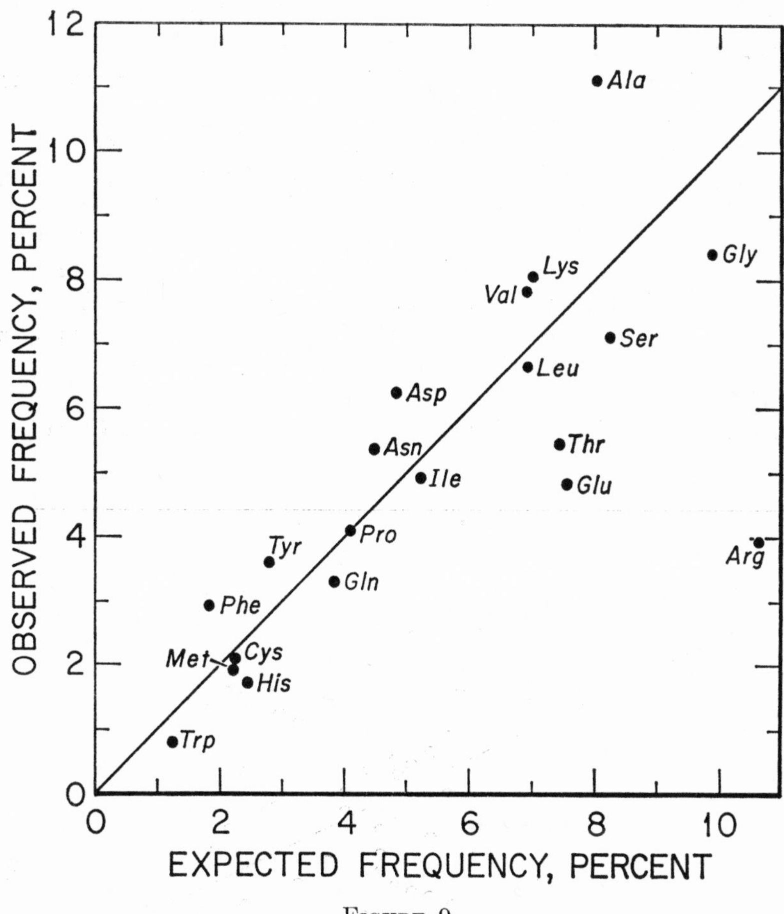

FIGURE 9

16 bacterial proteins, 2679 residues. Same as Figure 8, but for 16 bacterial and viral proteins. Note that nearly all amino acid frequencies that are greater than expected in Figure 8 are greater than expected in Figure 9. For all amino acid frequencies except arginine, $r = 0.87$.

frequencies depart. In view of the magnitude of the discrepancy, however, some documentation is in order.

Apparently mutations to arginine are selected against more often than are mutations to other amino acids because of its rather special chemical and structural properties, which make all arginine substitutions "radical." For substantiation of this hypothesis, one must turn to the only objective and quantitative measure of total physiochemical differences between amino acids, the difference index of Sneath [49]. Sneath classified the 20 amino acids according to 134 dichotomous categories of structural detail and physiochemical activity.

The difference index is the unweighted sum of all the categories not shared by a given pair of amino acids.

In Figure 10, adapted with permission from Clarke [5], all of the 75 amino acid substitutions that can be achieved by single nucleotide base changes are arranged according to two criteria: (1) Sneath's difference index (horizontal axis); and (2) Clarke's calculation of the log of the relative probability of evolutionary acceptance of the amino acid substitution (vertical axis). Clarke arrived at the latter values by dividing the number of such substitutions in several phylogenetic trees calculated by Dayhoff [8] by the relative probability for each substitution of arising by mutation (vertical axis). Clarke's regression clearly shows that the more different two amino acids are according to Sneath's index, the less likely is a mutational interchange between them to be accepted as an evolutionary event.

Interchanges involving arginine are indicated by open circles. These tend to

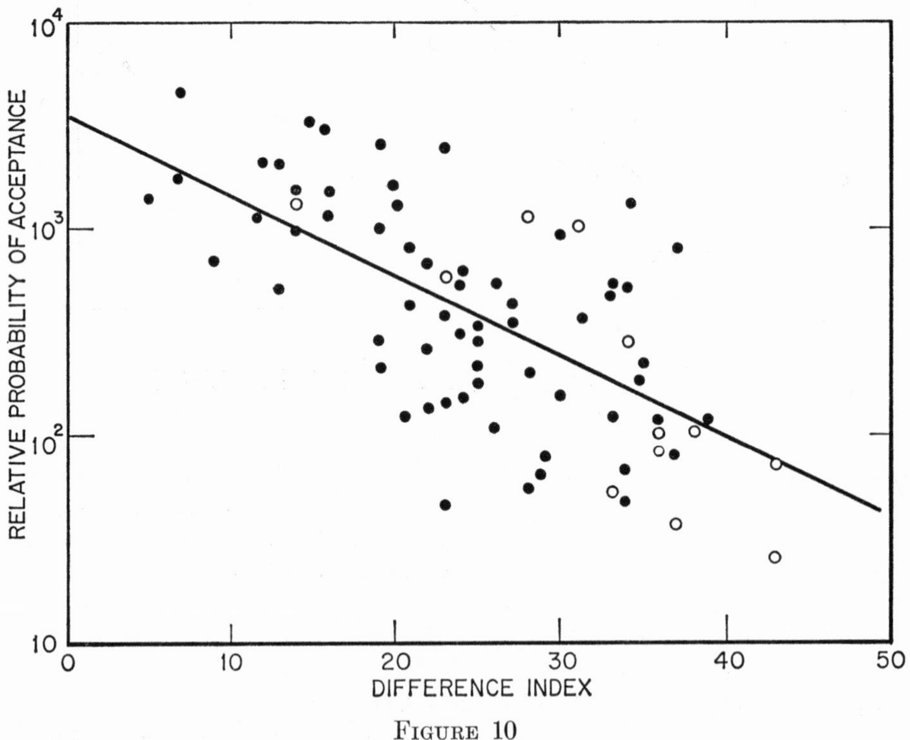

FIGURE 10

The relationship between the relative probability of evolutionary acceptance and Sneath's index of chemical difference for 75 single base amino acid interchanges. The regression indicates that substitutions involving dissimilar amino acids are relatively unlikely to become accepted as evolutionary events. Open circles indicate pairs involving arginine. Adapted with permission from Clarke [5].

cluster in the lower right. The mean index value for the 12 single base amino acid substitutions involving arginine is 33.0; the mean for the remaining 63 single base amino acid substitutions *not* involving arginine is 23.4 ($t = 3.14$, $p < 0.01$). According to Clarke's regression, this difference in mean index values (9.6 units) would predict that arginine substitutions are only about one tenth as likely to be evolutionarily acceptable as the average of all other amino acid substitutions, and the actual calculations confirm this (Figure 10).

REMARK. Incidental to the above analysis, but of considerable interest, is the observation that the mean difference index value is 25.0 for the 75 amino acid substitutions that can be achieved with a single nucleotide base change, while the mean index value for the remaining 115 amino acid substitutions, each requiring two or more nucleotide base changes, is 28.0. While this gives some quantitative support to the often reported observation that the genetic code itself appears to favor "conservative" mutational amino acid interchanges, the magnitude of the effect is rather small. The average difference value for 29 amino acid changes that are referable to nucleotide changes in the first position of the codon is 21.8; 42 interchanges can be achieved by second position changes and 7 by third position changes, with mean difference values of 26.5 and 25.1, respectively.

9.2. *First and second codon positions in structural DNA.* Ohta and Kimura [39] analyzed vertebrate *m*RNA composition in a manner similar to King and Jukes [24], by translating protein composition into implied *m*RNA codons and tabulating the first and second position nucleotide frequencies of the implied codons. They found that the nucleotide frequencies of the first position were very different from those of the second position (Table I).

I decided to test the hypothesis that the discrepancy is due to two known departures from randomness: (1) the chain terminating codons UAA, UAG or UGA were not present and (2) arginine is anomalously rare. Following the prodecure described by Ohta and Kimura, I constructed an iterative computer program in which the different codons assigned to a single amino acid would be apportioned according to their relative predicted frequencies; the sum of the frequencies of the codons coding for any one amino acid would be equal to the frequency of the amino acid in the proteins sampled. Real data was used for the frequencies of the 55 codons coding for 19 amino acids. The iterative program added *computed* frequencies for the six arginine codons and the three chain terminating codons, then analyzed all codon frequencies to derive new nucleotide frequencies before the next iteration. Analysis of the final composite of real and "expected" codon frequencies showed much smaller differences in nucleotide frequencies between the first and second codon positions, proving that most of the discrepancy was due to the difference between randomly expected and actual frequencies of arginine and chain terminating codons (Table I).

9.3. *The anomalous rarity of CpG and TpA DNA doublets.* CpG doublets occur in vertebrate DNA at about ten per cent of their expected frequency; TpA doublets are also found less frequently than expected, although the discrepancy

TABLE I

NUCLEOTIDE FREQUENCIES IN THE FIRST AND SECOND CODON POSITIONS OF IMPLIED *m*RNA

All figures are in per cent.

Amino acid composition of proteins can be translated into implied *m*RNA codon frequencies, which are then analyzed for nucleotide frequencies at the first and second positions of each codon. Ohta and Kimura [39] found a marked discrepancy between the frequency distributions of the first and second positions. This discrepancy can be shown to be due primarily to the known deficiencies of arginine and of the three chain terminating codons. U(1), C(1), A(1), G(1) are the frequencies of uracil, cytosine, adenine, and guanine in the first position; U(2), C(2), A(2), G(2) are the nucleotide frequencies in the second position in per cent.

17 mammalian proteins, Ohta and Kimura (actual gene composition) [39]	U(1) = 19.6 U(2) = 23.5	C(1) = 18.0 C(2) = 23.9	A(1) = 28.2 A(2) = 32.9	G(1) = 34.2 G(2) = 19.7
Absolute difference	3.9	5.9	4.7	14.5
Same data (corrected for deficiencies of Arg and chain terminating codons)	U(1) = 22.3 U(2) = 20.4	C(1) = 19.1 C(2) = 20.1	A(1) = 28.8 A(2) = 32.4	G(1) = 29.7 G(2) = 27.1
Absolute difference	1.9	1.0	3.6	2.6
29 unrelated mammalian proteins (corrected for deficiencies of Arg and chain terminating codons)	U(1) = 22.2 U(2) = 20.6	C(1) = 20.2 C(2) = 21.0	A(1) = 28.8 A(2) = 31.7	G(1) = 28.8 G(2) = 26.8
Absolute difference	1.6	0.8	2.9	2.0
16 bacterial and phage proteins (corrected for deficiencies of Arg and chain terminating codons)	U(1) = 19.2 U(2) = 20.9	C(1) = 16.9 C(2) = 20.9	A(1) = 30.8 A(2) = 32.4	G(1) = 33.1 G(2) = 25.8
Absolute difference	1.7	4.0	1.6	7.3

is not so marked. It seems highly likely that there is some connection between the rarity of CpG doublets and the rarity of arginine; four of the six arginine codons contain the doublet. What might be the cause and effect relationship? Selection against arginine would not directly lower the CpG doublets to such an extent; arginine is not that rare, and CpG doublets occur in other codons and presumably bridge adjacent codons. Furthermore, arginine is also rare in bacteria, and the CpG doublet deficiency does not occur in bacteria. On the other hand, the CpG doublet deficiency could cause the arginine deficiency in vertebrates, but then one would have to postulate two mechanisms for the rarity of arginine, one in vertebrates and one in microorganisms.

Elsewhere [23], I have suggested that the primary cause is the generally deleterious character of new arginine substitution mutations. These are selected against directly in microorganisms. They are also selected against directly in vertebrates, but in addition vertebrates have evolved a secondary mechanism that serves to protect them against these deleterious mutations. The patterns of mutation are determined, to some large extent, by the action of the DNA repair

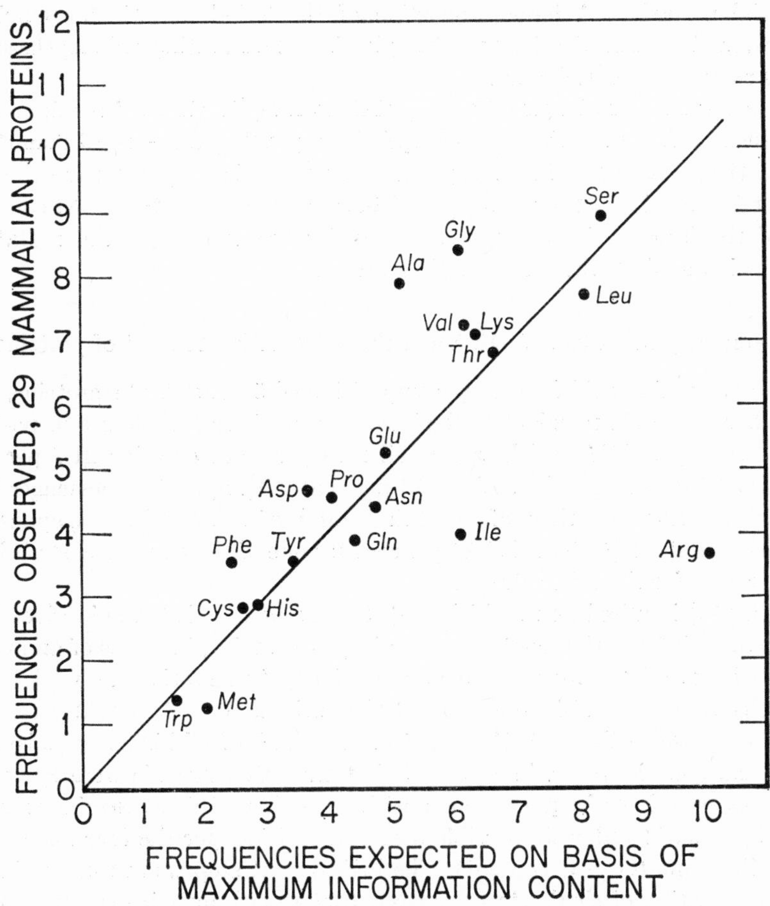

FIGURE 11

Expected frequencies of amino acids based on the nucleotide frequencies giving maximum information content to protein composition, and compared with frequencies observed in a sample of 29 unrelated mammalian proteins taken from Dayhoff's Atlas [8].

enzyme systems and also by DNA replication systems. Vertebrates have apparently evolved a means of editing out CpG doublets when they occur, or perhaps even long after they occur. The selective advantage of such a mechanism is obvious.

Chain termination mutations are even more likely to be deleterious than are arginine mutations. Two of the three *m*RNA chain terminating codons contain the doublet UpA, which corresponds to TpA doublets in both strands of the relevant DNA. The deficiency of TpA in vertebrate DNA is also probably due to secondary mechanisms that protect against harmful mutation while allowing beneficial mutation. A large proportion of the lethal mutations occurring in bacteria are base substitutions resulting in chain terminating codons; these may be less common in vertebrates.

Large mammalian viruses that bring their own replicating and repair enzymes with them do not show the CpG or TpA discrepancies, while small mammalian viruses that utilize the host's replicating and repair enzymes have the same doublet frequencies as their hosts [52]. This seems to be a rather good confirmation of the hypothesis that the discrepancies are caused by the mutational patterns imposed by these enzymatic systems.

10. Asymmetry in structural DNA and the information content of proteins

From the amino acid frequency composition of 53 vertebrate proteins, King and Jukes were able to calculate the base composition of the first two positions of the messenger RNA as follows: uracil 22.0 per cent, cytosine 21.7 per cent, adenine 30.3 per cent, and guanine 26.1 per cent [24]. Since the messenger RNA base frequencies reflect those of one of the strands of structural DNA, apparently the purine content (A + G) is 56.4 per cent for one of the two DNA strands and 43.6 per cent for the other strand.

Smith [48] inquired as to which G + C content of DNA would give rise to the maximal information content (diversity) of proteins, in terms of amino acid frequencies predicted by random permutations of nucleotides as read by the genetic code. He was able to show that the optimal G + C content was 41 per cent, very close to the G + C content of the DNA of all higher organisms. In his calculations, however, he assumed that the DNA strands were symmetric with respect to base frequency composition. Without this constraint there are three degrees of freedom in possible base composition (for the four nucleotides) rather than one (G + C content). The logical next step was to determine the base composition of single strand *m*RNA that would optimize the information content (diversity) of predicted amino acid frequencies as read by the genetic code.

Dr. Glenn Sharrock assisted me in devising a computer program that would determine which array of *m*RNA base frequencies would optimize the diversity of amino acids. The optimal frequencies of amino acids are 0.05 each, the true maximum of the Shannon index of diversity or information content. The base frequencies were allowed to vary so as to allow predicted amino acid frequencies to approach the optimal frequencies. The sum square deviations between calculated and optimal amino acid frequencies were minimized by the Gauss-Newton method on nonlinear regression through steepest descents (UCLA Biomed program $x - 85$) [14]. We decided to leave the frequencies of the chain terminating and arginine codons out of the maximization criterion since they are known to be subject to strong natural selection and/or to behave anomalously. Widely different initial values all converged on the same predicted optimal base frequencies (Table II).

TABLE II

BASE COMPOSITION OF mRNA
All figures are in per cent.

The highest information content (Shannon index of diversity) that can be achieved with random permutations of mRNA nucleotides, as translated by the genetic code, is achieved with the nucleotide frequencies listed in the first column. The second column gives mRNA frequencies previously calculated from protein composition. It appears that evolutionary mechanisms tend to maximize the diversity of new mutations within constraints set by the universal genetic code.

mRNA nucleotide	Optimal composition for maximum index of diversity for predicted amino acid frequencies	Composition calculated from amino acid content of 53 vertebrate proteins (King and Jukes [24])
Uracil	23.3	22.0
Cytosine	19.4	21.7
Adenine	32.3	30.3
Guanine	25.0	26.1

These values seem close enough to suggest that base frequencies are indeed under genetic and evolutionary control, through control of repair and replication systems that determine the rates and pattern of mutational substitutions; and that the evolutionarily determined pattern of mutation is such as to maximize the diversity of amino acid substitutions. Such a pattern may be adaptive because it maximizes the functional novelty of new mutations without increasing the total mutation rate.

There are other good reasons to believe that the base composition of DNA is under genetic control to some considerable extent. There is a marked deviation from expectation in the respective frequencies of each of the twelve different kinds of base substitutions, when one considers evolutionarily fixed changes, and these changes are markedly asymmetric with regard to DNA strands [9]. Quite likely the twelve kinds of base substitution differ similarly in direct mutational probability. In noninformational DNA, which is not subject to constraints of natural selection and which, in all likelihood, constitutes most vertebrate DNA, the equilibrium frequency of each nucleotide is determined by the equilibrium of forward and backward mutation rates. For example, the frequency of G will be determined by the rate of mutation to G divided by the sum of mutations to and from G. These rates in turn are influenced and controlled by the characteristics of the repair and replication enzymes: the affinity of each enzyme for each kind of nucleotide, the ability to recognize some kinds of errors more readily than others, and so on. The asymmetry of structural DNA is best explained by assuming that there are differences between the transcribed and nontranscribed strands that cause them to interact differently with these enzyme systems. This might be the same difference that enables RNA polymerase to recognize and transcribe one strand exclusively.

It appears that there is a very large element of randomness in adaptive as well as neutral evolution at the molecular level. The general statistical patterns of evolutionary molecular change are dictated primarily by mutation. At the same time, it is beginning to appear that there is a large and heretofore unexpected element of *non*randomness in mutation itself.

REFERENCES

[1] B. AMES, "The nature and frequency of spontaneous mutations," unpublished paper presented at the Sixth Berkeley Symposium on Mathematical Statistics and Probability, Conference on Evolution, April, 1971.

[2] F. J. AYALA, "Evolution of fitness. I. Improvement in the productivity and size of irradiated populations of *Drosophila serrata* and *Drosophila birchii*," *Genetics*, Vol. 53 (1966), pp. 883–895.

[3] T. BAYLEY, J. A. CLEMENTS, and A. J. OSBAHR, "Pulmonary and circulatory effects of fibrinopeptides," *Circ. Res.*, Vol. 21 (1967), pp. 469–485.

[4] B. BLOMBÄCK, M. BLOMBÄCK, P. OLSSON, L. SVENDSEN, and G. ABERG, "Synthetic peptides with anticoagulant and vasodilating activity," *Scand. J. Clin. Lab. Invest. Suppl.*, Vol. 107 (1969), pp. 59–64.

[5] B. CLARKE, "Selective contraints on amino-acid substitutions during the evolution of proteins," *Nature*, Vol. 228 (1970), pp. 159–160.

[6] J. F. CROW, "Darwinian and non-Darwinian evolution," *Proceedings of the Sixth Berkeley Symposium on Mathematical Statistics and Probability*, Berkeley and Los Angeles, University of California Press, 1972, Vol. 5, pp. 1–22.

[7] J. F. CROW, and M. KIMURA, *An Introduction to Population Genetics Theory*, New York, Harper and Row, 1970.

[8] M. O. DAYHOFF, *Atlas of Protein Sequence and Structure*, Vol. 4, Silver Spring, Md. National Biomedical Research Foundation, 1969.

[9] W. M. FITCH, "Evidence suggesting a non-random character to nucleotide replacements in naturally occurring mutations," *J. Mol. Biol.*, Vol. 26 (1967), pp. 499–507.

[10] W. M. FITCH and E. MARGOLIASH, "The usefulness of amino acid and nucleotide sequences in evolutionary studies," *Evol. Biol.* (edited by W. Steere, T. Dobzhansky, and M. K. Hecht), Vol. 4 (1971), to appear.

[11] W. M. FITCH and E. MARKOWITZ, "An improved method for determining codon variability in a gene and its application to the rate of fixation of mutations in evolution," *Biochem. Genetics*, Vol. 4 (1970), pp. 579–593.

[12] T. C. GIBSON, M. L. SCHEPPE, and E. C. COX, "On the fitness of an *E. coli* mutation gene," *Science*, Vol. 169 (1970), pp. 686–690.

[13] J. B. S. HALDANE, "The cost of natural selection," *J. Genet.*, Vol. 56 (1957), pp. 11–27.

[14] H. O. HARTLEY, "The modified Gauss-Newton method for fitting non-linear regression functions by least squares," *Technometrics*, Vol. 3 (1961), pp. 269–280.

[15] J. JOSSE, A. D. KAISER, and A. KORNBERG, "Enzymatic synthesis of deoxyribonucleic acid. VIII. Frequencies of nearest neighbor base sequences in deoxyribonucleic acid," *J. Biol. Chem.*, Vol. 236 (1961), pp. 861–875.

[16] B. P. KAUFMANN, "Spontaneous mutation rate in *Drosophila*," *Amer. Natur.*, Vol. 81 (1947), pp. 77–80.

[17] M. KIMURA, "Some problems of stochastic processes in genetics," *Ann. Math. Statist.*, Vol. 28 (1957), pp. 882–901.

[18] ———, "On the evolutionary adjustment of spontaneous mutation rates," *Genet. Res.*, Vol. 9 (1967), pp. 23–34.

[19] ———, "Genetic variability maintained in a finite population due to mutational production of neutral and nearly neutral isoalleles," *Genet. Res.*, Vol. 11 (1968), pp. 247–269.

[20] ———, "Evolutionary rate at the molecular level," *Nature*, Vol. 217 (1968), pp. 624–626.

[21] M. KIMURA, and T. OHTA, "Population genetics, molecular biometry and evolution," *Proceedings of the Sixth Berkeley Symposium on Mathematical Statistics and Probability*, Berkeley and Los Angeles, University of California Press, 1972, Vol. 5, pp. 43–68.

[22] J. L. KING, "Continuously distributed factors affecting fitness," *Genetics*, Vol. 55 (1967), pp. 483–492.

[23] ———, "The influence of the genetic code on protein evolution," *Biochemical Evolution and the Origin of Life* (edited by E. Schoffeniels), Amsterdam, North Holland, 1972, pp. 3–13.

[24] J. L. KING and T. H. JUKES, "Non-Darwinian evolution," *Science*, Vol. 164 (1969), pp. 788–798.

[25] D. E. KOHNE, "Evolution of higher-organism DNA," *Quart. Rev. Biophys.*, Vol. 3 (1970), pp. 327–375.

[26] D. E. KOHNE, J. A. CHISCON, and B. H. HOYER, "Evolution of mammalian DNA," *Proceedings of the Sixth Berkeley Symposium on Mathematical Statistics and Probability*, Berkeley and Los Angeles, University of California Press, 1972, Vol. 5, pp. 193–210.

[27] E. G. Leigh, Jr., "Natural selection and mutability," *Amer. Natur.*, Vol. 104 (1970), pp. 301–305.

[28] R. C. LEWONTIN and J. L. HUBBY, "A molecular approach to the study of genic heterozygosity in natural populations. II. Amount of variation and degree of heterozygosity in natural populations," *Genetics*, Vol. 54 (1966), pp. 595–609.

[29] S. E. LUNA and M. DELBRÜCK, "Mutations of bacteria from virus sensitivity to virus resistance," *Genetics*, Vol. 28 (1943), pp. 491–511.

[30] J. MAYNARD SMITH, " 'Haldane's dilemma' and the rate of evolution," *Nature*, Vol. 219 (1968), pp. 1114–1116.

[31] ———, "Natural selection and the concept of a protein space," *Nature*, Vol. 225 (1970), pp. 563–564.

[32] E. MAYR, *Animal Species and Evolution*, Cambridge, Harvard University Press, 1963.

[33] P. S. MOORHEAD and M. M. KAPLAN (editors), *Mathematical Challenges to the Neo-Darwinian Interpretation of Evolution*, Philadelphia, Wistar Institute Press, 1967.

[34] N. E. MORTON, J. F. CROW, and H. J. MULLER, "An estimate of the mutational damage in man from data on consanguineous marriages," *Proc. Nat. Acad. Sci. U.S.A.*, Vol. 42 (1956), pp. 855–863.

[35] A. G. Motulsky, "Some evolutionary implications of biochemical variants in man," *Proceedings of the Eighth Congress of Anthropological and Ethnological Sciences*, 1970, pp. 364–365.

[36] T. MUKAI, "The genetic structure of natural populations of *D. melanogaster*. I. Spontaneous mutation rate of polygenes controlling viability," *Genetics*, Vol. 50 (1964), pp. 1–19.

[37] H. J. MULLER, "Advances in radiation mutagenesis through studies on *Drosophila*," *Progress in Nuclear Energy, Series VI, Biological Sciences*, Vol. 2, London, Pergamon Press, 1959.

[38] A. NOVICK and L. SZILARD, "Genetic mechanisms in bacteria and bacterial viruses I. Experiments on spontaneous and chemically induced mutations of bacteria growing in the chemostat," *Cold Spring Harbor Symp. Quant. Biol.*, Vol. 16 (1953), pp. 337–344.

[39] T. OHTA and M. KIMURA, "Statistical analysis of the base composition of genes using data on the amino acid composition of proteins," *Genetics*, Vol. 64 (1970), pp. 387–395.

[40] ———, "On the constancy of the evolutionary rate of cistrons," *J. Molec. Evol.*, Vol. 1 (1971), pp. 18–25.

[41] P. A. PARSONS and W. F. BODMER, "The evolution of overdominance: natural selection and heterozygote advantage," *Nature*, Vol. 190 (1961), pp. 7–12.

[42] E. M. PRAGER and A. C. WILSON, "Multiple lysozymes of duck egg white," *J. Biol. Chem.*, Vol. 246 (1971), pp. 523–530.

[43] F. J. Ryan, "Spontaneous mutation in non-dividing bacteria," *Genetics*, Vol. 40 (1955), pp. 726–738.

[44] ———, "Natural mutation in nondividing bacteria. *Trans. N.Y. Acad. Sci. Ser. 2*, Vol. 19 (1957), pp. 515-517.

[45] F. B. Salisbury, "Natural selection and the complexity of the gene," *Nature*, Vol. 224 (1969), pp. 342–343.

[46] V. M. Sarich and A. C. Wilson, "Rates of albumin evolution in primates," *Proc. Nat. Acad. Sci. U.S.A.*, Vol. 58 (1967), pp. 142–148.

[47] ———, "Immunological time scale for hominic evolution," *Science*, Vol. 158 (1967), pp. 1200–1203.

[48] T. F. Smith, "The genetic code, information density, and evolution," *Math. Biosci.*, Vol. 4 (1969), pp. 179–187.

[49] P. H. A. Sneath, "Relations between chemical structure and biological activity in peptides," *J. Theor. Biol.*, Vol. 12 (1966), pp. 157–195.

[50] C. Stern, "The role of genes in differentiation," *Proc. Int. Genet. Symp.*, Tokyo, 1957, pp. 70–72.

[51] A. C. Stevenson and C. B. Kerr, "On the distributions of frequencies of mutation to genes determining harmful traits in man," *Mutat. Res.*, Vol. 4 (1967), pp. 339–352.

[52] J. H. Subak-Sharpe, "The doublet pattern of nucleic acids in relation to the origin of viruses," *Handbook of Molecular Cytology* (edited by A. Lima-de-Faria), Amsterdam, North Holland, 1969.

[53] J. A. Sved, "Possible rates of gene substitution in evolution," *Amer. Natur.*, Vol. 102 (1968), pp. 283–292.

[54] L. K. Wainright, "Spontaneous mutation in stored spores of a *Streptomyces*," *J. Gen. Microbiol.*, Vol. 14 (1956), pp. 533–544.

[55] H. J. Whitfield, Jr., R. G. Martin, and B. Ames, "Classification of aminotransferase (C gene) mutants in the histidine operon," *J. Molec. Biol.*, Vol. 21 (1966), pp. 335–355.

[56] S. Wright, "Random drift and the shifting balance thoeory of evolution," *Mathematical Topics in Population Genetics* (edited by K. Kojima), Berlin-Heidelberg-New York, Springer-Verlag, 1970.

COMPARISON OF POLYPEPTIDE SEQUENCES

THOMAS H. JUKES
UNIVERSITY OF CALIFORNIA, BERKELEY

1. Introduction

Evolution is able to take place because the inherited information in living organisms is subject to change. Most of this information is constant from generation to generation, but changes that take place in a very small part of it are necessary for evolution. Hereditary information is stored in the base sequences of DNA. Molecular evolution is, therefore, concerned with alterations in DNA. Changes in DNA are of three general types: *point mutations*, in which one base pair is substituted for another; *recombination*, in which two double strands of DNA cross over, so that there is a lengthening of one double strand and a shortening of the other; and *duplication*, proliferation, or multiplication, in which the amount of DNA per cell becomes increased. Any change of these three types is evolutionary if it is predominantly adopted into the genome of a species.

Part of the DNA, probably only a small fraction, consists of base sequences that provide information for the synthesis of proteins, and it is through proteins that most of the phenotypic expression of hereditary information takes place. This paper is concerned with evolutionary changes in proteins, as detected by analyzing proteins and by comparing the proteins of different species of living organisms with each other. Differences between two similar proteins, such as the hemoglobins obtained from man and monkey, will represent evolutionary changes in DNA that have entered into the makeup of the species. The last-mentioned two types of change, recombination and duplication, are infrequently adopted; for example, there seem to have been only six evolutionary events of recombination in the hemoglobins in about 500 million years. In contrast, base replacements take place incessantly. These produce the "evolutionary clock" that ticks slowly in proteins, independent of speciation, generation time or gene duplication, but *not* independently of the type of protein as defined by its essential function. We shall illustrate this by comparing the amino acid sequences of polypeptide chains, with the use of the amino acid code. A family of proteins, the globins, that occurs in all vertebrates, will be used as illustrative examples.

Supported by a grant from the National Aeronautics and Space Administration NgR 05-003-020 "The Chemistry of Living Systems" to the University of California.

Most of the studies on polypeptide chains have been carried out by protein chemists, and hence, biochemical viewpoints are quite prominent in this field. In considering amino acid substitutions during evolution, emphasis has been placed on the amino acids themselves, rather than on the nucleic acid code which is responsible for the changes. One result of this is a great underestimate of the number of evolutionary events during the divergence of two homologous proteins. Another result is that amino acids have been placed by several authors into so-called similar groups.

I shall include all and any base substitutions that are incorporated in the DNA of a species in the definition of *evolutionary events,* regardless of where the substitution is, or whether it produces a change in an amino acid residue. For example, there is an evolutionary substitution of CUA for UUA, both of which code for leucine, in the coat protein cistrons of the RNA viruses f2 and R17. Such neutral changes can affect the results produced by subsequent mutations.

Perhaps the first demonstration that amino acid replacements were an expression of evolutionary differentiation was the discovery by Sanger and co-workers [3], [13], [29] of differences between insulin A chains as shown in Table I. The remainders of the chains are identical.

TABLE I

Species	Residue No. 8 9 10
Cow	-Ala-Ser-Val-
Horse	-Thr-Gly-Ile-
Sheep	-Ala-Gly-Val-
Pig	-Thr-Ser-Ile-

By using the genetic code (Table II), the preceding series can be rewritten as alterations of a single strand of DNA (see Table III), where A, C, G, T are the four bases present in DNA, N is any one of them, and Y is C or T.

The molecular steps in this evolutionary process would seem to be simplicity itself, but this is not so. Consider the comparisons of the β chains of hemoglobins portrayed in Table IV.

The difference between the second and third lines is the result of a point mutation, as originally identified by Ingram [16]. There are more than 100 different identified point mutations in human hemoglobin, and each corresponds to a single base change in the genetic code (Table II). For this and other reasons, we conclude that the change from Glu to Val in hemoglobin S was caused by a point mutation from A to T in the codon for glutamic acid at site 6. But we do not know the steps in the evolutionary change from Gly to Pro at site 5 in lines

TABLE II

THE AMINO ACID CODE

Y = U or C; R = A or G; N = U, C, A or G.

Phenylalanine (Phe)	UUY	Serine (Ser)	UCN	Tyrosine (Tyr)	UAY	Cysteine (Cys)	UGY
Leucine (Leu)	UUR			Chain Termination (CT)	UAR	Chain Termination (CT)	UGA
						Tryptophan (Trp)	UGG
Leucine (Leu)	CUN	Proline (Pro)	CCN	Histidine (His)	CAY	Arginine (Arg)	CGN
				Glutamine (Gln)	CAR		
Isoleucine (Ile)	AUY, AUA	Threonine (Thr)	ACN	Asparagine (Asn)	AAY	Serine (Ser)	AGY
Methionine (Met)	AUG			Lysine (Lys)	AAR	Arginine (Arg)	AGR
Valine (Val)	GUN	Alanine (Ala)	GCN	Aspartic acid (Asp)	GAY	Glycine (Gly)	GGN
				Glutamic acid (Glu)	GAR		

TABLE III

Species	Base Sequence
Cow	-G-C-N-A-G-Y-G-T-Y-
Horse	-A-C-N-G-G-Y-A-T-Y-
Sheep	-G-C-N-G-G-Y-G-T-Y-
Pig	-A-C-N-A-G-Y-A-T-Y-

1 and 2. The interpretation, as Gatlin [12] says of matters related to the information in DNA, depends on the context. Indeed, the question of the extent to which two amino acids are interchangeable in a sequence depends on the context more than on their chemical structures. There may have been two changes, such as Gly to Ala and Ala to Pro, or three changes, such as Gly to Ala, Ala to Ser, and Ser to Pro, in the millions of years during which human beings and horses differentiated by separate pathways from a common ancestor.

TABLE IV

Horse hemoglobin	-Ser-Gly-Glu-Glu-Lys-
Human hemoglobin A	-Thr-Pro-Glu-Glu-Lys-
Human hemoglobin S	-Thr-Pro-Val-Glu-Lys-

If we next compare the same sequence in the β hemoglobin chain with the corresponding region in myoglobin, the difference is so great that no homology is perceptible (see Table V). Only by comparing the entire sequences and the tertiary structures of these two proteins can their homology be established. Their evolutionary separation took place probably more than 500 million years ago. We wonder if the glutamic acid at site 7 remained constant during this period, or if it perhaps changed, let us say, from Glu to Asp and back to Glu again, because Asp is found at this site in two other hemoglobin chains.

TABLE V

	4	5	6	7	8
β chain	-Thr-	Pro-	Glu-	Glu-	Lys-
Myoglobin	-Ser-	Asp-	Gly-	Glu-	Trp-

There is a way of measuring the rate of change in DNA molecules directly (Kohne [23]). It is only a rough measurement, and does not identify the changes. I will describe the principle of the method very simply. When two strands of DNA are separated by heating, they reassociate on cooling—the bases seek out their original partners. Most of the DNA in higher organisms consists of what is termed "unique sequence DNA," meaning that its sequence of bases is not repeated elsewhere in the chromosomes. If this fraction of DNA from human cells is heated, and mixed with the corresponding fraction from heated monkey DNA, the single strands of human DNA will form double strands with some of their corresponding partners from monkey DNA, during slow cooling. However, the partnership is imperfect because the human and monkey DNA, which have both descended from a common ancestor, have accumulated base changes during evolution so that the sequences are somewhat different from each other. The hybrid DNA has a lower melting point than the nonhybrid DNA, corresponding to a lowering of about 1°C for each 1.5 per cent of unmatched base sites. This quantitation may be used as an evolutionary clock. If we deduce from the fossil record that the lines of descent leading to human beings and green monkeys separated about 30 million years ago (Wilson and Sarich [31]), then there have been about 18×10^{-10} substitutions per nucleotide pair per year. Most proteins diverge during evolution at a rate lower than this. We deduce from comparing homologous proteins that Darwinian natural selection acts to reject some of the changes in them caused by differentiation of DNA. If we could separate and examine the region of DNA that codes for proteins, we should expect to find that it had been held more constant than other regions in DNA (except for the regions that are transcribed into specialized molecules of RNA, such as transfer RNA and ribosomal RNA). The majority of the unique sequence DNA apparently differentiates sufficiently fast [23], that, evidently, most of its base changes are neutral and are incorporated into the genomes of the species by genetic drift. These changes originate mainly as point mutations. When these occur in structural genes, many of them are translated into amino acid changes in proteins. The genetic drift of such changes occurs at different rates in different proteins, depending on how much the protein can change without losing its function. Cytochromes change slowly, hemoglobins fairly fast, and antibodies somewhat faster.

To study such changes adequately, it is necessary to use the genetic code, for changes in amino acid sequences are responses to changes in DNA. The genetic code is the alphabet of molecular evolution. Students of this subject should be so familiar with the code that when they see ACG on an automobile license plate they should immediately think of threonine.

2. Methods for comparing polypeptides

I shall now describe two of the models used in comparing homologous polypeptide sequences. The first of these ignores the code, and uses a first order decay curve to predict the number of evolutionary amino acid replacements that have occurred at each site. The model has been used by Margoliash and Smith [25], Zuckerkandl and Pauling [32], Dayhoff [5], Kimura [21], and King and Jukes [22]. It says that one or more hits (that is, base substitutions) at any one codon site produce an end result of substitution of the amino acid at that site. The changes are enumerated without reference to the genetic code. Kimura [21] used this procedure to estimate the number of evolutionary events (that is, amino acid replacements) per unit of time as follows. Let x be the average number of unchanged amino acid sites per site in the comparison of two polypeptide chains. For the human: carp α Hb comparison, $x = 0.514$. Then, from the Poisson distribution, $m = 0.665$. This is the number of evolutionary events per site. The evolutionary interval between human and carp is 2×375 million years $= 0.075 \times 10^{10}$ years, that is, twice the length of time since each has separated from a common ancestor. Therefore, evolutionary events in α Hb have occurred at the rate of $0.665 \div (0.075 \times 10^{10}) = 8.9 \times 10^{-10}$ per codon per year since the separation of the carp and human lines. This model makes the following assumptions:

(i) evolutionary amino acid replacements are distributed at random along the polypeptide chains;

(ii) invariable sites, such as the histidines at positions 63 and 121, are included as being subject to evolutionary change;

(iii) no replacements are considered to be revertants to an original amino acid residue.

All three assumptions are erroneous.

I have used a second model, based on minimum base differences per codon (MBDC) (Jukes [18]). The procedure makes no assumptions. It consists of counting all the visible base changes in the codons when two polypeptide chains are compared. Visible base changes are defined as the minimum number of base changes needed for an interchange between two amino acids which are compared. Using this procedure, the minimum number of evolutionary events since the separation of the human and carp lines is as given in Table VI. This corresponds

TABLE VI

Sites compared	Minimum base differences				Average per codon
	0	1	2	3	
140	72	41	27	0	0.68

to 9.1×10^{-10} per codon per year, which is quite close to the value calculated by Kimura's procedure. However, it is an obvious underestimate, for it is calculated in terms of *minimum* base replacements per codon. Assuming randomness, the value of 0.68 implies that there has been the Poisson distribution of events in

TABLE VII

0	1	2	3	4	5
71	48	16	4	0.6	0.1

140 loci shown in Table VII; that is, only 21 codons should have undergone more than one base replacement. The actual number is 27, which indicates that the number of unhit sites *plus* sites hit only once is not less than 113. Referring again to the Poisson distribution, a minimum of 0.805 evolutionary events per codon separate the carp and human α Hb lines, or 10.7×10^{-10} per codon per year. Yet this is still a gross underestimate. Evolutionary events must be measured in terms of base replacements in DNA. As shown by Holmquist [14], the number of primary mutagenic events may be as much as 3.5 times as great as the expected number of amino acid differences between two homologous polypeptide chains. To illustrate this matter, let us examine the effect of a series of base replacements on an individual codon. Single base substitutions in the threonine codon ACU, lead to the example in Table VIII.

TABLE VIII

UCU	CCU	GCU	AUU	AAU	AGU	ACC	ACA	ACG
Ser	Pro	Ala	Ile	Asn	Ser	Thr	Thr	Thr

The three changes to other Thr codons are not expressed, and are, therefore, not included in the next step. This examines the next series of changes, each attributable to two substitutions in the original codon (see Table IX). By

TABLE IX

UCU	to	CCU	ACU	GCU	UUU	UAU	UGU	UCC	UCA	UCG
CCU	to	UCU	ACU	GCU	CUU	CAU	CGU	CCC	CCA	CCG
GCU	to	UCU	CCU	ACU	GUU	GAU	GGU	GCC	GCA	GCG
AUU	to	UUU	CUU	GUU	ACU	AAU	AGU	AUC	AUA	AUG
AAU	to	UAU	CAU	GAU	AUU	ACU	AGU	AAC	AAA	AAG
AGU	to	UGU	CGU	GGU	AUU	ACU	AAU	AGC	AGA	AGG

following this procedure for each of the other three Thr codons the totals for changes of two bases per Thr codon are as shown in Table X, together with examples from other amino acids.

TABLE X

RESULTS OF TWO-BASE SUBSTITUTIONS IN CODONS FOR THREONINE GLYCINE, ASPARAGINE, AND METHIONINE

The results are derived from single base changes, that is, from the codons for amino acids that are mutations resulting from one base substitutions in the original sets of codons for threonine, glycine, asparagine, and methionine, respectively.

(a) Revertants to original amino acid.
(b) Two-base changes that simulate single base changes.
(c) Recognizable two-base changes.

	From threonine (ACN)			From glycine (GGN)	
(a)	(b)	(c)	(a)	(b)	(c)
Thr-24	Ala- 20	Asp- 4	Gly-23	Arg-28	His- 4
	Pro- 20	Cys- 4		Ser- 17	Gln- 4
	Ser- 30	Gln- 4		Trp- 4	Pro- 8
	Lys- 10	Glu- 4		Cys- 8	Leu-11
	Asn- 10	Gly- 8		Glu-10	Met- 2
	Arg- 18	His- 4		Asp-10	Ile- 6
	Ile- 15	Leu- 12		Ala- 20	Thr- 8
	Met- 5	Phe- 4		Val- 20	Lys- 4
		Trp- 2			Asn- 4
		Tyr- 4			Phe- 4
		Val- 8			Tyr- 4
Totals 24	128	58	23	117	59
	From asparagine (AAY)			From methionine (AUG)	
(a)	(b)	(c)	(a)	(b)	(c)
Asn-20	Asp- 6	Ala- 4	Met-9	Leu- 12	Ala- 2
	Tyr- 6	Arg- 12		Val- 8	Asn- 4
	His- 6	Cys- 4		Lys- 4	Gln- 2
	Ser- 10	Gln- 8		Thr- 8	Glu- 2
	Thr- 14	Glu- 8		Arg- 6	Gly- 2
	Ile- 12	Gly- 4		Ile- 6	Phe- 4
	Lys- 4	Leu- 2			Pro- 2
		Met- 2			Ser- 6
		Phe- 2			Trp- 2
		Pro- 4			
		Val- 4			
Totals 20	58	54	9	44	26

The percentage distribution of these changes is seen in Table XI. Extended calculations along these lines show that, roughly speaking, 19 recognizable two-base changes correspond to 54 codons that have been hit more than once. The remaining 35 codons either show no change in amino acid assignment, or show changes that could correspond to a single base replacement.

I shall not pursue the intricacies of the comparison as regards two-base changes any further. Holmquist [14] has presented some mathematical treatment of this question in an accompanying paper. However, the existence of the above relationships must be considered when homologous proteins are compared. For example, the β hemoglobin chains of human and horse differ at 25 loci, and six

TABLE XI

	Thr	Gly	Asn	Met
(a) Revertants	11	11	13	11
(b) Two-base changes simulating single base changes	61	59	43	56
(c) Recognizable two-base changes	28	30	44	33

of the differences correspond to changes that are recognizable as resulting from a minimum of two base replacements in a codon. We shall, for convenience, term these "two-base changes." On the basis of randomization, the number of such changes should be about 2.3. Therefore, there must be some channeling of evolutionary changes in a nonrandom manner, as noted in the cytochromes *c* by Fitch and Margoliash [8] and developed further by Fitch and Markowitz [9]. This channeling seems to take place without regard to much of the often emphasized chemical similarity between certain pairs of amino acids, such as the presence of hydrophobic side chains. The recognizable two-base interchanges in the comparison of human and horse β chains are between Pro and Gly; Thr and Leu; Gly and His; Ala and His; Cys and Val; Pro and Glu. The members of these pairs of amino acids do not show a close chemical resemblance to each other. Evidently protein molecules pay very little attention to the rules laid down by some biochemists as to which amino acids are suitable for evolutionary replacements.

Evolutionary changes in proteins, in addition to replacements of one amino acid by another, include duplication, lengthening and shortening. All these are reflections of corresponding changes in DNA. Lengthening and shortening are produced by genetic recombination. Shortening results in gaps that are found by comparing two polypeptide sequences; for example, there is a gap of five amino acids in the Gun Hill mutant of the β chain of human hemoglobin (Bradley, Wohl, and Rieder [2]). Note (Table XII) that the gap may be written in any of three ways, because of the repetition of Leu-His. When a deletion becomes incorporated in one member of a homologous pair of proteins, the subsequent differentiation of the polypeptide chains can make it very difficult to locate the gap, or even to justify the assumption that it has occurred. The matter was discussed and analyzed by Cantor [4]. The problem is particularly acute if the

TABLE XII

Normal	-Ser	-Glu	-Leu	-His	-Cys	-Asp	-Lys	-Leu	-His	-Val	-Asp	-Pro-
Mutant	-Ser	-Glu	-Leu	-His						-Val	-Asp	-Pro-
or	-Ser	-Glu	-Leu						-His	-Val	-Asp	-Pro-
or	-Ser	-Glu-						-Leu	-His	-Val	-Asp	-Pro-

gap is near the end of a polypeptide sequence, where the homology on one side of the gap can almost never be good enough to justify the gap. As Cantor points out, "When two relatively dissimilar sequences are compared, it is almost always possible to improve the homology by the insertion of one or more gaps $\cdots$. In general, adding gaps increases the number of comparisons by about L^{N_g} where N_g is the number of gaps and L is the number of amino acid pairs so that the extra homology introduced by a gap must more than compensate for the increased number of comparisons."

The measurement of such homology is facilitated by using the genetic code. Cantor's method for searching for gaps is to introduce gaps into computerized comparisons of sequences of amino acids. These comparisons depend on minimum base differences per codon (Table XIII), and include the use of a computer program as described by Fitch [10]. Cantor's procedure, which is illustrated in Figure 1, enables the gap to be placed at a point where the difference between the two polypeptide chains is minimized.

The locating of gaps is helped by the availability of a large number of sequences from a family of homologous proteins. The hemoglobins are unusually useful for this procedure, because gene duplications are superimposed on speciation. This makes it possible to locate a gap of five consecutive residues at positions 53 to 57 in all the α chains by comparing them with the myoglobin and β chains, none of which contain the gap. A large gap is in lamprey hemoglobin between the G and H helical regions (Fujiki and Braunitzer [11], Li and Riggs [24]). Carp and lamprey hemoglobins contain extra Met residues, inserted between positions 70 and 71, and 85 and 86, respectively. The gaps and insertions in the globins are summarized in Table XIV.

Figure 2, modified from the proposal by Ingram [17], shows how the different globin chains have evolved by duplication and differentiation from a common ancestral gene. Below the diagram are arrows indicating the order at which various lines of descent have branched off as compared with the order in which the events of duplication occurred. The time relationship shows that the lamprey should have only α type chains, because its line of descent separated before the origin of β chains, that is, before the appearance of tetrameric hemoglobins, which are not present in lampreys. The carp should have both α and β chains, but not the γ chain, because on the line of descent leading to human γ this duplication apparently originated at about the time of the marsupial-eutherian divergence. This was estimated as about 130 million years ago (Air, Thompson, Richardson, and Sharman [1]). A comparison of the human β and γ chains shows about the same divergence as that between the human and kangaroo β chains (Table XV). This would place the γ chain as having appeared at about the same time as the placental circulatory system, and after the origin of amphibians and birds. The embryonic frog and chicken hemoglobins (which have not been "sequenced") should, therefore, have diverged prior to the separation of the mammalian β and γ chains.

TABLE XIII

Minimum Base Differences per Codon for the 61 Codons for the Amino Acids

The assignments of the codon are in Table I.

		GCN Ala	CGN Arg	AAY Asn	GAY Asp	UGY Cys	CAR Gln	GAR Glu	GGN Gly	CAY His	AUY Ile	AUA Ile	CUN Leu	AAR Lys	AUG Met	UUY Phe	CCN Pro	UCN Ser	ACN Thr	UGG Trp	UAY Tyr	GUN Val	AGY Ser	AGR Arg	UUR Leu
GCN	Ala	0	2	2	1	2	2	1	1	2	2	2	2	2	2	2	1	1	1	2	2	1	2	2	2
CGN	Arg	2	0	2	2	1	1	2	1	1	2	2	1	2	2	2	1	2	2	1	2	2	1	1	2
AAY	Asn	2	2	0	1	2	2	2	2	1	1	2	2	1	2	2	2	2	1	3	1	2	1	2	3
GAY	Asp	1	2	1	0	2	2	1	1	1	2	3	2	2	3	2	2	2	2	3	1	1	2	3	3
UGY	Cys	2	1	2	2	0	3	3	1	2	2	3	2	3	3	1	2	1	2	1	1	2	1	2	2
CAR	Gln	2	1	2	2	3	0	1	2	1	3	2	1	1	2	3	1	2	2	2	2	2	3	2	2
GAR	Glu	1	2	2	1	3	1	0	1	2	3	2	2	1	2	3	2	2	2	2	2	1	3	2	2
GGN	Gly	1	1	2	1	1	2	1	0	2	2	2	2	2	2	2	2	2	2	1	2	1	1	1	2
CAY	His	2	1	1	1	2	1	2	2	0	2	3	1	2	3	2	1	2	2	3	1	2	2	3	3
AUY	Ile	2	2	1	2	2	3	3	2	2	0	1	1	2	1	1	2	2	1	3	2	1	1	2	2
AUA	Ile	2	2	2	3	3	2	2	2	3	1	0	1	1	1	2	2	2	1	3	3	1	2	1	1
CUN	Leu	2	1	2	2	2	1	2	2	1	1	1	0	2	1	1	1	2	2	2	2	1	2	2	1
AAR	Lys	2	2	1	2	3	1	1	2	2	2	1	2	0	1	3	2	2	1	2	2	2	2	1	2
AUG	Met	2	2	2	3	3	2	2	2	3	1	1	1	1	0	2	2	2	1	2	3	1	2	1	1
UUY	Phe	2	2	2	2	1	3	3	2	2	1	2	1	3	2	0	2	1	2	2	1	1	2	3	1
CCN	Pro	1	1	2	2	2	1	2	2	1	2	2	1	2	2	2	0	1	1	2	2	2	2	2	2
UCN	Ser	1	2	2	2	1	2	2	2	2	2	2	2	2	2	1	1	0	1	1	1	2	2	2	1
ACN	Thr	1	2	1	2	2	2	2	2	2	1	1	2	1	1	2	1	1	0	2	2	2	1	1	2
UGG	Trp	2	1	3	3	1	2	2	1	3	3	3	2	2	2	2	2	1	2	0	2	2	2	1	1
UAY	Tyr	2	2	1	1	1	2	2	2	1	2	3	2	2	3	1	2	1	2	2	0	2	2	3	2
GUN	Val	1	2	2	1	2	2	1	1	2	1	1	1	2	1	1	2	2	2	2	2	0	2	2	1
AGY	Ser	2	1	1	2	1	3	3	1	2	1	2	2	2	2	2	2	2	1	2	2	2	0	1	3
AGR	Arg	2	1	2	3	2	2	2	1	3	2	1	2	1	1	3	2	2	1	1	3	2	1	0	2
UUR	Leu	2	2	3	3	2	2	2	2	3	2	1	1	2	1	1	2	1	2	1	2	1	3	2	0

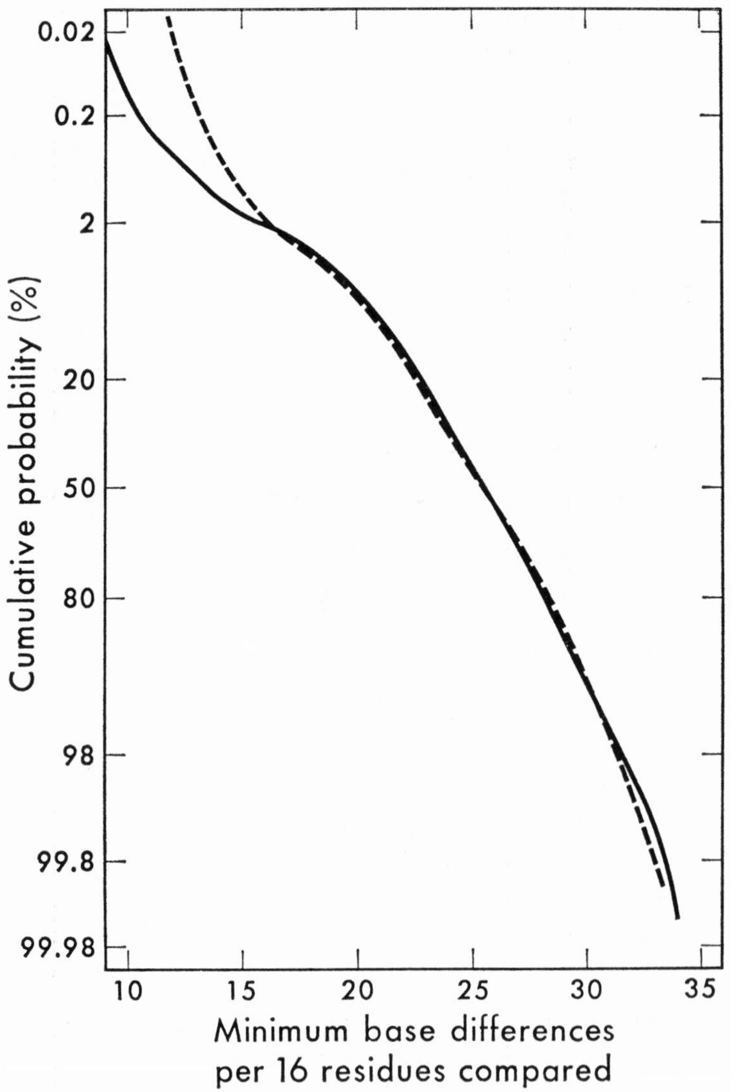

FIGURE 1

Computer comparisons of the first 40 amino acids of α and β human hemoglobin chains including a gap of two residues on the β chain (solid line); and on the α chain (broken line). The shift in the β distribution is evidence for a gap of two on the β chain. (From Jukes and Cantor [19].)

TABLE XIV

EVOLUTIONARY CHANGES IN LENGTH IN GLOBIN CHAINS

Globin	Nature of change (Residues numbered to include gaps)	Time of occurrence
β, γ and δ hemoglobins	Addition of one residue at NH_2-terminus Deletion of 19 and 20	Before origin of birds
α hemoglobin of terrestrial species	Deletion of 48	Subsequent to divergence of bony fishes
α hemoglobin of bony fishes	Deletion of 64. Addition of 70.5	Subsequent to divergence of bony fishes
α hemoglobin	Deletion of 53–57	Subsequent to divergence of lamprey
Lamprey hemoglobin	Addition of 9 residues to N-terminus. Addition of 78.5. Deletion of residues 110–119 and 148	Subsequent to divergence of lamprey
Frog β hemoglobin	Deletion of 1–6	After divergence of amphibians

The duplication of the α chain to give rise to the α and β chains enables comparisons to be made between the α and β chains of any two groups of animals that appeared subsequently. Such comparisons are not possible with the cytochromes *c* because these have evolved throughout the vertebrate line without

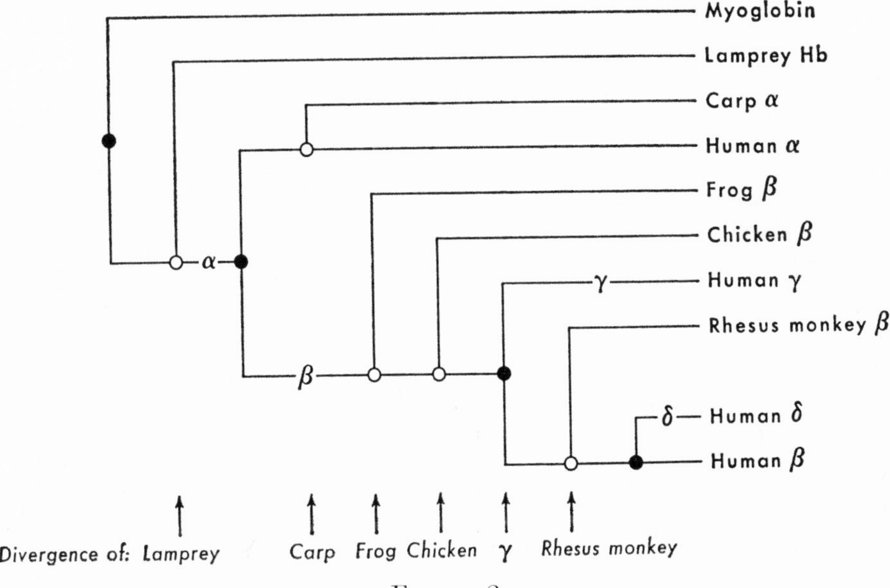

FIGURE 2

Divergence of globin chains by gene duplication (dots) and speciation (circles).

TABLE XV

COMPARISONS OF VARIOUS HEMOGLOBIN (Hb) AND MYOGLOBIN (Mb) AMINO ACID SEQUENCES IN TERMS OF AMINO ACID DIFFERENCES (AAD) AND MINIMUM BASE DIFFERENCES PER CODON (MBDC)

	Sites compared	Minimum base differences per codon 0	1	2	3	Per site AAD	MBDC
(i) α chains							
Human with Horse	141	123	14	4	0	0.13	0.16
Bovine	141	124	14	3	0	0.12	0.14
Mouse	141	125	13	3	0	0.11	0.13
Rabbit	141	116	22	3	0	0.18	0.20
Horse with Bovine	141	125	11	5	0	0.16	0.15
Mouse	141	119	17	5	0	0.16	0.19
Rabbit	141	116	19	6	0	0.18	0.22
Bovine with Mouse	141	122	17	2	0	0.13	0.15
Rabbit	141	116	22	3	0	0.18	0.20
Mouse with Rabbit	141	118	19	4	0	0.16	0.19
Carp with Human	140	72	41	27	0	0.49	0.68
Rhesus	140	72	39	29	0	0.49	0.69
Horse	140	72	39	29	0	0.49	0.69
Bovine	140	74	38	28	0	0.47	0.67
Rabbit	140	70	46	24	0	0.50	0.67
Mouse	140	72	46	22	0	0.49	0.64
Chicken	140	68	44	28	0	0.51	0.71
(ii) Mammalian α and β chains							
Human α with Human β	139	64	53	22	0	0.54	0.70
Rhesus β	139	64	50	25	0	0.54	0.72
Horse β	139	64	52	23	0	0.54	0.70
Rabbit β	139	61	55	23	0	0.56	0.73
Rabbit α with Human β	139	75	53	21	0	0.53	0.68
(iii) Lamprey with some Hb chains							
Lamprey with Human α	130	45	57	28	0	0.65	0.87
Rabbit α	130	42	54	34	0	0.68	0.94
Carp α	130	39	53	38	0	0.70	0.99
Chicken α	130	36	59	34	1	0.72	1.00
(iv) Human β with Horse β	146	122	18	6	0	0.16	0.20
Human γ	146	107	29	10	0	0.27	0.34
Kangaroo β	146	108	24	14	0	0.26	0.36
Frog β	140	80	39	20	1	0.43	0.59
(v) Myoglobins (Mb) with Hemoglobins							
Horse Mb with Horse α	141	36	55	50	0	0.74	1.10
Human α	141	39	54	47	1	0.72	1.04
Mouse α	141	36	62	43	0	0.75	1.05
Carp α	141	35	66	39	1	0.75	1.05
Horse β	145	35	63	45	2	0.76	1.10
Human β	145	38	56	49	2	0.74	1.10
Frog β	140	36	52	51	1	0.74	1.13
Lamprey	135	33	60	41	1	0.75	1.09

indications of duplication. Comparisons of groups of α hemoglobin chains with groups of β chains are shown in Table XV. There are fewer two base changes in the α:β comparison than in the carp α: mammalian α comparison, although the total difference is greater in the α:β comparison. This points to different evolutionary patterns of amino acid replacements in the two cases. A possible explanation is that the α:β divergence is modified by the fact that the two chains have to fit each other to form an α:β dimer (Perutz [27]), while the comparisons of α chains with each other may show a pattern depending on the fact that mammalian α chains must form dimers with both β and γ chains.

A comparison of horse myoglobin with various hemoglobin chains shows a surprising constancy of difference, clearly indicating divergent rather than convergent evolution. Lamprey hemoglobin has apparently differentiated from the common ancestor of the globins (Figure 2) at the same rate as the other hemoglobins. The "primitive" anatomy and physiology of the lamprey have led to its being regarded as a "living fossil." Nevertheless, lamprey hemoglobin seems to have changed during evolution to about the same extent as the other hemoglobins.

3. Silent changes in codons

Changes in the third base of codons for amino acids will in most cases not produce a change in the amino acid assignment, so that in such cases they are, therefore, selectively neutral. Evidence has recently come to light that such "silent" changes take place during evolution. Three small RNA viruses that grow in *E. coli* are f2, MS2, and R17. The coat proteins of these contain a sequence of 129 amino acids. R17 and MS2 coat proteins are identical and f2 differs only by having leucine instead of methionine at residue 88. Recently the base sequences of large portions of these viruses have been determined, including those portions that code for the coat proteins. There are six "silent" base differences in cistrons for the identical coat proteins of MS2 and R17. Other "silent" changes include UCC(f2) *versus* UCU(R17); CUA(f2) *versus* UUA(R17); and GGC(f2) *versus* GGU(R17).

These are evidently neutral changes that have occurred by mutation and have been incorporated by genetic drift. Subsequent changes in such codons may cause marked differences in two lines, as shown by the hypothetical example in Figure 3.

Doubtless many other silent changes in DNA will soon come to light. There could well be between 10 and 15 such silent differences in the genes for the identical molecules of human and chimpanzee hemoglobins, as calculated from the rate of evolutionary separation of the hemoglobins and the five million years estimated for the chimpanzee-human divergence by Wilson and Sarich [31].

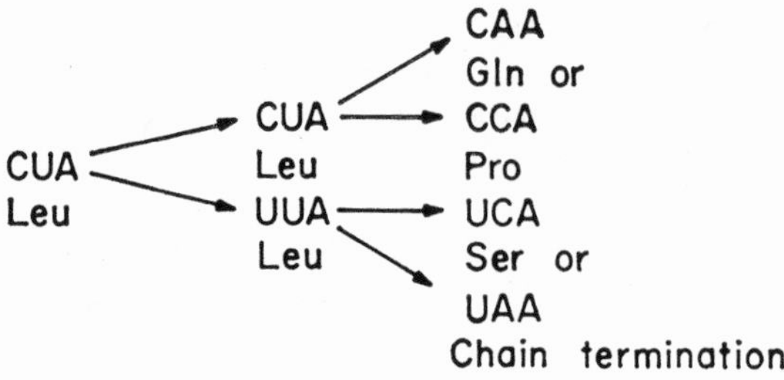

FIGURE 3

4. The disappearance of homology during evolution

The evolutionary differentiation of two polypeptide chains can eventually proceed to a point where their homology can no longer be perceived. This is a gradual process, and, as a result, some sequences are at the borderline of disappearing homology. The process of divergence may be followed phylogenetically by comparing the complete sequences of various globins, but it is perhaps more interesting to observe it in internal duplications in proteins, because the homology can then be followed to its point of disappearance.

The best example of internal repetition is human haptoglobin α-2 (Table XVI). In this protein segment 2, of 59 residues, 13 to 71, is repeated exactly at residues 72 to 130 except for one (single base) interchange, Glu to Lys. This may be diagrammed as shown in Figure 4. Moreover, the same segment occurs once, that is, without repetition, in haptoglobin α-1, residues 13 to 71 (see Figure 5). Evidently this repetition in haptoglobin α-2 is a recent evolutionary event, and, obviously, partial duplication of the gene has taken place. This example establishes the existence of the phenomenon of internal duplication in a protein molecule.

The next example is clostridial ferredoxin, for example, that of *Clostridium butyricum* (Table XVI). Using a similar diagram, the duplication is represented in Figure 6. Segments 1 and 1′ are, respectively, 25 and 26 amino acid residues in length; 1′ has an additional amino acid at the fifth residue. After adjusting for this, segments 1 and 1′ are identical in 14 out of 25 residues at corresponding sites. The existence of the internal duplication is clear, but the event took place so long ago in evolutionary time that considerable differentiation has taken place, and four of the eleven differing residues are separated by two-base codon changes.

The third example (see Figure 7) is the A and B chains of insulin. These occur in the single molecule of proinsulin. The B chain comes first. Residues 1 to 20 repeat at 65 to 84; if three gaps are inserted in each of the two regions (Table XVI) nine out of the 17 pairs of homologous residues are identical and four of the eight differing residues are separated by two or more base changes per codon.

The next two examples illustrate the eventual disappearance of homology.

TABLE XVI

Examples of Internal Repetition

Abbreviations:

A	Ala	G	Gly	M	Met	S	Ser
C	Cys	H	His	N	Asn	T	Thr
D	Asp	I	Ile	P	Pro	V	Val
E	Glu	K	Lys	Q	Gln	W	Trp
F	Phe	L	Leu	R	Arg	Y	Tyr
				—	Gap	Z	= Glx

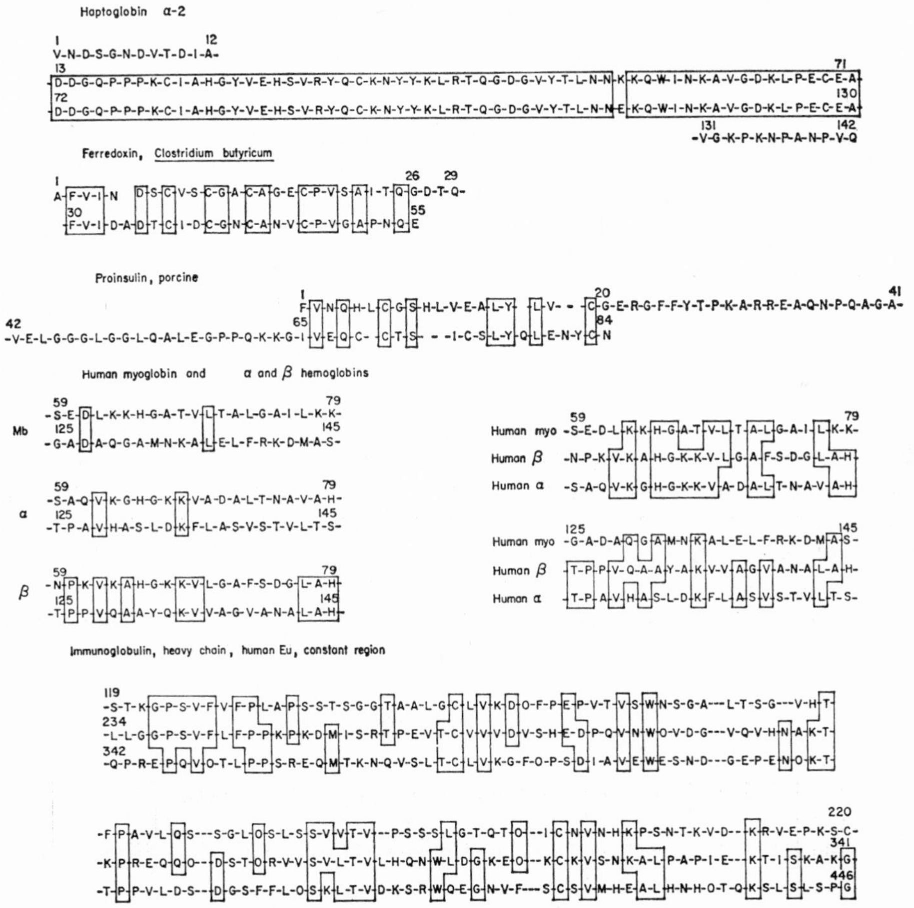

Haptoglobin α-2

```
1                      12
V-N-D-S-G-N-D-V-T-D-I-A-
13                                                                                                  71
-D-D-G-Q-P-P-P-K-C-I-A-H-G-Y-V-E-H-S-V-R-Y-Q-C-K-N-Y-Y-K-L-R-T-Q-G-D-G-V-Y-T-L-N-N-K-K-Q-W-I-N-K-A-V-G-D-K-L-P-E-C-E-A-
72                                                                                                 130
-D-D-G-Q-P-P-P-K-C-I-A-H-G-Y-V-E-H-S-V-R-Y-Q-C-K-N-Y-Y-K-L-R-T-Q-G-D-G-V-Y-T-L-N-N-E-K-Q-W-I-N-K-A-V-G-D-K-L-P-E-C-E-A-
                                                                                   131                     142
                                                                                   -V-G-K-P-K-N-P-A-N-P-V-Q
```

Ferredoxin, Clostridium butyricum

```
1                                               26  29
A-F-V-I-N     D-S-C-V-S-C-G-A-C-A-G-E-C-P-V-S-A-I-T-Q-G-D-T-Q-
 30                                              55
 -F-V-I-D-A-D-T-C-I-D-C-G-N-C-A-N-V-C-P-V-G-A-P-N-Q-E
```

Proinsulin, porcine

```
                                           1                                  20                                              41
                                           F-V-N-Q-H-L-C-G-S-H-L-V-E-A-L-Y-  -L-V- - -C-G-E-R-G-F-F-Y-T-P-K-A-R-R-E-A-Q-N-P-Q-A-G-A-
42                                       65                                   84
-V-E-L-G-G-G-L-G-G-L-Q-A-L-E-G-P-P-Q-K-K-G-I-V-E-Q-C-  -C-T-S- - -I-C-S-L-Y-Q-L-E-N-Y-C-N
```

Human myoglobin and α and β hemoglobins

```
    59                                   79
    -S-E-D-L-K-K-H-G-A-T-V-L-T-A-L-G-A-I-L-K-K-
Mb  125                                  145
    -G-A-D-A-Q-G-A-M-N-K-A-L-E-L-F-R-K-D-M-A-S-

    59                                   79
    -S-A-Q-V-K-G-H-G-K-K-V-A-D-A-L-T-N-A-V-A-H-
α   125                                  145
    -T-P-A-V-H-A-S-L-D-K-F-L-A-S-V-S-T-V-L-T-S-

    59                                   79
    -N-P-K-V-K-A-H-G-K-K-V-L-G-A-F-S-D-G-L-A-H-
β   125                                  145
    -T-P-P-V-Q-A-A-Y-Q-K-V-V-A-G-V-A-N-A-L-A-H-
```

```
             59                                        79
Human myo    -S-E-D-L-K-K-H-G-A-T-V-L-T-A-L-G-A-I-L-K-K-
Human β      -N-P-K-V-K-A-H-G-K-K-V-L-G-A-F-S-D-G-L-A-H-
Human α      -S-A-Q-V-K-G-H-G-K-K-V-A-D-A-L-T-N-A-V-A-H-

             125                                       145
Human myo    -G-A-D-A-Q-G-A-M-N-K-A-L-E-L-F-R-K-D-M-A-S-
Human β      -T-P-P-V-Q-A-A-Y-A-K-V-V-A-G-V-A-N-A-L-A-H-
Human α      -T-P-A-V-H-A-S-L-D-K-F-L-A-S-V-S-T-V-L-T-S-
```

Immunoglobulin, heavy chain, human Eu, constant region

```
119
-S-T-K-G-P-S-V-F-V-F-P-L-A-P-S-S-T-S-G-G-T-A-A-L-G-C-L-V-K-D-O-F-P-E-P-V-T-V-S-W-N-S-G-A---L-T-S-G---V-H-T-
234
-L-L-G-G-P-S-V-F-L-F-P-P-K-P-K-D-M-I-S-R-T-P-E-V-T-C-V-V-V-D-V-S-H-E-D-P-Q-V-N-W-O-V-D-G---V-Q-V-H-N-A-K-T-
342
-Q-P-R-E-P-Q-V-O-T-L-P-P-S-R-E-Q-M-T-K-N-Q-V-S-L-T-C-L-V-K-G-F-O-P-S-D-I-A-V-E-W-E-S-N-D---G-E-P-E-N-O-K-T-

                                                                                                            220
-F-P-A-V-L-Q-S---S-G-L-O-S-L-S-S-V-V-T-V--P-S-S-S-L-G-T-Q-T-O--I-C-N-V-N-H-K-P-S-N-T-K-V-D--K-R-V-E-P-K-S-C-
                                                                                                            341
-K-P-R-E-Q-Q-O--D-S-T-O-R-V-V-S-V-L-T-V-L-H-Q-N-W-L-D-G-K-E-O--K-C-K-V-S-N-K-A-L-P-A-P-I-E--K-T-I-S-K-A-K-G
                                                                                                            446
-T-P-P-V-L-D-S--D-G-S-F-F-L-O-S-K-L-T-V-D-K-S-R-W-Q-E-G-N-V-F---S-C-S-V-M-H-E-A-L-H-N-H-O-T-Q-K-S-L-S-L-S-P-G
```

The first of these considers two cytochromes *c*: those of *Neurospora* and tuna. Residues 20 to 34 in *Neurospora* are a repetition of 5 to 19. There has been much evolutionary differentiation but the repetition is still distinct (Table XVII). When the same sequences are compared in tuna, there is no sign of homology

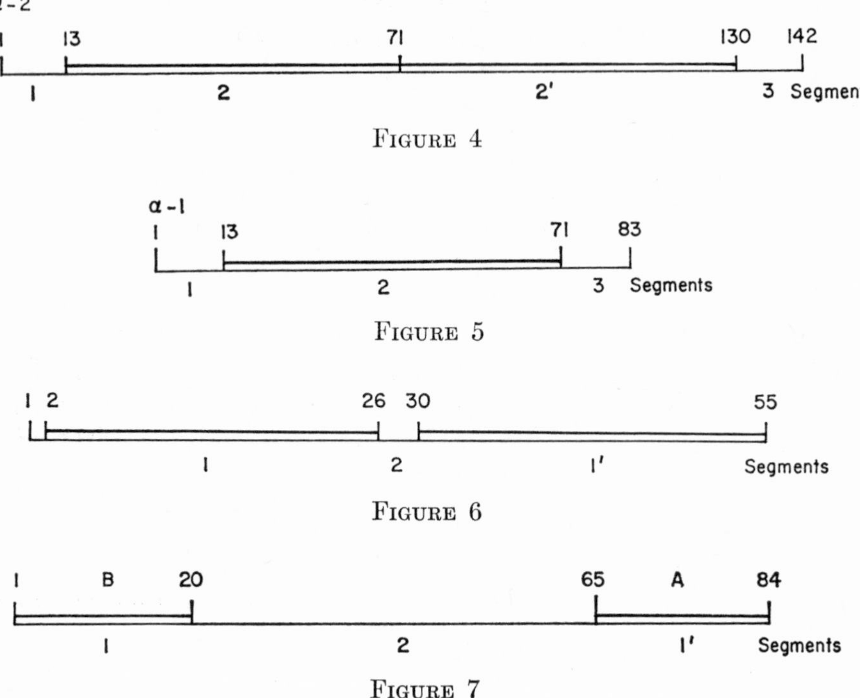

FIGURE 4

FIGURE 5

FIGURE 6

FIGURE 7

(Table XVIII). However, when the segments of *Neurospora* and tuna are compared directly with each other, the homology of 5 to 19 *Neurospora* with 5 to 19 tuna, is obvious, and so are the homologies of the two 20 to 34 regions, as seen in Table XIX. The tuna internal comparison (5 to 19 *versus* 20 to 34) is a clear case of complete evolutionary erosion of homology.

The same phenomenon has occurred in hemoglobin (Table XVI) except that the homologous segments do not occur consecutively, but are separated by

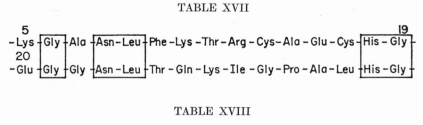

TABLE XVII

5	-Lys	**Gly**	Ala	**Asn**	**Leu**	Phe	Lys	Thr	Arg	Cys	Ala	Glu	Cys	**His**	**Gly**- (19)
20	-Glu	**Gly**	Gly	**Asn**	**Leu**	Thr	Gln	Lys	Ile	Gly	Pro	Ala	Leu	**His**	**Gly**-

TABLE XVIII

5	-Lys	Gly	Lys	Lys	Thr	Phe	Val	Gln	Lys	Cys	Ala	Gln	Cys	His	Thr- (19)
20	-Val	Glu	Asn	Gly	Gly	Lys	His	Lys	Val	Gly	Pro	Asn	Leu	Trp	Gly- (34)

TABLE XIX

	5														19
N.	Lys	Gly	Ala	Asn	Leu	Phe	Lys	Thr	Arg	Cys	Ala	Glu	Cys	His	Gly
T.	Lys	Gly	Lys	Lys	Thr	Phe	Val	Gln	Lys	Cys	Ala	Gln	Cys	His	Thr

	20														34
N.	Glu	Gly	Gly	Asn	Leu	Thr	Gln	Lys	Ile	Gly	Pro	Ala	Leu	His	Gly
T.	Val	Glu	Asn	Gly	Gly	Lys	His	Lys	Val	Gly	Pro	Asn	Leu	Trp	Gly

about 45 residues. Residues 59 to 79 and 125 to 145 in the human β chain are clearly homologous (Fitch [10], Jukes and Cantor [19]) because eight of the 21 pairs of corresponding amino acids are identical. The corresponding regions in horse myoglobin show no significant homology with each other (only two identities in 21 pairs, with an expectation of one identity in 20 random sequences of amino acids).

The same is true of the human α chain. Direct comparisons of corresponding sequences with each other, however (Table XVI), show that regions 59 to 79 are homologous and 125 to 145 are homologous in myoglobin, α Hb, and β Hb. The similarity of the myoglobin sequences to the hemoglobins is, as to be expected, less than the similarity of the α and β hemoglobin regions to each other. For some reason, the internal homology in the β chain, but not in the other two, has been retained. This is unexpected in view of the fact that the β chains of various species of vertebrates are diverging from each other more rapidly than is the case with either the myoglobins or the α chains that have been examined.

To facilitate further examination and because of the interest in this particular comparison, the internal homology in the hemoglobins is shown in more detail in Table XX. The prototype is derived by comparing all the codons for the amino acid residues in a vertical row, and assigning the predominant nucleotides to the codon for the amino acid in the prototype. The polypeptide sequences for residues 59 to 79 and 125 to 145 are both from helical regions (E and H) in which the hydrophobic side chains point inwards. This would tend to favor homology, whether or not the two regions had a common origin. However, other helical regions such as A, F, and G, which also have hydrophobic, inner directed, side chains do not show homology, in the β and γ chains, with residues 59 to 79 and 125 to 145. Further discussion of this homology in the globins is in Jukes and Cantor [19].

There are three possible reasons for homology in two polypeptide sequences: it may be by chance, by convergent evolution, or by divergent evolution. Chance homology is examined in terms of probability. The decision between convergent and divergent evolution is aided by probability, molecular structure, and phylogeny. As an example, let us examine the sequences 125–145 in various

TABLE XX

Comparisons of Sequences in Polypeptide Chains of Certain Globins

Mb = human myoglobin; α = α chain of human hemoglobin; β = β chain; γ = γ chain; Lamprey = *Petromyzon* hemoglobin. The numbering system is identical with that in Table XVI. The underlined residues are identical with residues in the chain above them.

	50 53 58
Mb	-Leu-Lys-Ser - -Glu-Asp-Glu-Met-Lys-Ala-
	59 79
	-Ser -Glu-Asp-Leu-Lys-Lys-His-Gly-Ala-Thr-Val -Leu-Thr-Ala-Leu-Gly-Gly- Ile -Leu-Lys-Lys-
	125 145
	-Gly-Ala-Asp-Ala-Gln-Gly-Ala-Met-Asn-Lys-Ala -Leu-Glu-Leu-Phe-Arg-Lys-Asp-Met-Ala -Ser
	50 58
α	-Leu-Ser -His — — — — — — Gly-
	59 79
	-Ser -Ala-Gln-Val-Lys-Gly-His-Gly-Lys-Lys-Val -Ala-Asp-Ala-Leu-Thr-Asn-Ala -Val -Ala -His-
	125 145
	-Thr-Pro-Ala-Val-His-Ala-Ser-Leu-Asp-Lys-Phe-Leu-Ala -Ser -Val-Ser -Thr-Val -Leu-Thr-Ser
	50 58
β	-Leu-Ser -Thr- -Pro-Asp-Ala -Val -Met-Gly-
	59 79
	-Asn-Pro-Lys-Val-Lys-Ala-His-Gly-Lys-Lys-Val -Leu-Gly-Ala-Phe-Ser -Asp-Gly-Leu-Ala -His-
	125 145
	-Thr-Pro-Pro-Val-Gln-Ala-Ala-Tyr-Gln-Lys-Val -Val-Ala -Gly-Val-Ala-Asn-Ala -Leu-Ala -His
	50 58
γ	-Leu-Ser -Ser - -Ala-Ser -Ala - Ile -Met-Gly-
	59 79
	-Asn-Pro-Gly-Val-Lys-Ala-His-Gly-Lys-Lys-Val -Leu-Thr-Ser -Leu-Gly-Asp-Ala - Ile -Lys-His-
	125 145
	-Thr-Pro-Glu-Val-Gln-Ala-Ser-Trp-Gln-Lys-Met-Val-Thr-Gly-Val-Ala-Ser -Ala -Leu-Ser -Ser
	50 58
Lamprey	-Leu-Thr-Thr- -Ala-Asp-Gln-Leu-Lys-Lys-
	59 79
	-Ser -Ala-Asp-Val-Arg-Trp-His-Ala -Glu-Arg- Ile -Ile -Asn-Ala-Val-Asn-Asp-Ala -Val -Ala -Ser-
	145
	— — — — — — Gly-Phe-Glu-Lys-Leu-Ser -Met-Cys-Ile -Ile -Leu-Met-Leu-Arg-Ser
Prototype	-Ser -Pro-Asp-Val-Lys-Ala-His-Gly-Lys-Lys-Val -Leu-Thr-Ala-Val-Ala-Asp-Ala -Leu-Ala -Ser-

human globins (Table XX). There are only two differences between the β and δ chains, at positions 127 and 128 and the homology seems apparent by probability alone. However, there are 8 differences between β and γ, 13 between β and α, and 17 between β and Mb (Table XX). Therefore, in terms of probability, there is no visible homology between these regions of β and Mb. Despite this, the globins are regarded as a family of proteins that have divergently evolved by gene duplication followed by differentiation, and, by this argument, the sequences 125–145 are all homologous. This conclusion is supported by an examination of the tertiary structures of the globins; these structures are homologous. Furthermore, a progressive divergence can be seen when the sequences of the α Hb chains of vertebrates are compared. The amino acid differences between human and *Rhesus* average 0.03 per site; human and horse, 0.13; human and chicken, 0.25; human and carp, 0.49; human and lamprey, 0.65. The human: lamprey difference reaches the edge of probability, yet the homology is accepted because the progressive phylogenetic divergence mirrors the zoological and fossil comparisons. This pattern of thought favors divergent, rather than convergent, evolution as an explanation for the homology in this series. This example is perhaps the most elegant illustration that has come to light of the disappearance of homology as a result of evolutionary erosion.

Repetitive sequences in polypeptide chains must be examined by the same criteria for homology: probability, convergence, and divergence. But in this case, phylogeny is not available for support, and tertiary structure can actually work against us. We are thrown back on probability. The homology between 59–79 and 125–145 in the β Hb chains or between the two halves of clostridial ferredoxin (Table XVI) must be evaluated by probability alone. If two such sequences are in unrelated regions of the tertiary structure, the credibility of the proposal that they are a repetition may be questioned. But it should be remembered that, after such an internal repetition, a protein may evolve towards a more "useful" tertiary structure and, during this evolution, the homology of the repeated regions may change rapidly. Therefore, an existing homology between two separated portions of a single polypeptide chain may indeed be indicative of an evolutionary duplication.

The establishment of cases of convergent evolution, on the other hand, would seem to be entirely dependent upon similarities in tertiary structure, such as recently noted for subtilisin and trypsinogen. In such cases, evidence of strong homology in comparisons of the amino acid sequences would probably override convergent evolution in favor of divergent evolution. A case in point is chymotrypsinogen and trypsinogen, in which the amino acid difference is only 0.44 per site when the appropriate gaps are inserted. The similarity of these two sequences is, on this basis, too great to be attributed to convergent evolution.

Homology may still be detectable in proteins when it is not possible to demonstrate it by a comparison of the amino acid sequence. An example is the similarity of the tertiary structures of *Chironomus* and mammalian hemoglobins, which have similar configurations (Huber, Formanek, and Epp [15]) when compared by

X-ray methods. In such a case it may be difficult to decide whether the homology is the result of convergent as opposed to divergent evolution.

4.1. *Immunoglobulins.* These proteins, the antibodies, contain repetitious sequences of about 110 amino acid residues. The IgG immunoglobulins, for example, consist of two "light" chains, each with two such sequences and two "heavy" chains, each with four such sequences. The N-terminal sequences of both the light and heavy chains are highly variable with respect both to length and sequences. The constant regions are easier to compare with each other. The internal repetition of a constant region in the heavy chain of the Eu (γ type) immunoglobulins is shown in Table XVI. The homology of the three repetitions is clearly visible and the alignment is made easy by the presence of 13 identical amino acids at corresponding sites in all three sequences. These sequences have diverged from each other at a constant rate as estimated by amino acid differences and minimum base differences per codon.

The tertiary structure of immunoglobulins is unknown, but the repeating sequences evidently have similar secondary structures, characterized by disulfide bridges (Putnam [28]). These connect pairs of cysteine residues shown at positions 144 and 200; 259 and 321; 367 and 425 in Table XVI. Most of the residues are needed to preserve the structure and function of the molecule. Some of these residues are common also to the constant regions of the light chain and to the variable regions of light and heavy chains (Jukes and Holmquist [20]).

5. Fibrinopeptides

During blood clotting, which typically takes place outside the circulatory system of animals, a complex series of reactions leads to the conversion of prothrombin to thrombin. This then reacts with fibrinogen, splitting off two fibrinopeptides (A and B) and producing fibrin. The two fibrinopeptides markedly differ from each other in the same animal. Each of them shows great evolutionary diversity, so that closely related species have different fibrinopeptides A and the same is true for fibrinopeptide B (Table XXI).

The enzyme thrombin, however, is quite nonspecific, and the thrombins of all mammalian species so far examined will cross-react with any mammalian fibrinogen (Doolittle [6]). The point of attack is the first Arg-Gly linkage in each of the A and B chains. Evidently the adjoining amino acids have no measurable effect on the affinity of thrombin for the Arg-Gly peptide bond, for none of the A chains in Table XXI have the same amino acid preceding this bond as is found in the corresponding B chain. Thrombin, which is a protease, is so nonspecific that it is standardized by its reaction with a methyl ester of tosyl-blocked arginine. Nonspecificity is typical of proteases.

The marked differences between the A and B chains in the same animal may be a clue to the reason for nonspecificity of thrombin, because the same thrombin molecule releases two different fibrinopeptides A and B. Obviously, if the enzyme were more specific, for example, tailored to fit a special sequence of amino acids,

TABLE XXI

SEQUENCES OF FIBRINOPEPTIDES

Abbreviations as in Table XVI. Z = Glx, a derivative of glutamic acid. The asterisks indicate invariant residues.

(1) 39 Fibrinopeptides A

```
                                              *       *
Human   – – – A-D-S-G – E-G-D-F-L-A-E-G-G-G-V-R
Others  A D G S N T K D S S E S I T A   A T G
        E T D T K P A T G – T   E D D     A I
              P E V E   K   S     S G     V
              G – A D   T         H       H
              V   G V   D         G
              –   – Q   –         P
                    P             E
                    –
```

(2) 30 Fibrinopeptides B

```
                                                  *
Human   – – – – – – – – Z-G-V-N-D-N-E-E-G-F-F-S-A-R
Others  Z H A L D D D D E E E E E R A K V H L D G
          F S D Y Y Y H T V Q G G G P V L T V G
          G P I   S     – D D D N – V   S R P
          P L A           – – – –   D   D G
          A Y E                     I     L
          S I H                     T     P
            T T                     K     I
            F                             V
                                          S
```

the evolutionary rate of differentiation of the fibrinopeptides would be much slower than it is, because each change in a fibrinopeptide would have to be accompanied by a corresponding change in the thrombin molecule. An analogy is the slow rate of evolutionary differentiation of cytochrome *c*. The biological function of cytochrome *c* necessitates its interaction with two other proteins, cytochrome c_1 and cytochrome oxidase. It is postulated that this circumstance helps to slow down the rate of evolution of cytochrome *c*.

The amino acid residues at the various fibrinopeptide sites are shown in Table XXI. Comparisons of the sequences have been made by various authors in order to relate differences in the fibrinopeptides to phylogeny and evolutionary rate of change. There are, however, certain peculiarities in the fibrinopeptides that make such comparisons difficult to evaluate. First, the sequences are very short so that not many amino acid residues can be compared, and, second, the chains vary in length, and it is difficult to align them without using arbitrary gaps.

It is obvious, however, that the fibrinopeptides change more rapidly in evolution than any of the polypeptide sequences that have so far been compared, with the possible exception of the calcitonins (Milhaud [26]). It seems likely that most of the changes in the fibrinopeptides are neutral for reasons described above.

6. Discussion

The examples in Table XVI show that internal repetitions are common rather than exceptional in protein evolution. Proteins are large molecules, usually containing more than one hundred amino acid residues. The probability of an existing sequence having emerged by chance from the superastronomical number of possibilities for permuting twenty variables (the different amino acids) into a sequence of more than one hundred units is almost infinitely small. It seems more likely that a short sequence of amino acids having some catalytic activity could be formed during the early stages of the emergence of life. Repeated end to end duplication of such a sequence could perhaps preserve and enhance its activity. Meanwhile, numerous point mutations would occur, and some of these, in nonfunctional regions of the molecule, might favorably influence the development of an advantageous tertiary structure which would be preserved by natural selection.

It is axiomatic that any mutation that changes an amino acid residue simultaneously alters the chemical properties of a protein. We have proposed (King and Jukes [22]) that many of such alterations are selectively neutral. The opposite conclusion is that any alteration of properties has an adaptive effect which enters into the process of natural selection, so that all changes which enter into the makeup of an organism during evolution are incorporated on a Darwinian basis. These two contrary viewpoints, in various guises, have been debated at this Symposium.

In support of the neutral concept, I wish to advance the suggestion that alterations in the properties of an enzyme may often be selectively neutral. It is well known in biochemistry that wide "margins of safety" exist in many physiological processes. The capacity of an enzymatic process in a living organism may far exceed the demands that are ever placed upon it. Söderqvist and Blombäck [30] state that, "A species-specific thrombin has varying clotting-times with different fibrinogens while a species-specific fibrinogen gives fairly constant clotting-times with different thrombins. The results suggest that mammalian thrombins have changed slowly while corresponding fibrinogens have undergone a more pronounced change during mammalian evolution. The differences in clotting-times are as great as ten times and cause no obvious negative adaptive effects."

Blood clotting is, of course, an extremely complex phenomenon involving a multitude of chemical reactions. Blood clotting within the circulatory system can be fatal to an animal. Speed of clotting may, therefore, be disadvantageous; the important feature is that the clot arrest the flow of blood from a ruptured vessel within a reasonably prompt time. Söderqvist and Blombäck conclude that some variation in this time can occur without selective advantage.

I conclude that it is possible that amino acid differences, between two homologous proteins in similar organisms may well be the result of a selectively neutral divergence. An example might be horse and bovine myoglobins, which differ in 17 of 153 amino acid residues.

It is to be expected that supporters of the neo-Darwinian viewpoint will continue to fit all molecular changes in evolution to the Procrustean bed of pan-selectionism. Time will not permit the resolution of the argument, for we vanish from the scene much more rapidly than evolutionary changes take place. We can, however, take advantage of modern scientific methods to study the result of evolutionary change at the molecular level. Much new information is available as a result of determinations of the sequences of amino acids in proteins, thus permitting new theoretical insights into molecular evolutionary processes.

7. Summary

7.1. Polypeptide sequences in families of homologous proteins may be used to study molecular evolution. When two such sequences are compared, amino acid differences are usually found at many corresponding sites. These differences are considered to be the result of replacements of nucleotides in DNA during divergent evolution.

7.2. It is possible to interpret the differences in terms of the amino acid code, and such interpretations show that the rate of nucleotide replacement is more rapid than is usually estimated on the basis of amino acid differences.

7.3. Evolutionary changes in amino acid sequences produce changes in the properties of proteins. Such changes may be deleterious, adaptive, or neutral in their phenotypic effects.

7.4. The detectable homology between two polypeptides with a common origin may disappear during evolution. This disappearance is the result of progressive evolutionary differentiation following gene duplication. Examples of the phenomenon are given.

Note added in proof. Since this manuscript was written, McLachlan [33], [34] has discussed at length methods for searching for repetitive homology in globin and other polypeptide chains.

The author thanks Doctors Richard Holmquist, Lila Gatlin, and Jack King for suggestions, and Mrs. Regina Cepelak for preparing the manuscript.

REFERENCES

[1] G. M. Air, E. O. P. Thompson, B. J. Richardson, and G. B. Sharman, "Amino-acid sequences of kangaroo, myoglobin and haemoglobin and the date of marsupial-eutherian divergence," *Nature*, Vol. 229 (1971), pp. 391–394.

[2] T. B. BRADLEY, JR., R. C. WOHL, and R. F. RIEDER, "Hemoglobin Gun Hill: deletion of five amino acid residues and impaired hemoglobin binding," *Science*, Vol. 157 (1967), pp. 1581–1583.

[3] H. BROWN, F. SANGER, and R. KITAI, "The structure of pig and sheep insulins," *Biochem. J.*, Vol. 60 (1955), pp. 556–565.

[4] C. R. CANTOR, "The occurrence of gaps in protein sequences," *Biochem. Biophys. Res. Commun.*, Vol. 31 (1968), pp. 410–416.

[5] M. O. DAYHOFF, *Atlas of Protein Sequence and Structure*, Vol. 4, Silver Spring, Md., National Biomedical Research Foundation, 1969.

[6] R. DOOLITTLE, Personal communication, 1971.

[7] G. M. EDELMAN, B. A. CUNNINGHAM, W. E. GALL, P. D. GOTTLIEB, U. RUTISHAUSER, and M. J. WAXDAL, "The covalent structure of an entire γG immunoglobulin molecule," *Proc. Nat. Acad. Sci. U.S.A.*, Vol. 63 (1969), pp. 78–85.

[8] W. M. FITCH and E. MARGOLIASH, "A method for estimating the number of invariant amino acid coding positions in a gene using cytochrome *c* as a model case," *Biochem. Genet.*, Vol. 1 (1967), pp. 65–71.

[9] W. M. FITCH and E. MARKOWITZ, "An improved method for determining codon variability in a gene and its application to the rate of fixation of mutations in evolution," *Biochem. Genet.*, Vol. 4 (1970), pp. 579–593.

[10] W. M. FITCH, "The relation between frequencies of amino acids and ordered trinucleotides," *J. Mol. Biol.*, Vol. 16 (1966), pp. 1–27.

[11] H. FUJIKI and G. BRAUNITZER, "The primary structure of lamprey haemoglobin and its contribution to the evolutionary aspects of haemoglobin molecules," *Folia Bioch. et Biol. Graeca*, Vol. 7 (1970), pp. 68–73.

[12] L. L. GATLIN, "Evolutionary Indices," *Proceedings of the Sixth Berkeley Symposium on Mathematical Statistics and Probability*, Berkeley and Los Angeles, University of California Press, 1972, Vol. 5, pp. 277–296.

[13] J. I. HARRIS, F. SANGER, and M. A. NAUGHTON, "Species differences in insulin," *Arch. Biochem. Biophys.*, Vol. 65 (1956), pp. 427–438.

[14] R. HOLMQUIST, "Theoretical foundations of paleogenetics," *Proceedings of the Sixth Berkeley Symposium on Mathematical Statistics and Probability*, Berkeley and Los Angeles, University of California Press, 1972, Vol. 5, pp. 315–350.

[15] R. HUBER, H. FORMANEK, and O. EPP, "Kristallstrukturanalyse des Met-Erythrocruorins bei 5, 5Å Auflösung," *Naturwissenschaften*, Vol. 55 (1968), pp. 75–77.

[16] V. M. INGRAM, "Gene mutations in human haemoglobin: the chemical difference between normal and sickle cell haemoglobin," *Nature*, Vol. 180 (1957), pp. 326–328.

[17] ———, "Gene evolution and the haemoglobins," *Nature*, Vol. 189 (1961), pp. 704–708.

[18] T. H. JUKES, "Some recent advances in studies of the transcription of the genetic message," *Advan. Biol. Med. Phys.*, Vol. 9 (1963), pp. 1–41.

[19] T. H. JUKES and C. R. CANTOR, "Evolution of protein molecules," *Mamm. Prot. Metab.*, Vol. 3 (1969), pp. 21–132.

[20] T. H. JUKES and R. HOLMQUIST, "Estimation of evolutionary changes in certain homologous polypeptide chains," *J. Mol. Biol.*, Vol. 64 (1972), pp. 163–179.

[21] M. KIMURA, "The rate of molecular evolution considered from the standpoint of population genetics," *Proc. Nat. Acad. Sci. U.S.A.*, Vol. 63 (1969), pp. 1181–1188.

[22] J. L. KING and T. H. JUKES, "Non-Darwinian evolution," *Science*, Vol. 164 (1969), pp. 788–798.

[23] D. E. KOHNE, "Evolution of higher organism DNA," *Quart. Rev. Biophys.*, Vol. 3 (1970), pp. 327–375.

[24] S. L. LI and A. RIGGS, "The amino acid sequence of hemoglobin V from the lamprey *Petromyzon marinus*," *J. Biol. Chem.*, Vol. 245 (1970), pp. 6149–6169.

[25] E. MARGOLIASH and E. L. SMITH, "Structural and functional aspects of cytochrome *c* in

relation to evolution," *Evolving Genes and Proteins* (edited by V. Bryson and H. J. Vogel), New York, Academic Press, 1965, pp. 221–242.

[26] G. MILHAUD, Personal communication, 1971.

[27] M. F. PERUTZ, "A model of oxyhaemoglobin," *J. Mol. Biol.*, Vol. 13 (1965), pp. 646–668.

[28] F. W. PUTNAM, "Immunoglobulin structure: variability and homology," *Science*, Vol. 163 (1969), pp. 633–644.

[29] F. SANGER and E. O. P. THOMPSON, "The amino-acid sequence in the glycyl chains of insulin. 1. The identification of lower peptides from partial hydrolysates," *Biochem. J.*, Vol. 53 (1953), pp. 353–374.

[30] T. SÖDERQVIST and B. BLOMBÄCK, "Fibrinogen structure and evolution," *Naturwissenschaften*, Vol. 58 (1971), pp. 16–23.

[31] A. C. WILSON and V. M. SARICH, "A molecular time scale for human evolution," *Proc. Nat. Acad. Sci. U.S.A.*, Vol. 63 (1969), pp. 1088–1093.

[32] E. ZUCKERKANDL and L. PAULING, "Evolutionary divergence and convergence in proteins," *Evolving Genes and Proteins* (edited by V. Bryson and H. J. Vogel), New York, Academic Press, 1965, pp. 97–166.

Added in proof.

[33] A. D. MCLACHLAN, "Tests for comparing related amino-acid sequences. Cytochrome c and cytochrome c_{551}," *J. Mol. Biol.*, Vol. 61 (1971), pp. 409–424.

[34] ———, "Repeating sequences and gene duplication in proteins," *J. Mol. Biol.*, Vol. 64 (1972), pp. 417–438.

STUDIES OF ENZYME EVOLUTION BY SUBUNIT HYBRIDIZATION

ROSS J. MACINTYRE
CORNELL UNIVERSITY

1. Introduction

Like many evolutionists in this era of molecular biology, I have been trying to assess the importance of natural selection at the molecular level.

Surely no one approach or single technique, or, in fact, no series of investigations which approaches the problem from just one direction will settle the dispute between the Darwinian and non-Darwinian camps. The final verdict will be pieced together from many experimental facts and theoretical insights. In this paper, I will present a progress report on a technique which will not by itself answer that outstanding question but which will, I hope, add to the evidence which will eventually decide the issue.

While initial results have been encouraging, there are still some outstanding questions. Specifically, does the technique measure what I think it does? And, do the differences in protein structure which it detects have any functional (and, thus, selective) significance?

2. Some background

Before I describe the technique, analyze the preliminary results, and discuss these questions, I will try to explain why I presently prefer the biochemical approach to the question of the importance of natural selection.

I began my research career as an experimental population geneticist interested in the adaptive significance of enzyme polymorphisms. My organism of choice was and still is, *Drosophila melanogaster*. However, I became rather profoundly pessimistic about my own or indeed anyone's ability to measure selective differences between the carriers of different electrophoretic variants of enzymes, or allozymes as they have been called (Prakash, Lewontin, and Hubby [17]). If, in fact, differences exist, we may not be able to measure them with present techniques. Why is this so? If one chooses to work with *Drosophila,* he may want to use population cages, the classical tool of the experimental population geneticist, to detect these selective differences in laboratory populations. Available evidence from population cages leads to a rather unsatisfactory conclusion, namely, if selective differences exist between different allozyme carriers, they must be very small.

For example, we have measured gene frequency changes in several allozyme systems of *D. melanogaster* and *D. simulans* over the last few years (MacIntyre and Wright [12] and O'Brien and MacIntyre [14]). In some instances, after the first few generations where complex chromosomal interactions may change single gene frequencies rather rapidly, a rather definite although slow change in gene frequencies may be seen. In others, no change is readily discernable. I have calculated selection coefficients in these cages under a simple model of selection against one homozygote. Table I shows the calculations and some of

TABLE I

SUMMARY OF GENE FREQUENCY ESTIMATES IN SEVERAL LONG TERM *Drosophila* POPULATION CAGES

The first seven cases are from MacIntyre and Wright [12] and the last two from O'Brien and MacIntyre [14].

$$\Delta\bar{p} = [s\bar{p}^2(1 - \bar{p}^2)]/(1 - s\bar{p}^2).$$

Cage	Enzyme	Number of generations	$\overline{\Delta p}$	s
3	esterase-6	28	.002	.017
4	esterase-6	28	.004	.03
5	esterase-6	27	.007	.05
6	esterase-6	30	.012	.11
7	esterase-6	32	.0007	.008
8	esterase-6	32	.002	.02
1	acid phosphatase-1	84	.0004	.003
1	leucine aminopeptidase-D	100	.003	.02

the background information. Data were taken from seven cages. In these populations, the number of generations was large enough to give us some confidence in the reality of any change in gene frequency on the one hand or in any equilibrium on the other. Three gene-enzyme systems were studied and in each, so-called "fast" and "slow" electrophoretic variants were the phenotypes used to determine gene frequencies.

In each cage, the mean change in allele frequency per generation is derived from p_0 (the "fast" allele frequency at apparent initial equilibrium) $-p_e$ (its frequency at end of the experiment)/number of generations and $\bar{p}$ (mean gene frequency of favored allele) is equal to p_0 (at initial apparent equilibrium) $+p_e$ (at end of experiment)/2. Finally, s (the selection coefficient) is calculated by rearranging the familiar formula in Table I. Note the very small selection coefficients ranging from 0.003 to 0.11. Both the reality of the gene frequency changes and the model of selection against only one homozygote may be properly questioned. What I want to do, however, is emphasize the very small changes in gene frequencies that characterize many population cage experiments with electrophoretic enzyme variants. We conclude simply that if the changes are real, the selection coefficients are small. A recent and detailed study on a gene-enzyme system in *D. pseudoobscura* fully supports this conclusion (Yamazaki [19]).

But what does this imply with regard to meaningfully answering the question about the adaptive significance of enzyme polymorphisms in population cages? One thing it may mean is that each cage may represent a unique event and that results will not be repeatable. Laird and McCarthy have estimated that the haploid genome of *D. melanogaster* contains about 90,000 unique cistrons. Taking this at face value, and dividing by the approximately 300 map units of the genome, we get an estimate of 300 genes per map unit! While this may be an overestimate, there is no compelling reason to reject it out of hand (see also, Davidson and Hough [3]). It is fair to say, I believe, that most population geneticists have built their models with many fewer genes in mind. There is at least one implication in this that does not bode well for the investigator setting up his populations in order to study gene frequency changes at an allozyme locus. Remember that the selection coefficients associated with the genotypes are likely to be very small. Now, suppose there are 600 genes within one map unit of his locus. In order to define the problem, let us make the additional (and undoubtedly unrealistic) assumption that selection dictates that each of these 600 loci should be occupied by a wild type dominant allele. That is, at each there is selection for the wild type homozygote. Yet spontaneous mutations to deleterious alleles will have been occurring in the population. Each locus will in fact be polymorphic in the sense that rare mutant alleles will be present in low frequencies.

TABLE II

PROBABILITY OF SELECTING A "WILD TYPE" CHROMOSOME OVER A TWO MAP UNIT REGION UNDER SEVERAL MODELS OF SELECTION AT EVERY LOCUS AND A MUTATION RATE OF 1×10^{-6}

s	h	q when $\Delta q = 0$	Probability of selecting a wild type chromosome
0.1	0	3×10^{-3}	$(.997^{600}) = .17$
0.01	0	10^{-2}	$(.99^{600}) = .0024$
0.001	0	3×10^{-2}	$(.97^{600}) = 1.5 \times 10^{-7}$
0.1	0.1	10^{-4}	$(.9999^{600}) = .95$
0.1	0.01	10^{-3}	$(.999^{600}) = .55$
0.01	0.001	10^{-1}	$(.90^{600}) = 3.5 \times 10^{-28}$

Table II shows what some of these frequencies might be given a mutation rate per locus of 1×10^{-6}. The table is set up so that at each of the 600 loci, selection can act either against only the recessive homozygote or against both the recessive homozygote and the heterozygote. Column 3 in Table II contains the mutation-selection equilibrium frequency of the rare mutant allele in each situation. The frequencies, especially in the absence of selection against the heterozygote, can be rather high. The important calculation is in the right hand column. This is the probability, given the various mutant allele frequencies at the 600 loci, that the investigator will choose a chromosome *free from mutant alleles* in the two map unit region when he chooses the founders for his experimental population. As you can see with selection coefficients associated with the deleterious mutant alleles of around 0.01, it is almost certain that he will choose chromosomes with some "genetic junk" closely associated with the allozyme locus. Note that each chromosome in the founders will almost always be unique with regard to its "genetic junk." Several investigations indicate that new deleterious mutations are associated with selection coefficients of 0.01 (for example, Kenyon [7]). Thus, if an investigator starts two populations each with different founders, the chromosomes may look like something I have drawn in Figure 1. In this figure it is assumed that the founders of a cage will contain four possibly different chromosomes. The area surrounding the allozyme locus represents one map unit on each side and is delimited by the number of genes (600) in the interval. Thus, in chromosome number one in cage one, the "fast" allele is linked to a mutant, deleterious allele at locus No. 33. The fourth chromosome for cage 2 has a "slow" allele linked to three different mutant genes. The dashes in Figure 1 simply indicate wild type alleles at the other loci. Notice that while the founders have the same initial frequencies of the "fast" allele of the enzyme locus, the other associated genes are very different.

In this model, I have assumed that the type of selection against each mutant allele at the 600 loci is the same, that is, the mutant alleles are completely recessive. This is, of course, quite unrealistic. In reality, if one can even think of selection acting in such an independent way, there may be selection against heterozygotes and mutant homozygotes at some loci, heterotic selection at others, and so on. The point I wish to make is that even in an oversimplified model, there can be tremendous variation in several samples of chromosomes. The rather variable results from duplicated experiments designed to detect single gene frequency changes in experimental populations, I think, support this idea.

Franklin and Lewontin [4] have put this concept in a sharper definition by equating the "unit of selection" as a correlated block of genes, that is, a small area of the chromosome in which the genes may be out of linkage equilibrium. Unfortunately, we have almost no experimental information about the extent of linkage disequilibria in natural populations. Nor are we, in my view, likely to be able to obtain meaningful data on this point in the near future. We simply don't have enough closely linked genes identified. Also, it is not very satisfying to have to replace the single locus, whose products we can directly examine, with the "correlated block" of genes whose boundaries may be always shifting and whose allelic contents may be largely unknown. The real danger here is that the gap between the new theory and feasible experimentation may be unbridgeable for the foreseeable future.

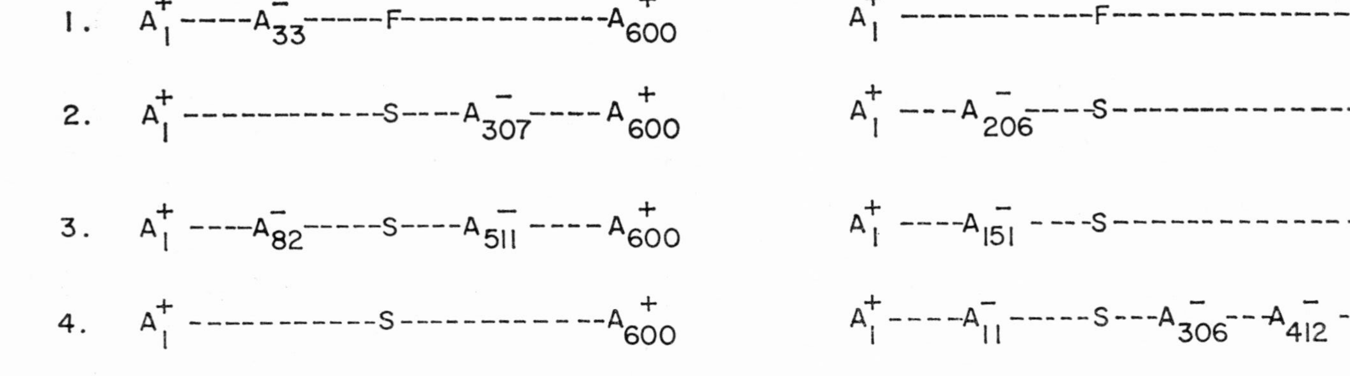

FIGURE 1

Chromosomes of hypothetical founders of two experimental populations. F and S refer to alleles specifying detectable electrophoretic variants of an enzyme. A_1–A_{600} refer to the 600 loci within one map unit of the gene controlling the allozymes.

Because of these considerations, I was left, in my own mind at least, with only one alternative—to approach the problem as a developmental biologist or a physiological geneticist rather than a population geneticist.

The basic difference between the neo- and non-Darwinians here can be simply stated: if allozymes or homologous enzymes in different species are selectively neutral, then they are functionally equivalent.

But how does one demonstrate the functional equivalence of two enzymes? One way is the so-called "shotgun" approach; that is, to measure and compare properties such as heat stabilities, inhibitor sensitivities, K_m's, and so on. But, if differences aren't found, it may simply be that the tests run weren't exhaustive, or if differences are found, they may not be relevant to the action of natural selection.

It is better, it seems to me, to proceed from that great truism of evolutionary biology, that natural selection acts on phenotypes and ask what is the phenotype of the enzyme? In this context, it must be its tertiary, and if it is an oligomeric enzyme, its quarternary structure. Thus, an enzyme must be folded properly to have a "wild type" active site or a "wild type" allosteric site or a "wild type" membrane attachment site. But, short of X-ray crystallography, how does one compare two enzymes with regard to their three dimensional configurations? Immunological tests are one way, but differences in antigenic sites may not always be accompanied by meaningful differences in enzyme structure.

In recent years, it has become evident that many, if not most enzymes, are composed of subunits (see Klotz and Darnall [8]). The bonding between these subunits is critical for the activity and perhaps the regulation of the activity of most of these enzymes (Cook and Koshland [2] and Iwatzuki and Okazaki [6]). It is, in fact, very fashionable as well as attractive to invoke subunit interactions in the control of cellular metabolism (Noble [13] and Haber and Koshland [5]). Hard evidence for this idea is rather sparse, however.

Nevertheless, studies on hemoglobin, the only protein for which the number and kind of amino acid residues involved in the quarternary structure are known, attest to the importance of this property in its basic function. Some 26 α chain amino acids contact the β chain and 27 amino acids in that latter subunit contact the α chain (Perutz, Miurhead, Cox, and Goaman [15]). So in each polypeptide chain almost twenty per cent of the amino acids are involved in the maintenance of a proper quarternary structure. The evolutionary conservatism of these amino acids is as pronounced as that of the amino acids involved in haem contact (see MacIntyre [11] for the details of this comparison). Furthermore, mutations in human hemoglobins affecting subunit interfaces, lead, in the majority of cases, to clinically detected anemias (Perutz and Lehmann [16]).

Thus, considering the all too meager definitive information from just one protein and much speculation, it can be stated at least as a working hypothesis that if differences in subunit affinities between allozymes or homologous enzymes can be demonstrated, then the enzymes should not be functionally equivalent.

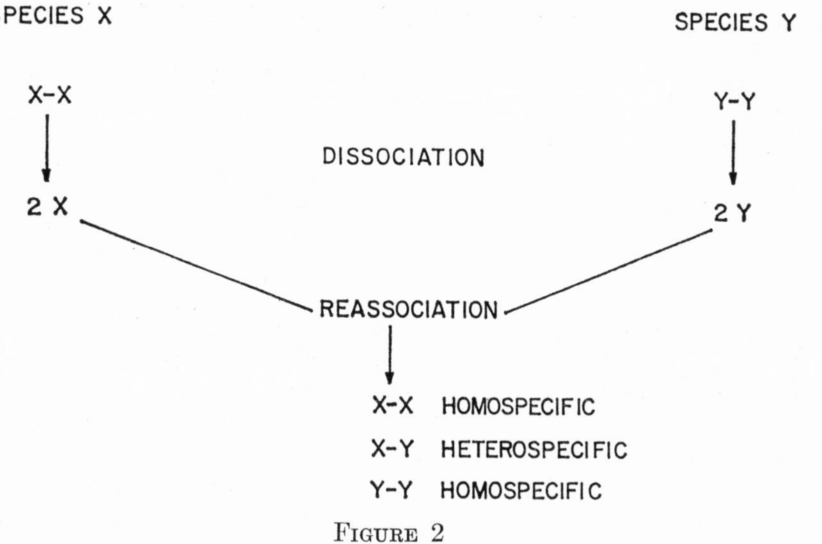

FIGURE 2

Outline of the experiments designed to detect interspecific differences in enzyme subunit affinities.

Observed homospecific:heterospecific ratio = enzyme activity of

$$[X\text{-}X + Y\text{-}Y]/(X\text{-}Y).$$

Expected homospecific:heterospecific ratio $= (p^2 + q^2)/2pq$,

where $p = [X\text{-}X + \frac{1}{2}(X\text{-}Y)]/[X\text{-}X + X\text{-}Y + Y\text{-}Y]$
and $q = [Y\text{-}Y + \frac{1}{2}(X\text{-}Y)]/[X\text{-}X + X\text{-}Y + Y\text{-}Y]$.

3. Experimental outline and details

In order to detect these differences, we have followed the simple rationale outlined in Figure 2. An enzyme from two species, which is assumed to be a dimer, is dissociated into its constituent subunits. These are mixed and the relative activities of the two homospecific (or homodimeric) and the heterospecific (or heterodimeric) enzymes are determined. The observed amounts can be expressed as an observed homospecific:heterospecific enzyme ratio, and compared to a ratio expected if subunit association is random. The data are frequently expressed as the *difference* between the two ratios, one observed and the other calculated.

At this point, and this will be discussed at greater length below, a difference implies either that the subunits do not randomly associate, or that the enzymes have different substrate turnover numbers.

We have made our most extensive tests to date with enzymes of different species because, initially, it wasn't clear if the method would distinguish between the undoubtedly more subtle allozymic differences. I will report only on the interspecific comparisons in this paper.

The enzyme I chose to work with is an acid phosphatase, by far the most prominent phosphatase after electrophoretic separation of extracts from *D. melanogaster* adults. The allozymic forms are shown in Figure 3. This is a starch gel pattern or zymogram developed for acid phosphatase activity of the extracts of four single flies (MacIntyre [12]). Flies monomorphic for "slow" and "fast" forms of the enzyme are designated as *AA* or *BB* homozygotes, respectively. The two heterozygous or *AB* flies in the middle of the gel show a three band pattern, characteristic of a dimeric enzyme (Shaw [18]). The gene had been mapped using electrophoretic variants shown in Figure 3.

We chose this enzyme for two reasons. First, there is ample interspecific variation in the electrophoretic mobility of this enzyme. Figure 4 shows the electropherogram of eight *Drosophila* species. *D. melanogaster* and *D. simulans* both have distinct allozymic forms of this enzyme. The enzymes from *D. emarginata* and *D. willistoni* have distinct subbands in acrylamide gels. This leads to some problems in the interpretation of the densitometrically delineated peaks of enzyme activity when the mixtures of subunits, after reassociation, are subjected to electrophoresis and staining. Many of these species are very closely related, for example, *D. willistoni* and *D. paulistorum*, *D. emarginata* and *D. sturtevanti*, and of course *D. melanogaster* and *D. simulans*. Their enzymes, however, are electrophoretically separable. This is fortunate since the quantitation of the homospecific and heterospecific enzymes involves either elution of

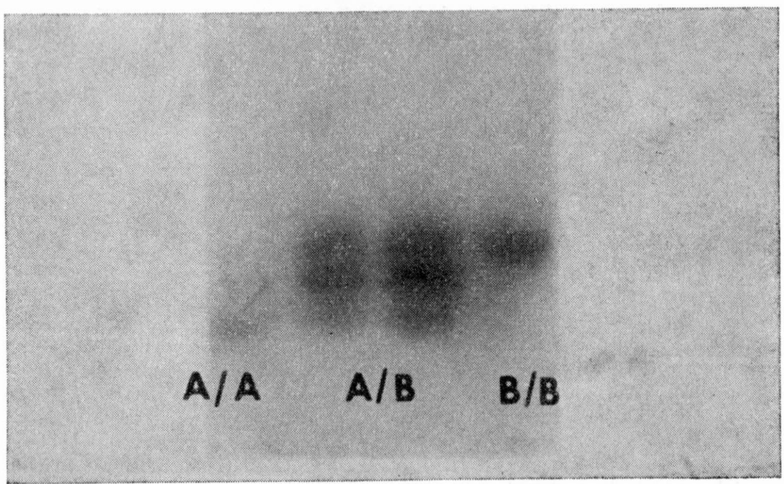

FIGURE 3

A starch gel with allozymic acid phosphatase patterns characteristic of "slow" or *AA* homozygotes, "slow"/"fast" or *AB* heterozygotes, and "fast" or *BB* homozygotes.

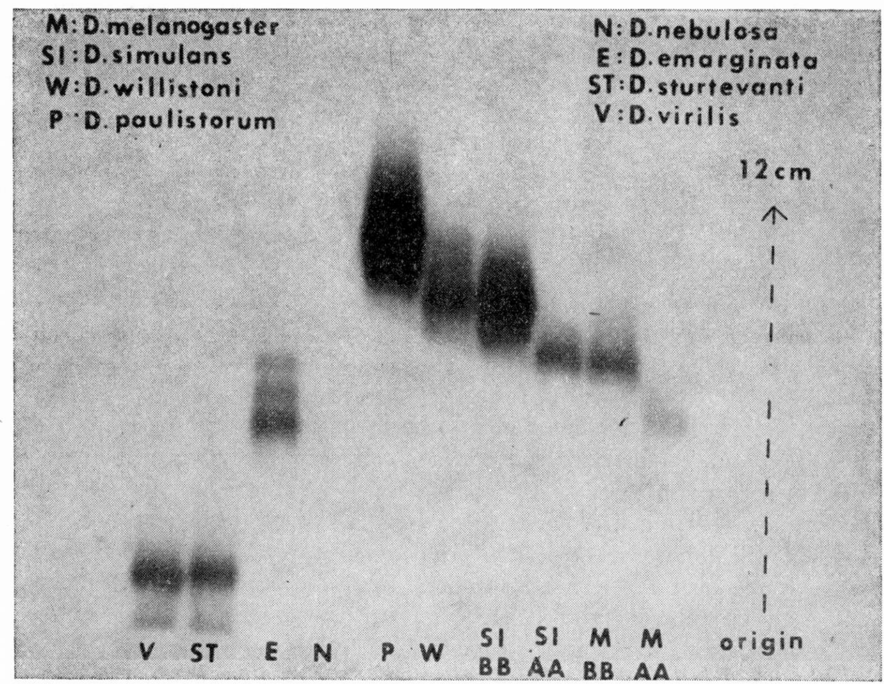

FIGURE 4

Acrylamide gel showing positions of the homospecific enzymes from eight *Drosophila* species. The *AA* and *BB* allozymes of *D. melanogaster* and *D. simulans* are also shown.

the precipitated dye from the gels or densitometry. On the other hand, the enzymes from the rather distantly related species *D. virilis* and *D. sturtevanti* are electrophoretically indistinguishable. Note that these prominent adult phosphatases are only assumed to be homologous with the one from *D. melanogaster*. Better evidence for this will be presented below.

The second reason this enzyme was chosen, was that in addition to the electrophoretic pattern of the heterozygote, we have other evidence indicating that this acid phosphatase is a dimer. This evidence will be published elsewhere (MacIntyre [9]) so I will simply summarize it here. First, the molecular weight of the enzyme from both *D. melanogaster* and *D. virilis*, the two most distantly related species in this survey, is 100,000, a value compatible with other enzymes known to be dimers (Klotz and Darnall [8]). Secondly, the subunit produced by the dissociating treatment we employ is almost exactly one half the size of the native enzyme. These subunits migrate to a single position in acrylamide gels where they reassociate to form active enzymes. This observation almost certainly rules out a heteromultimeric structure for the enzyme since it is improbable that the two different subunits would have the same net charge and size. Finally, we have chemically induced 15 mutations which eradicate or alter the activity of the enzyme. A complementation analysis with all 15 mutants produced results entirely compatible with the interpretation that a single structural gene codes for this enzyme (Bell and MacIntyre [1]).

4. Preliminary experiments

I do not intend to go into extensive detail about certain preliminary information. (See MacIntyre [9], [10] for the details.) We assay the enzyme's activity by coupling α-naphthol phosphate with the diazonium salt, Fast Red TR. The red complex is soluble in detergent and glacial acetic acid. Thus, we can use the same substrate for both test tube and gel assays. Furthermore, we have correlated the amount of dye eluted from gels with densitometric tracings. Use of the latter technique allows more data to be taken in a shorter period of time. Enzyme preparations used in the experiments are only partially purified, about ten fold over crude supernates, after dialysis at low pH and ammonium sulfate fractionation. The preparations are free from all other phosphatase activity, however. I will discuss below certain problems of interpretation which the use of rather crude enzyme preparations raises. The pH optimum for the enzyme, for all species examined so far, is 5.1.

Many accepted dissociating treatments were tried on the enzyme. The most effective, in terms of good recovery of enzyme activity, is exposure of the enzyme to high pH levels. Figure 5 shows the inactivation which takes place over rather narrow pH intervals. The inactivation profiles of the enzymes from three species *D. melanogaster*, *D. virilis*, and *D. simulans* are shown in this figure. Dilute NaOH was added to extracts and the pH was recorded. Aliquots were assayed for enzyme activity at various pH's. The data are plotted in terms of per cent of control activity remaining *versus* pH. The arrows at the top of this figure indicate the pH at which complete inactivation just occurs for each species' enzyme. Thus, the acid phosphatase from *D. melanogaster* is completely inactivated at pH 10.7, *D. simulans*' enzyme at pH 10.9 and *D. virilis*' at 11.0. These differences are repeatable. I should point out that care must be taken not to raise the pH too far beyond the level at which inactivation is complete. If the pH is raised too high, too little enzyme activity is regained after adjustment of the pH back to neutrality.

From 20 to 85 per cent of the initial activity can be regained if the pH inactivated preparations are dialyzed back against Tris-maleate buffers. Curiously enough, the enzymes from different species may differ in the pH optimum for reassociation. This is shown in Figure 6. In this experiment, aliquots of completely inactivated extracts were dialyzed against Tris-maleate buffers of pH's ranging from 4.9 to 8.2. The activity regained by each aliquot, here expressed as optical density at 540 millimicrons, was then plotted against the pH of the buffer against which it was dialyzed. Note that the three species' enzymes have different pH optima for reassociation.

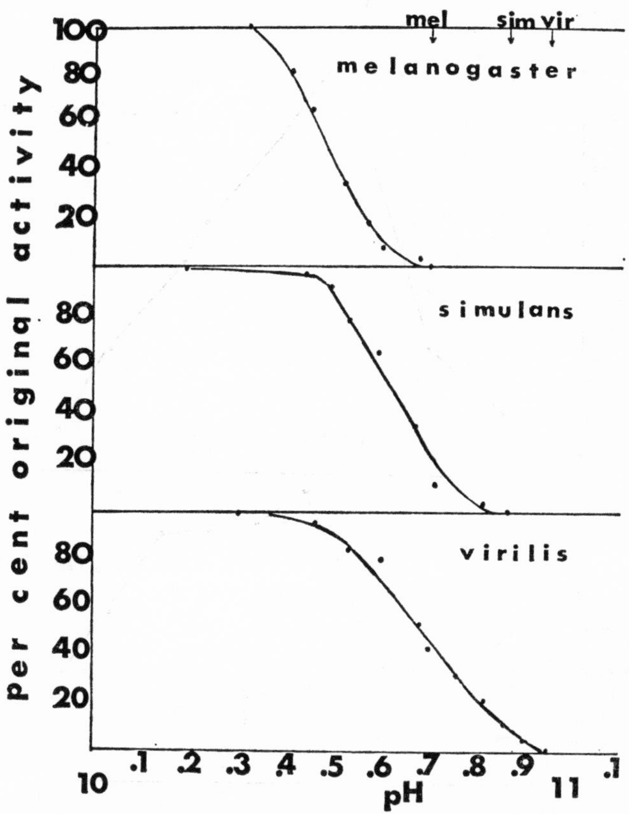

FIGURE 5

Inactivation curves of acid phosphatase-1 from *D. melanogaster*, *D. simulans* and *D. virilis*.

0.1 N NaOH was added to 5.0 ml of a partially purified enzyme preparation in 0.05 M NaCl at 25°C. After mixing, the pH was measured with a combination microelectrode. At the pH's indicated on the graph, 0.2 ml aliquots were removed and immediately assayed for acid phosphatase activity. Activities were corrected for the volume of NaOH added. The pH at which 100 per cent inactivation occurs for each species' enzyme is indicated on the top line.

In several preliminary subunit hybridization experiments with enzymes from *D. melanogaster*, *D. virilis*, and *D. simulans*, we tested for several things: (1) Does exposure to high pH dissociate the enzyme? (2) Do the subunits reassociate in all possible combinations during dialysis against the Tris-maleate buffers? (3) Can the electrophoretic patterns of the reassociated enzymes be reliably quantitated? (4) Are the results repeatable? Figure 7 shows that in the three tests, the enzyme dissociates at high pH and the subunits do reassociate

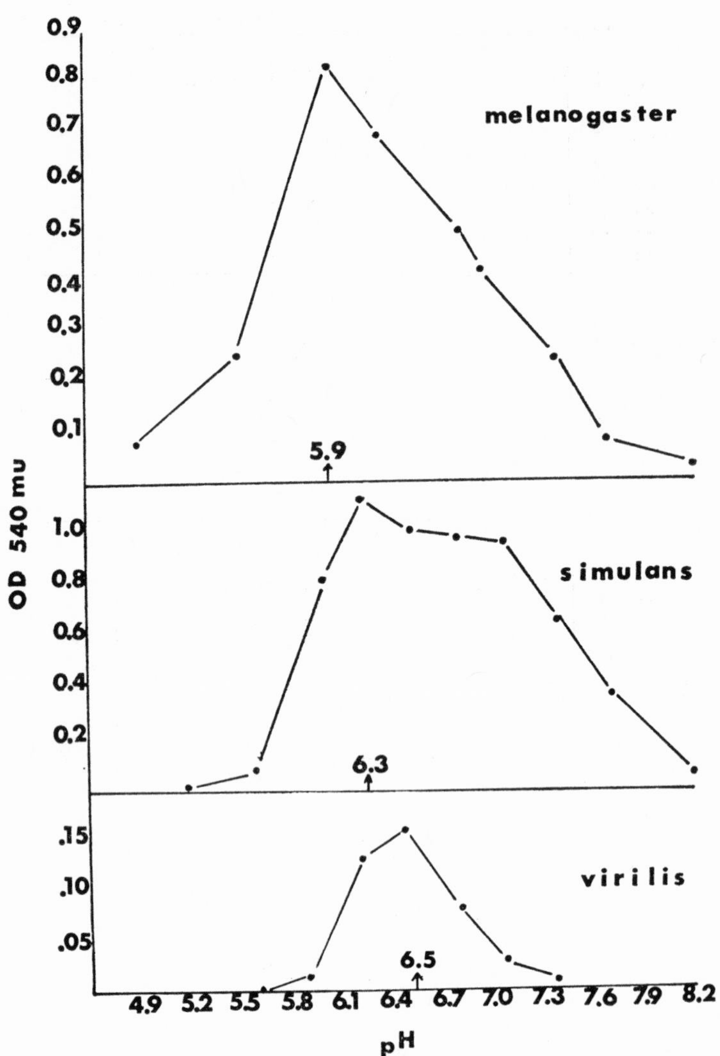

FIGURE 6

pH optima for reassociation of subunits of acid phosphatase-1 from *D. melanogaster*, *D. simulans* and *D. virilis*.

2.0 ml aliquots of pH inactivated acid phosphatase-1 from each of the three species (*D. melanogaster*; pH 10.7, *D. simulans*, pH 10.8, *D. virilis*, pH 11.0) were dialyzed against 0.4 Tris-maleate buffers at the indicated pH's for 72 hours at 4°C. The aliquots were then assayed for acid phosphatase activity. Apparent pH optima for each species are indicated on the abscissae.

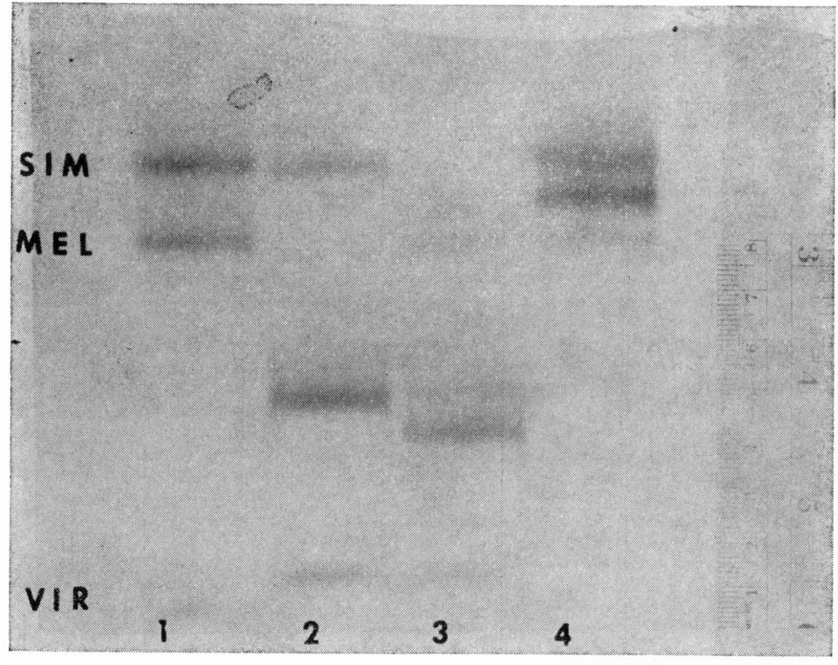

FIGURE 7

Electropherogram of acid phosphatase-1 from *D. virilis* (VIR), *D. simulans* (SIM) and *D. melanogaster* (MEL)(slot 1), the reassociated subunits from *D. simulans* and *D. virilis* (slot 2), from *D. melanogaster* and *D. virilis* (slot 3), and *D. melanogaster* and *D. simulans* (slot 4). In each mixture, the heterospecific enzyme is in the middle of each pattern.

to form *active* heterospecific enzymes during dialysis against the Tris-maleate buffer. In slot 1, the positions in the gel of the undissociated homospecific enzymes for each of the three species are shown. Above slot 2 is the pattern, after electrophoresis, of the reassociated mixture of subunits from *D. simulans* and *D. virilis*. The intermediate zone of activity represents the heterospecific enzyme. Slots 3 and 4 show the patterns of reassociated subunits from *D. melanogaster* and *D. virilis* (slot 3) and *D. melanogaster* and *D. simulans* (slot 4). Note that a heterospecific enzyme formed in each mixture of subunits from two different species. The presence of the active heterospecific enzyme almost certainly establishes the homologies between the enzymes. In Table III are the detailed quantitative data for one of the tests, *D. virilis* and *D. melanogaster*. Each experiment was repeated and the subunits reassociated at each species' pH optimum for reassociation, that is, at pH 5.9, the pH optimum for reassociation of the enzyme from *D. melanogaster* and at pH 6.5, the pH optimum for the *D. virilis* enzyme. The fourth column indicates how much activity was

TABLE III

RESULTS OF EXPERIMENTS INVOLVING REASSOCIATION OF ACID PHOSPHATASE-1 SUBUNITS OF *D. melanogaster* AND *D. virilis*

Ranges for the per cent of total acid phosphatase in each zone and for the homospecific:heterospecific ratios are given in parentheses. Number of determinations: experiment 1 (pH 5.9)-8, (pH 6.5)-7; experiment 2 (pH 5.9)-5, (pH 6.5)-4. The expected ratio is calculated as follows: p (*mel* subunits) = mean proportion of *melanogaster* homospecific enzyme + $\frac{1}{2}$ mean proportion of heterospecific enzyme; q (*virilis* subunits) = $1 - p$; expected ratio = $(p^2 + q^2)/2pq$. For example, in experiment 1 at pH 5.9: p (*melanogaster* subunits) = $(9 + \frac{1}{2}70)/100 = 0.44$; q (*virilis* subunits) = $1.00 - 0.44 = 0.56$; expected ratio = $[(0.44)^2 + (0.56)^2]/2(0.44)(0.56) = 1.00$.

pH of reassociation	Experiment no.	Subunits	Per cent reactivation after dialysis	Per cent of total acid phosphatase in each zone: *Melanogaster* homospecific	Heterospecific	*Virilis* homospecific	Homospecific:heterospecific ratio	Expected ratio	Difference
5.9	1	*melanogaster* *virilis* mixture	31 34 52	9 (7–11)	70 (67–76)	21 (17–22)	0.43 (0.32–0.49)	1.00 (1.00–1.04)	−0.57 (0.51–−0.68)
5.9	2	*melanogaster* *virilis* mixture	33 35 29	12 (11–13)	70 (68–74)	18 (15–19)	0.42 (0.35–0.47)	1.00 (1.00)	−0.58 (−0.53–−0.65)
6.5	1	*melanogaster* *virilis* mixture	22 49 38	10 (9–12)	67 (64–69)	23 (21–25)	0.50 (0.45–0.56)	1.00 (1.00–1.04)	−0.50 (−0.44–−0.57)
6.5	2	*melanogaster* *virilis* mixture	22 47 30	9 (8–10)	71 (70–71)	20 (19–22)	0.41 (0.41–0.43)	1.00 (1.00)	−0.59 (−0.59–−0.61)

regained, both by subunits from one species alone and when mixed in equal proportions. The relative activities of the three zones in the electrophoretic pattern of the reassociated mixture of subunits are in the next three columns. The ranges for each set of determinations (usually from 4–8) are in parentheses. If experiment number 1 (at the reassociation pH of 5.9) is taken as an example, then the average relative activity of the *D. melanogaster* homospecific enzyme is 9 per cent. The *D. virilis* homospecific enzyme contributes 21 per cent to the total. The heterospecific enzyme activity represents 70 per cent of the activity in the reassociated mixture of subunits. The observed homospecific:heterospecific activity ratio in this example is $(9 + 21)/70$ or 0.43. At the bottom of the table, the expected ratio for this particular experiment is calculated. The frequency of *D. melanogaster* subunits which reassociated to form active enzymes is $(9 + \frac{1}{2}70)/100$ or 0.44. The frequency of effectively reassociating *D. virilis* subunits, then, is $1 - 0.44$ or 0.56. The expected ratio is then equal to $(0.56)^2 + (0.44)^2/2(0.44)(0.56)$ or approximately 1. Then, finally, the difference between the observed and expected ratio in this example is $0.43 - 1.00$ or -0.57.

TABLE IV

SUMMARY OF RESULTS FROM ACID PHOSPHATASE-1 SUBUNIT REASSOCIATION EXPERIMENTS INVOLVING *D. melanogaster*, *D. simulans*, AND *D. virilis*

The ratios are the results of two experiments for each interspecific test.

Interspecific test	Reassociated at pH	Observed homospecific:heterospecific enzyme ratio	Difference from expected ratio
melanogaster	5.9	(1) .43 (2) .42	−.57 −.58
×			
virilis	6.5	(1) .50 (2) .41	−.50 −.59
simulans	6.3	(1) .60 (2) .79	−.40 −.21
×			
virilis	6.5	(1) .79 (2) .75	−.21 −.25
melanogaster	5.9	(1) 1.44 (2) 1.38	+.44 +.38
×			
simulans	6.3	(1) 1.50 (2) 1.33	+.42 +.33

Table IV summarizes the results for the three pairwise tests involving *D. melanogaster*, *D. virilis*, and *D. simulans*. Reported in this table are both the observed homospecific:heterospecific ratios and the differences of these from the ratios expected if subunit association were random and enzyme activities were equal. The correspondence between the duplicate experiments is good in every case but that one involving *D. virilis* and *D. simulans* at a reassociation pH of 6.3. We have repeated this test several more times, however, and more consistently find a difference of about −0.30. Note the independence of the results from the pH optimum of reassociation. The data are expressed ultimately as a difference between observed and expected ratios. The sign of the difference is important. A minus indicates more than expected heterospecific enzyme activity is present. A plus sign indicates more than expected homospecific enzyme activity has been measured.

The use of *D. melanogaster* and *D. simulans* allowed us to directly test another assumption implicit in the methodology, namely, that the methods used in purification, dissociation, and reassociation do not impair the functional integrity of the subunits. Specifically, with these two species we can make several *in vitro versus in vivo* comparisons, since they both have electrophoretic variants and will form viable interspecific hybrids. Thus, we determined the differences between observed and expected homospecific:heterospecific ratios in three tests and compared them with those obtained from the corresponding interspecific hybrids or intraspecific heterozygotes. The data are summarized in Table V.

TABLE V

COMPARISON OF *in vitro* AND *in vivo* RESULTS WITH REGARD TO REASSOCIATION OF ACPH-1 SUBUNITS

Difference between observed and expected homospecific/heterospecific ratio ± s.d.

Test	Difference ± S.D.	Probability
(A) *D. melanogaster*		
AA × *BB* (*in vitro*)	+0.31 ± 0.20	$t = 0.48$
Acph-1^A/*Acph-1*^{B} heterozygotes (single flies)	+0.22 ± 0.19	$P > 0.5$
(B) *D. simulans*		
AA × *BB* (*in vitro*)	+0.23 ± 0.06	$t = 1.96$
Acph-1^A/*Acph-1*^{B} heterozygotes (mass homogenates)	+0.32 ± 0.12	$0.10 > P > 0.05$
(C) Interspecific hybrids		
BB (*melanogaster*) × *BB* (*simulans*) (*in vitro*)	+0.32 ± 0.03	$t = 0.22$ $P > 0.5$
D. melanogaster ♀♀ × *D. simulans* ♂♂ (single flies)	+0.33 ± 0.12	
D. simulans ♀♀ × *D. melanogaster* ♂♂ (single flies)	+0.28 ± 0.04	$t = 2.22$ $0.05 > P > 0.01$

In the first case, the electropherograms of *D. melanogaster* heterozygotes for the alleles specifying the "slow" and "fast" electrophoretic variants were compared with the pattern obtained when partially purified, dissociated and reassociated "slow" and "fast" forms of the enzyme were used as the sources of the mixed subunits. In the second case, *D. simulans* heterozygote patterns are compared in a similar way with the corresponding dissociated and reassociated "slow" and "fast" enzymes from this species. Also, reciprocal interspecific hybrids are compared with the *in vitro* results from the *D. melanogaster* × *D. simulans* test. The results agree well in every case except one, but this is close to the acceptable level. The *in vitro* results appear to faithfully reflect *in vivo* enzyme subunit associations.

5. Interspecific tests

The initial tests with *D. melanogaster*, *D. simulans*, and *D. virilis* gave me enough confidence to go ahead and survey several selected species from the genus.

TABLE VI

SUMMARY OF THE PHYLOGENETIC RELATIONSHIP AND PRELIMINARY INFORMATION ABOUT THE ACID PHOSPHATASE-1 ENZYMES FROM THE ELEVEN SPECIES USED IN THE SUBUNIT HYBRIDIZATION TESTS

The asterisk indicates that the *BB* enzyme of *D. melanogaster* = 1.00. *Simulans*, *melanogaster*, and *nebulosa* have electrophoretic variants.

Species	Species group	Subgenus	Electrophoretic position*	pH optimum	pH of dissociation	pH optima for reassociation
D. melanogaster	*melanogaster*	*Sophophora*	0.83, 1.00, 1.07	5.1	10.3–10.7	5.9
D. simulans	*melanogaster*	*Sophophora*	1.03, 1.17, 1.31	5.1	10.5–10.9	6.2
D. willistoni	*willistoni*	*Sophophora*	1.24	5.1	10.8–11.2	6.4
D. paulistorum	*willistoni*	*Sophophora*	1.38	5.1	10.8–11.2	6.9
D. nebulosa	*willistoni*	*Sophophora*	1.41, 1.69	5.1	10.2–10.7	6.5
D. emarginata	*saltans*	*Sophophora*	0.76	5.1	10.2–10.7	6.5
D. sturtevanti	*saltans*	*Sophophora*	0.28	5.1	10.3–10.9	6.1
D. virilis	*virilis*	*Drosophila*	0.28	5.1	10.4–11.0	6.5
D. mulleri	*repleta*	*Drosophila*	0.28	5.1	10.0–10.7	6.9
D. mercatorum	*repleta*	*Drosophila*	0.28	5.1	9.9–10.6	6.7
D. immigrans	*immigrans*	*Drosophila*	0.69	5.1	10.2–10.8	7.2

Table VI summarizes the phylogenetic relationships of the species and necessary preliminary information about each which must be obtained before the experiments can be conducted. Eleven species which were used in all the possible pairwise tests are listed in the left column of the table. The relative electrophoretic positions of the homospecific enzymes (and allozymes where these have been found) are given in column 4. The "fast" or *BB* form of the *D. melanogaster* enzyme is set at 1.00. Note the interspecific variation in the pH inactivation interval and in the pH optimum for reassociation. In Figures 8 and 9 are the gel patterns obtained for some of the possible tests, that is, those tests in which the homospecific and heterospecific enzymes can be well enough

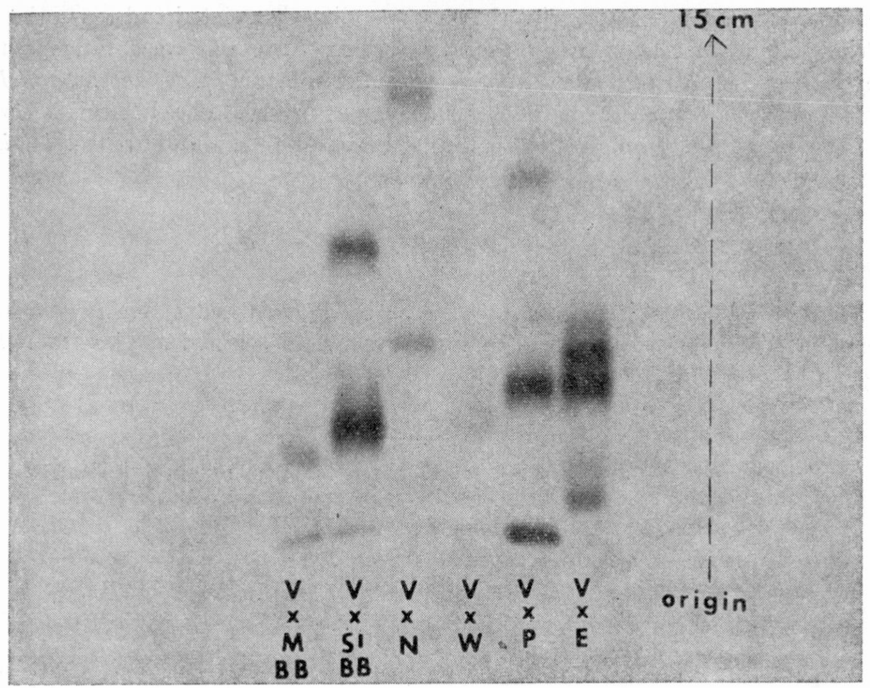

FIGURE 8

Acrylamide gel showing the six patterns of homospecific and heterospecific enzymes which formed during reassociation of subunits from the *D. virilis* acid phosphatase with subunits from the enzymes of six other *Drosophila* species. See Figure 4 for species designations.

separated by electrophoresis for an adequate densitometric analysis. Figure 8 shows eight tests with *D. virilis* as one common species. It can be seen in both figures, that in every case, a heterospecific enzyme forms, indicating that these are homologous enzymes. The subbanding in some patterns is also evident, but these can be correctly resolved if electrophoresis is carried out for a long enough time.

As you might expect, the quantitative data are massive. Figures 10 and 11 show them in graphic form. Given for each test is the mean difference between the observed and expected homospecific:heterospecific ratio and the 95 per cent confidence interval. Six to ten determinations were made on each test. Note that if the 95 per cent confidence interval overlaps the line of zero in the center, either subunit association is random or the enzymes have equal activities. In tests which fall on the right side of the line, the homospecific enzymes are preferentially formed or are more active. On the left side of the line, the heterospecific enzyme is disproportionally represented. The species pairs are roughly arranged, top to bottom, with increasing phylogenetic distance.

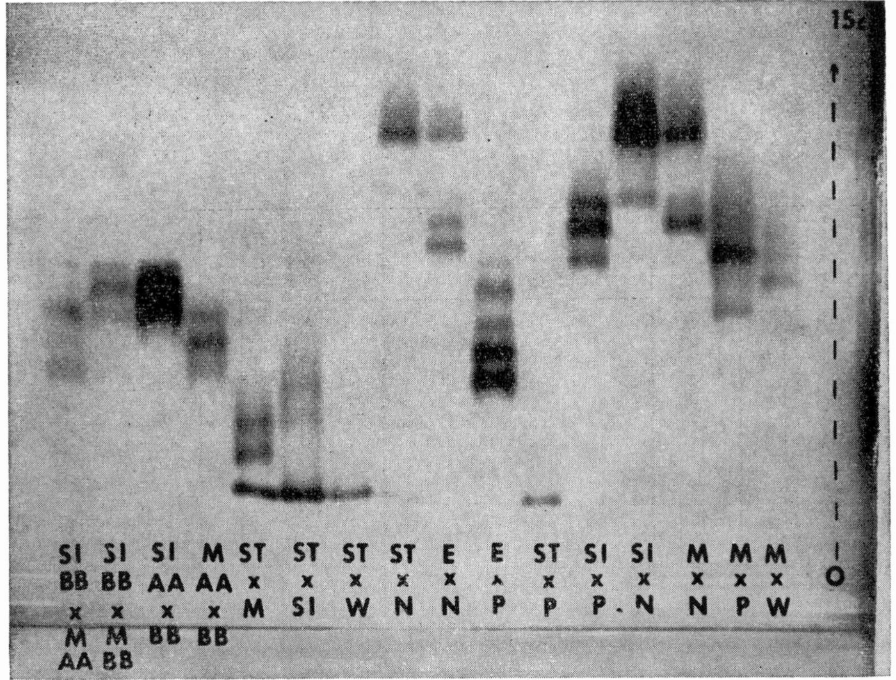

FIGURE 9

Acrylamide gel showing all possible subunit reassociation patterns involving the following species: *D. melanogaster* (M) with *AA* and *BB* allozymes, *D. simulans* (SI) also with *AA* and *BB* allozymes, *D. willistoni* (W), *D. paulistorum* (P), *D. nebulosa* (N), *D. emarginata* (E) and *D. sturtevanti* (ST).

The general conclusion to be drawn from these results is that there is variation between the homologous enzymes of closely related species that is detectable with the technique of subunit hybridization.

Many other observations can be made from the data presented in Figures 10 and 11. First of all, the intraspecific tests and the interspecific tests between species in the same species groups, without exception, have differences that indicate preferential formation or activity of the homospecific enzymes. In tests between species groups, however, both plus and minus differences were found. The meaning of this is not evident.

Secondly, the test appears to be extremely sensitive. A selected group of data was abstracted in Table VII. Reported here are the differences between observed and expected ratios in the tests involving *D. virilis* and *D. sturtevanti* with six other species from the subgenus *Sophophora*. Note that the enzymes of *D. melanogaster* and *D. simulans* are clearly different both in the tests with *D. virilis* and in the tests with *D. sturtevanti*. On the other hand, the enzymes from

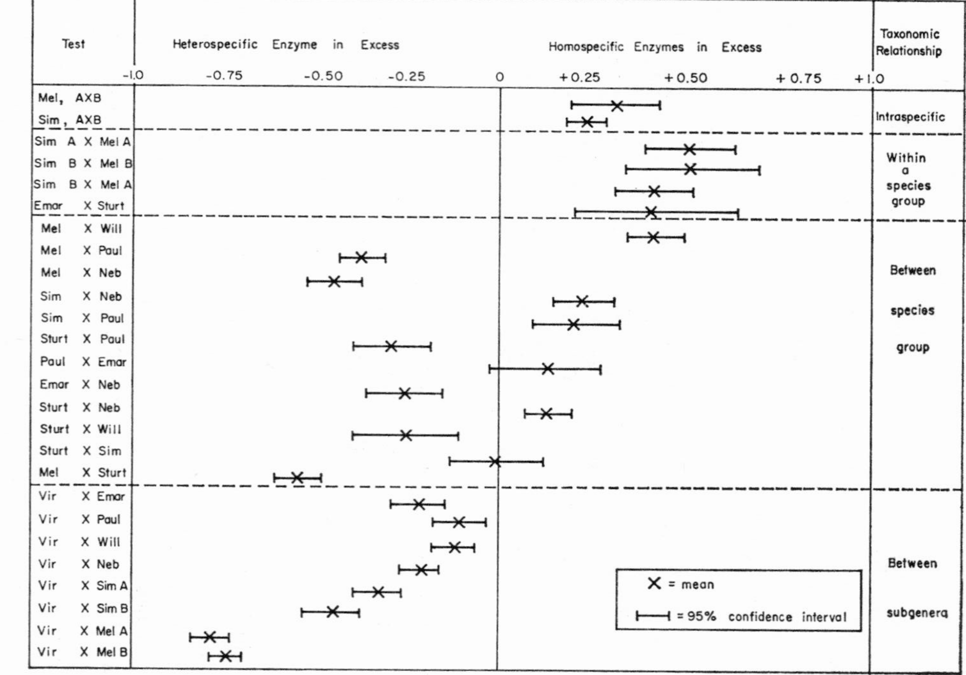

FIGURE 10

Differences of observed homospecific:heterospecific acid phosphatase activity ratios from those expected if subunit association were random, including all possible subunit hybridization tests involving the species listed in Figures 8 and 9. Code: Mel = *D. melanogaster*; Sim = *D. simulans*; Emar = *D. emarginata*; Sturt = *D. sturtevanti*; Will = *D. willistoni*; Paul = *D. paulistorum*; Neb = *D. nebulosa*; Vir = *D. virilis*.

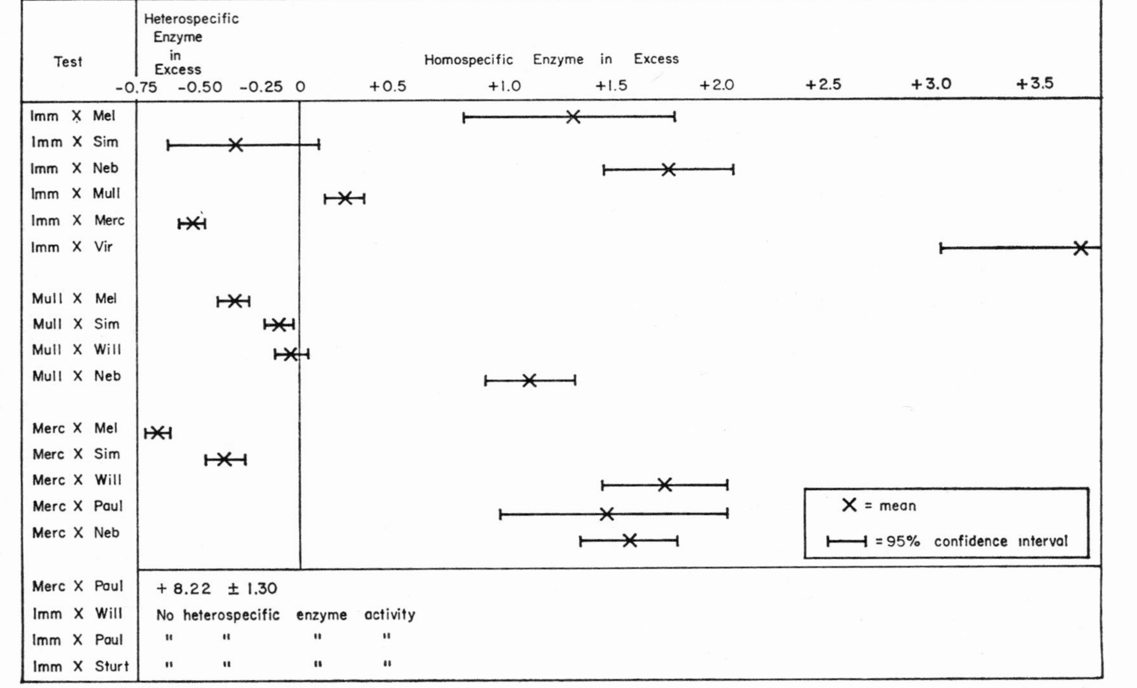

FIGURE 11

Differences of observed homospecific:heterospecific acid phosphatase activity ratios from those expected if subunit association were random, including all possible subunit hybridization tests involving *D. mulleri*, *D. mercatorum*, and *D. immigrans* with themselves and the other eight species listed in Table VI. Code: Imm = *D. immigrans*; Mull = *D. mulleri*; Merc = *D. mercatorum*; see Figure 10 for other species.

TABLE VII

RESULTS OF TESTS INVOLVING *D. virilis* AND *D. sturtevanti*
DIFFERENCE OF OBSERVED HOMOSPECIFIC:HETEROSPECIFIC RATIO FROM THAT EXPECTED IF SUBUNIT ASSOCIATION WERE RANDOM

Standard errors are shown with ± sign. Only *D. simulans* compared with *D. sturtevanti* is not significantly different from zero.

	D. virilis	*D. sturtevanti*
D. melanogaster	−0.74 ± .02	−0.54 ± .02
D. simulans	−0.46 ± .03	−0.01 ± .05
D. willistoni	−0.12 ± .02	−0.25 ± .06
D. paulistorum	−0.11 ± .03	−0.29 ± .04
D. nebulosa	−0.22 ± .02	+0.12 ± .03
D. emarginata	−0.23 ± .03	+0.42 ± .08

D. paulistorum and *D. willistoni,* two closely related species, are not differentiated by either test. Yet, their enzymes are separable by electrophoresis. *D. nebulosa,* a close relative of *D. willistoni* and *D. paulistorum* does appear to have an acid phosphatase which shows quite different reassociation patterns. Even more striking is the difference between the electrophoretically identical enzymes of *D. sturtevanti* and *D. virilis.* In every test except the one with *D. willistoni,* these two enzymes respond differently.

A third point to be made from the data in Figures 10 and 11 is that no two enzymes, when all possible pairwise tests are considered, are unequivocally the same in the properties measured by this technique. Note especially the differences between the enzymes of *D. mulleri* and *D. mercatorum.* Even the enzymes from *D. willistoni* and *D. paulistorum,* which appear to be the most similar, give very different results when tested with the dissociated enzyme from *D. melanogaster.* Finally, the subunits from *D. virilis* in almost every test preferentially associate (or form more active enzymes) in heterospecific combinations. On the other hand, the subunits from *D. immigrans* show little, if any, tendency to hybridize with subunits from the enzymes of different species. A complete analysis of these data will be published elsewhere.

6. Discussion

Some mention must be made at this point of the still unanswered questions about this method of detecting evolutionary changes in homologous enzymes. One obvious problem is that the proteins are not pure. In other words, could extrinsic factors and not the structures of the subunits be responsible for these patterns? Obviously, only with pure preparations can this problem really be solved. However, we have further purified the enzymes from *D. melanogaster* and *D. virilis* to about 80 fold by phosphocellulose chromatography. The very same homospecific:heterospecific enzyme ratio is obtained when these substantially purer preparations are used as the source of the subunits. In addition, we have obtained uncontaminated preparations of just the *D. melanogaster*-

D. virilis heterospecific enzyme. When it is dissociated and the subunits allowed to reassociate, both the two homospecific and the heterospecific enzymes form. The homospecific:heterospecific enzyme ratio in this experiment is again exactly the same as the ratio obtained when *D. melanogaster* and *D. virilis* homospecific enzymes are dissociated. (See MacIntyre [9] for details.)

If the differences between observed and expected ratios are due to intrinsic properties of the subunits, that is, differences in amino acid sequences, then another unanswered question arises. Specifically, do the differences reflect changes in amino acids affecting subunit interfaces or the enzymatic activity of the protein? Thus, in the test of *D. melanogaster* and *D. virilis* subunits, the homospecific:heterospecific ratio was in the range 0.3 to 0.4. Does this mean almost three times as much of the *D. melanogaster-D. virilis* heterospecific enzyme formed during reassociation? Or perhaps the expected amount forms but this novel phosphatase might turn over three times as many molecules of substrate as either homospecific enzyme.

In these experiments, there have not been measurable amounts of protein in the electrophoretically separated zones of acid phosphatase activity. We could, of course, start with pounds of flies rather than grams in the initial extraction of the enzyme, but as long as the preparations are only partially purified, one cannot be sure that measurements of specific activities would provide definitive information. We have some indirect evidence which suggests, in the *D. melanogaster* $\times$ *D. virilis* test, that the homospecific:heterospecific enzyme ratio reflects actual amounts of enzyme formed during subunit reassociation. Specifically, (1) the K_m's of all three enzymes are identical, and (2) the initial rate of increase of enzyme activity during reassociation is more rapid when *D. melanogaster* and *D. virilis* subunits are mixed than when only one kind of subunit is allowed to reassociate. (See MacIntyre [11] for details.) We hope to probe this question further by purifying the enzymes from *D. melanogaster*, *D. simulans*, and *D. virilis* as much as possible and repeating the tests, this time measuring specific activities. Immunological assays may be another way to measure the amount of acid phosphatase protein formed during reassociation.

If, in fact, the technique measures variation in subunit affinities, then what kind and how many amino acid substitutions are responsible for the differences? Obviously, only when amino acid sequences are determined for the acid phosphatases will this question be answered. One could, however, it seems to me, study the relationships with hemoglobin relatively easily. Besides affording an empirical basis for evolutionary comparisons which use the technique of subunit hybridization, such a comparison should also provide useful information to the protein chemist interested in quarternary structure.

There is still another explanation for the results of the interspecific tests which can be ruled out by a closer examination of the data. It might be argued that the degree of the difference between observed and expected homospecific: heterospecific enzyme ratios depends only upon the net charge of the dissociated subunits. That is, if the difference in net charge between the subunit from species X and species Y is substantial, then more heterospecific enzymes will form because of the greater electrostatic attraction between the unlike subunits. If this were strictly true, however, one should never see a homospecific:heterospecific enzyme ratio greater than the one expected if subunit association were random. Furthermore, as shown in Figure 12, when the differences between

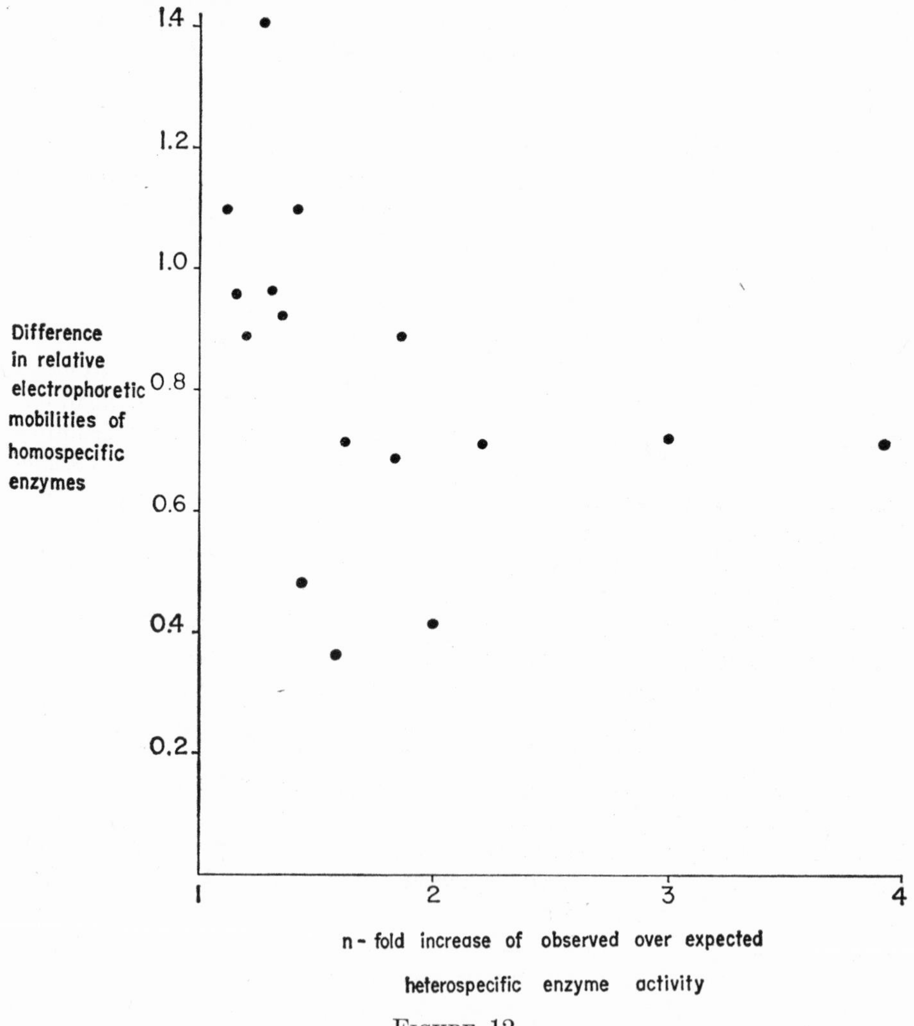

FIGURE 12

Plot of subunit affinity as measured by n-fold increase of observed over expected heterospecific enzyme activity against the difference between the electrophoretic mobilities of the participating homospecific enzymes. Data were taken only from those tests in which the difference between observed and expected homospecific:heterospecific enzyme activities had a minus sign.

observed and expected homospecific:heterospecific enzyme ratios (from the tests in which the heterospecific enzyme activity is greater than expected) are plotted against the distance between the two homospecific enzymes after electrophoretic separation, no definite correlation can be seen.

In order to make Figure 12 understandable, I will discuss how one of the points was plotted. In the *D. melanogaster* × *D. virilis* test, the difference between the two ratios is about 0.7 when the enzyme activities are determined by densitometry. Thus, the actual increase of observed over expected heterospecific enzyme activity in this test is about 3.8. This is plotted on the ordinate. The relative electrophoretic mobilities of the two homospecific enzymes are 1.00 (*D. melanogaster*) and 0.28 (*D. virilis*). The difference of 0.72 is plotted on the abscissa. The point plotted from the coordinates is then determined, and in Figure 12, is the point at the far right on the graph. All the other points representing the other tests in which excessive heterospecific enzyme activity was measured were determined in a similar fashion. Note that if the excess heterospecific enzyme activity were due only to the electrostatic attraction between unlike subunits, the data should plot as a straight line originating at zero.

Despite certain unanswered questions, the technique appears to be quite sensitive in its ability to detect differences between homologous enzymes, even those from sibling species. We were encouraged enough by this conclusion to begin a rather extensive survey of natural populations of *D. melanogaster*, *D. simulans*, and *D. nebulosa*. We hope to detect differences in electrophoretically identical allozymes from different populations, that is, to see if we can extend our present estimates of gene-enzyme variability in natural populations. Also, we hope to determine if the subunits of polymorphic electrophoretic variants of acid phosphatase differ significantly in their ability to form homo- and heterodimeric enzymes. If this could be shown, then it would suggest that the mutational differences between the alleles are not selectively neutral.

7. Summary

Let me conclude with a purposefully overstated summary. I have outlined a technique that allows us to detect differences in the homologous enzymes of closely related species and perhaps even in the same enzyme from different populations within a species. These differences are due to amino acid substitutions affecting subunit association and/or the activity of the enzyme. It is unlikely that either kind of difference would be selectively neutral.

REFERENCES

[1] J. Bell and R. MacIntyre, "An analysis of induced null mutants of acid phosphatase-1 in *Drosophila melanogaster*," *Isozyme Bull.*, Vol. 4 (1971), p. 19.

[2] R. A. Cook and D. E. Koshland, "Specificity in the assembly of multisubunit proteins," *Proc. Nat. Acad. Sci. U.S.A.*, Vol. 64 (1969), pp. 247–252.

[3] E. H. Davidson and B. R. Hough, "Genetic information in oocyte RNA," *J. Mol. Biol.*, Vol. 56 (1971), pp. 491–506.

[4] J. Franklin and R. C. Lewontin, "Is the gene the unit of selection?," *Genetics*, Vol. 65 (1970), pp. 707–734.

[5] J. E. Haber and D. Koshland, "Relation of protein subunit interactions to the molecular species observed during cooperative binding of ligands," *Proc. Nat. Acad. Sci. U.S.A.*, Vol. 58 (1967), pp. 2087–2092.

[6] N. Iwatzuki and R. Okazaki, "Mechanism of regulation of deoxythymidine kinase of *E. coli*. I. Effect of regulatory deoxynucleotides on the state of aggregation of the enzyme," *J. Mol. Biol.*, Vol. 29 (1967), pp. 139–156.

[7] A. Kenyon, "Comparison of frequency distributions of viabilities of second with fourth chromosomes from caged *Drosophila melanogaster*, "*Genetics*, Vol. 55 (1967), pp. 123–130.

[8] J. Klotz and D. W. Darnall, "Subunit structured proteins; A table," *Science*, Vol. 166 (1969), pp. 126–128.

[9] R. MacIntyre, "Evolution of Acid phosphatase-1 in the genus Drosophila as estimated by subunit hybridization. I. Methodology," *Genetics*, Vol. 68 (1971), pp. 483–508.

[10] ———, "A method for measuring activities of acid phosphatases separated by acrylamide gel electrophoresis," *Biochem. Genet.*, Vol. 5 (1971), pp. 45–56.

[11] ———, "Multiple alleles and gene divergence in natural populations," *Brookhaven Symp. Biol.*, Vol. 23 (1971), in press.

[12] R. MacIntyre and T. R. F. Wright, "Responses of Esterase-6 alleles of *Drosophila melanogaster* and *D. simulans* to selection in experimental populations," *Genetics*, Vol. 53 (1966), pp. 371–387.

[13] R. W. Noble, "Relation between allosteric effects and changes in the energy of bonding between subunits," *J. Mol. Biol.*, Vol. 39 (1969), pp. 479–489.

[14] S. J. O'Brien and R. J. MacIntyre, "Transient linkage disequilibrium in Drosophila," *Nature*, Vol. 230 (1971), pp. 335–336.

[15] M. F. Perutz, H. Miurhead, J. M. Cox, and L. C. G. Goaman, "Three dimensional Fourier synthesis of horse oxyhemoglobin at 2.8 Å resolution: The atomic model," *Nature*, Vol. 219 (1968), pp. 131–139.

[16] M. F. Perutz and H. Lehmann, "Molecular pathology of human haemoglobin," *Nature*, Vol. 219 (1968), pp. 902–909.

[17] S. Prakash, R. C. Lewontin, and J. L. Hubby, "A molecular approach to the study of genic heterozygosity in natural populations. IV. Patterns of genic variation in central, marginal and isolated populations of *Drosophila pseudoobscura*," *Genetics*, Vol. 61 (1969), pp. 841–858.

[18] C. R. Shaw, "The use of genetic variation in the analysis of isozyme structure," *Brookhaven Symp. Biol.*, Vol. 17 (1964), pp. 117–130.

[19] T. Yamazaki, "Measurement of fitness at the esterase-5 locus in *Drosophila pseudoobscura*," *Genetics*, Vol. 67 (1971), pp. 579–603.

THE EVOLUTION OF POLARITY RELATIONS IN GLOBINS

HELMUT VOGEL and EMILE ZUCKERKANDL
CENTRE NATIONAL DE LA RECHERCHE SCIENTIFIQUE, MONTPELLIER

1. Introduction

The well known correlation between the hydrophobicity of amino acid residues and their position in the interior of protein molecules was predicted long ago (Kauzmann [9]) and was more recently verified by X-ray diffraction studies (Kendrew [10]).

At the Rutgers conference in 1964, the examination of substitution patterns in globins showed that "there are more sites that seem to specialize in carrying residues fit for apolar bonding than any other sites at which the residues found are limited to one given chemical category." Apolar bonding, it was said, "may be the most specifically determined business of molecular sites in globular proteins" and, further: "we may venture the generalization that the outside of the globin molecule, and perhaps of globular proteins in general, is more variable than the inside" (Zuckerkandl and Pauling [19]).

This presumption was based in part on the already available knowledge that the majority of the polar amino acid residues are on the outside of the globins [10] and in part on the observation [19] that charged sites and other polar sites are more variable, on the average, than apolar sites, with the exception notably of glycine and of alanine sites.

In 1967, it was shown, by counting the minimal number of base substitutions on a deduced molecular phylogenetic tree, that maximum variability of sites coincided mostly with exteriority of the sites. This was established for a stretch representing two thirds of the globin chain (Derancourt, Lebor, and Zuckerkandl [5]).

On the basis of their study of the structural and functional implications of amino acid substitutions found in abnormal human hemoglobins, Perutz and Lehmann concluded in 1968 [12] that, whereas the conditions of amino acid substitutions are functionally restrictive in the interior of the hemoglobin molecule, they are liberal at its surface. The surface should be more variable than the interior, as Epstein [7] has also shown.

It was therefore unexpected to find (Zuckerkandl, Derancourt, and Vogel [18]), on the basis of an inventory of the different types of probable amino acid

This work was supported in part by grants from the Délégation Générale de la Recherche Scientifique et Technique and from the Fondation pour la Recherche Médicale Française.

substitutions during evolution, that in globins, and also in cytochrome *c*, apolar residues are lost and gained practically as frequently as polar residues. Alanine and glycine residues were not classified as apolar and were considered separately, as by Epstein [6], because they are found frequently at the surface as well as in the interior of globular proteins (Perutz, Kendrew, and Watson [11]). If these results are correct, then something must be incorrect about the assumption that a good correlation prevails between polarity, variability, and exteriority.

To reinvestigate this question, data on variability at each molecular site during globin evolution were obtained with the help of a Beckman 816 computer for a set of 39 chains. They include chains such as the frog β chain and the *Chironomus* (insect) chain that give to the comparisons and deductions a somewhat wider or better established evolutionary scope than was obtained heretofore.

Data on exteriority are readily available through the list of "internal" residues given by Perutz, Kendrew, and Watson in their 1965 paper (Table 4), and in the sinusoidal curve in Figure 1 of the same paper. Data on the sites involved in interchain contacts in oxy- and in deoxyhemoglobin are, in turn, available in recent papers by Perutz and Lehmann [13], and Bolton and Perutz [1]. These sites are exterior from the point of view of the tertiary globin structure, but interior (if the notion is used loosely) with respect to the quaternary structure of a tetrahemic hemoglobin molecule. Perutz's results are valid for the case of horse hemoglobin. It may be that during the evolution of tetrahemic hemoglobins the contacts between chains did not consistently involve the same sites. From the evolutionary point of view, whatever information is derived from the data on interchain contacts in horse hemoglobin is therefore only a probable approximation.

For an evaluation of polarity we adopted Woese's [15] polarity scale, based on his "polar requirement" index.

The present discussion is based largely on the results of Perutz and co-workers from which most advances in the field are derived and to whom we are so greatly indebted.

The list of globin chains used in the present computations is given in the legend to Table II.

In earlier work (Zuckerkandl and Pauling [19]), as in part of this paper, a concept of "variability" is used, that may be termed "comparative variability," in contrast to another concept, "evolutionary variability," that will be used mainly in the second part of this paper.

Comparative variability refers to a set of sites characterized as "sites for amino acid i" or as "sites for a subset A of amino acids." A site is called "site for amino acid i" if and only if amino acid i is found there in at least one known chain of the class of proteins under consideration. Evidently, a given site can thus be assigned to several amino acids. A site is called "site for the subset A" if and only if it is a site for at least one amino acid of the subset A. For instance, a site is called a "charged site" if it is a site for aspartic acid, glutamic acid, arginine, *or* lysine. (Histidine may or may not be counted among the charged amino acids). Comparative variability then refers to the number of different types of amino acids that occur at a site for amino acid i or at a site for a subset A, averaged over all those sites, and decremented by one.

Evolutionary variability refers only to a given site and is defined as the number of times that site was subject to amino acid substitution during evolution of the class of proteins under consideration. It has been customary to try to minimize the hypothetical element in this concept by using the minimum number of those substitutions that can be reconciled with the phylogenetic tree as determined from the topological analysis using the chains as a whole.

In the ideal case (no back mutations, only one step mutations resulting in amino acid substitutions), both variabilities should be numerically identical. Both back mutations and the attempt to account for two and three step mutations by counting them correspondingly often increase evolutionary variability. Since the tendency for back mutations and perhaps also for more-step mutations might vary from site to site, one cannot expect too close a correspondence between the two concepts of variability.

After presenting an overall picture of the variations of polarity in globins, we shall compare contemporary globin chains among themselves for numbers of types of substitutions (what and how many kinds of amino acids can be accommodated at different sites) and in a later section use the figures for the presumed evolutionary variability at each site as deduced from molecular phylogenetic trees (Zuckerkandl, Derancourt, and Vogel [18]) (number of times each residue has been substituted within the sector of evolution under consideration).

2. Variations of polarity in globins

The overall compositional polarity (the mean polarity per amino acid residue) seems to vary very little among globular proteins. For the sample given in Table I, this value is confined between 7.0 and 7.9, that is, between the values for alanine and glycine.

The constancy in the overall polarity of all hemoglobins (Table II) hides a considerable variation that becomes apparent as different corresponding stretches of different globin molecules are compared among themselves. Such variations relate to means obtained over stretches as long as, for instance, helix G (19 amino acids). When means characterizing stretches of significant length vary considerably, it would be surprising that the constancy of mean polarity of whole chains be a random effect and natural selection as the cause of such constancy is more probable. The existence of a tendency to preserve overall polarity in different groups of globin chains has been demonstrated (Epstein [7], H. Vogel [15]).

The modulation of the sectional mean polarities forms a "melody" charac-

TABLE I

MEAN POLARITY PER MOLECULAR SITE FOR DIFFERENT GLOBULAR PROTEINS

The polarity index used is Woese's [15] "polar requirement."

	$\bar{P}$
Hemoglobin	
human α chain	7.44
human β chain	7.53
Myoglobin, sperm whale	7.70
Ferredoxin	7.43
Ribonuclease, bovine	7.71
Tobacco mosaic virus, strain *Bulgare*	7.43
Chymotrypsinogen A, bovine	7.64
Glucagon, bovine	7.71
Cytochrome *c*, human	7.92

teristic of each type of chain. In Figure 1 the same values, as well as those for some individual chains, are represented in a notation in which the height of each note is proportional to the average "polar requirement" value and its duration is in rough relation to the length of the molecular section.

If the overall polarity of chains is considered, myoglobins are only slightly more polar than other globin chains (Tables I and II). Major differences appear however in individual helical and nonhelical regions. In comparison with α and non-α chains of hemoglobins, myoglobins are more polar along regions C, CD, D, EF, and H. Consideration of the "melodies" of Figure 1 shows that α and non-α chains differ from each other slightly, but significantly, in some sections. The patterns of the β chains and of the other non-α chains of mammalian hemoglobins are very similar. The lemur β chain is an exception. It is peculiar in many respects and in part quite different from both the human β and γ chains. It may well be the result of a distinct gene duplication (Zuckerkandl [17]). Carp α and frog β chains have a polarity line different from that of the corresponding mammalian chains, though the frog chain is recognizable as a β chain in terms of the mammalian pattern. For part of its polarity pattern, the dog α chain also is peculiar, whereas the dog β chain has a normal β chain pattern.

The stretches with highest polarity are seen to occur in the globin molecules that do not associate to form tetrahemic structures, namely in the myoglobins, in the lamprey chain, and in the *Chironomus* III chain. The *mean* polarity of the lamprey and *Chironomus* chains, however, does not differ from that of other hemoglobin chains.

The polarity of helices A and B is relatively stable, A being nearly always more polar than B. Helix C has a low polarity except in the monohemic globins. The evolution of tetrahemic globins may thus have required a lowering of the polarity of helix C. Helices E and F enclose the heme group. The polarity of E is not highly variable, that of F more so. Helix E is more polar than F, except

TABLE II

VARIATIONS IN POLARITY ALONG THE GLOBIN CHAINS

Polarity indices as in Table I. Helical and interhelical sections as defined by Perutz, Kendrew, and Watson [11]. Except when indicated otherwise, the values represent the means of all chains of the type considered. The following globin chains were used in the present work: (1) α *chains*—human, *Rhesus*, dog (Jones [8]), mouse, rabbit, horse, pig, bovine, sheep A, goat A, llama, kangaroo, carp; (2) β *chains*—human, *Rhesus*, lemur, dog (Jones [8]), rabbit, horse, pig, llama, bovine, sheep A, sheep B, sheep C, barbary sheep, goat A, kangaroo, frog; (3) *others*—human γ, human δ, sheep fetal, bovine fetal, lamprey, *Chironomus* III (Buse, Braig, and Braunitzer [2]), myoglobins of sperm whale, horse, cattle, and kangaroo.
Sources: if not stated otherwise, Dayhoff [4].

Molecular section	NA	A	B	C	CD	D	E	EF	F	FG	G	GH	H	HC	
Number of residues per section in human β chain	3	16	16	7	8	7	20	8	9	5	19	5	24		Total
α chains	5.25	8.00	7.70	6.74	7.11	8.01	7.72	8.19	7.37	7.32	6.82	7.30	7.11		7.4 ± 0.1
β chains	6.00	7.90	7.46	6.54	7.75	7.45	7.91	8.29	6.98	8.15	7.29	7.69	7.31		7.5 ± 0.1
γ, δ, "fetal β"	5.90	7.85	7.43	6.58	7.73	7.13	7.97	8.45	6.95	8.27	7.63	7.70	7.30		7.5
Myoglobins	6.00	7.87	7.73	8.80	8.23	9.03	7.53	9.60	7.20	8.00	6.93	7.63	7.97	7.80	7.8 ± 0.2
Lamprey	7.0	7.4	7.3	7.6	7.0	9.3	7.9	8.9	8.2	7.3	6.4	7.4	7.1		7.5
Chironomus III	4.9	~8.3	7.0	7.8	7.2	8.7	~6.9	10.3	8.3	8.4	7.9	8.2	6.4		7.5

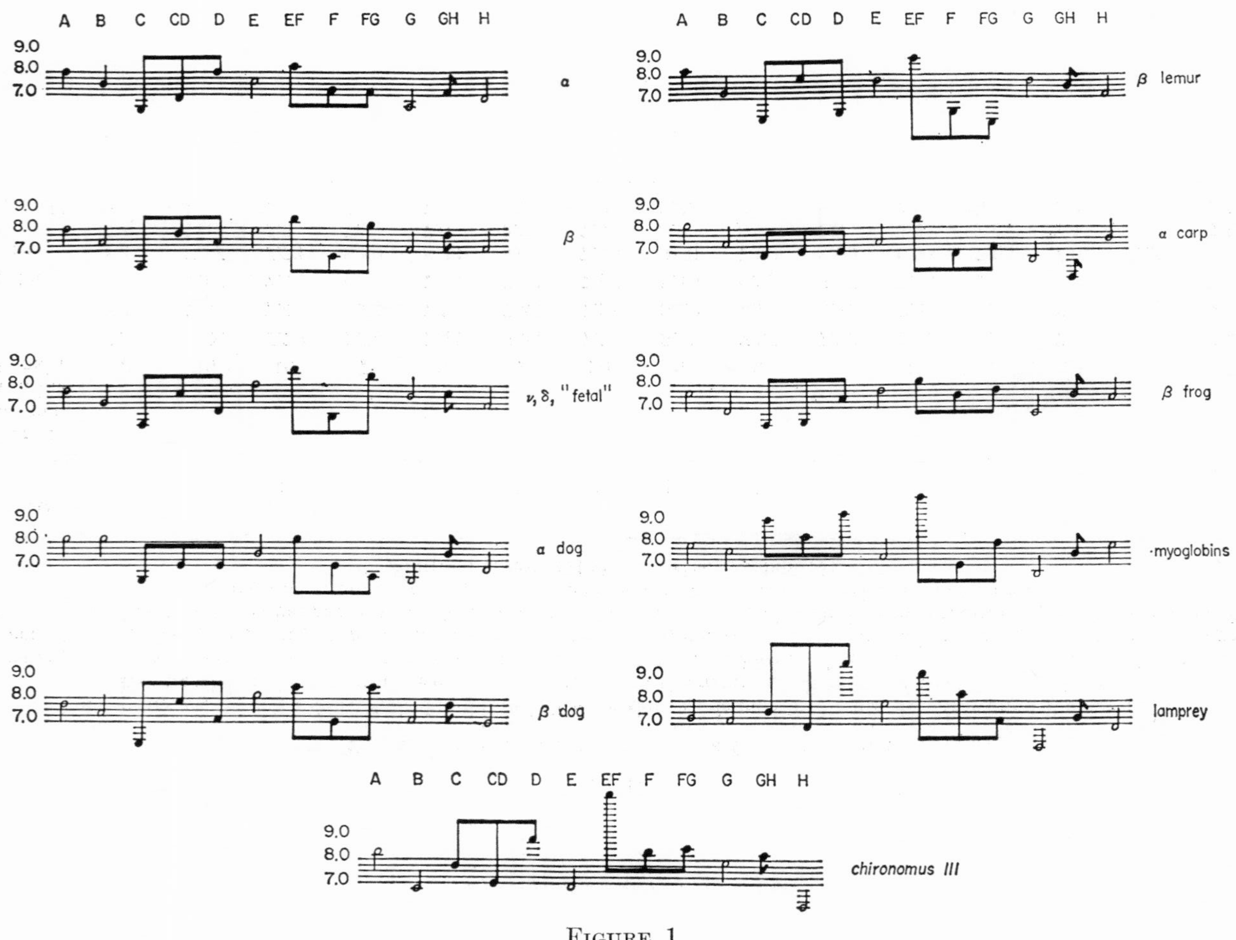

FIGURE 1

Change in "polar requirement" along globin chains (see Table II).

in lamprey and *Chironomus*, and G is nearly always less polar in α chains than in non-α chains, less also in the lamprey chain. Helix H displays a very stable polarity, that is, relatively high only in the myoglobins and low in the *Chironomus* chain. With the exception of the frog β chain, H is more polar than G in α chains, but not in non-α chains. The myoglobins and the lamprey chain behave in this respect like α chains.

As to interhelical sections, the polarity of the CD segment is lower in α chains, higher in β chains, the β chain of frog being an exception. The lamprey and *Chironomus* chains are in this respect like an α chain, the myoglobins like β chains. Section EF always has a relatively high polarity, but in the monohemic globins higher than in the others.

On the whole, the polarity line of the lamprey chain resembles that of an α chain more than that of a non-α chain. That of myoglobin is quite distinct from either.

Further data are needed for accurately picturing such apparent evolutionary trends and their functional significance.

3. Types of amino acids accommodated at different molecular sites

3.1. *"Internal" sites.* Perutz and his co-workers [11] listed 33 internal sites in globin, defined as sites cut off from the surrounding water. Of the 33, only three were then known to accommodate not only apolar, but also polar residues, namely serine or threonine. Internal sites able to accommodate polar residues now number fourteen, on the assumption that sites listed as internal by Perutz are so in all globin chains and that the homologies, taking into account gaps, have been correctly established. *Thus, 40 per cent or more of the "internal" sites in the globin molecule do not exclude polar residues at one time or another.* In nine of the cases, the polar amino acid found is serine or threonine. Other polar amino acids include glutamic acid or glutamine, aspartic acid, asparagine, and histidine (Table III).

A given chain usually accommodates no more than one polar residue in its interior. When an alignment of residues is such that a number of polar amino acids fall on molecular positions listed as interior, the alignment is likely to be faulty, such as that given in the *Handbook of Biochemistry* [14] for the lamprey chain, where at least nine polar residues coincide with "internal" sites. Apolarity at a maximum of internal sites in a given globin chain, and probably in globular proteins in general, may thus be used as a supplementary criterion for the correctness of amino acid alignments.

Internal polar sites are mostly those at which also either glycine, or alanine, or both have been found (9 cases out of 14; random expectation 5.1). Exceptions include *Chironomus* three times, which again raises the question of the strict applicability of the horse data. Conversely, at internal sites limited to apolar residues, glycine or alanine usually do not occur. (They do so only in 3 cases out of 19; random expectation 6.9).

TABLE III

"Comparative Variability" at Internal Globin Sites

Site		Val	Leu	Ile	Met	Phe	Tyr	Trp	Ala	Gly	Cys	Others	Organisms in which the other residues occur
(a) Sites not Known to Accommodate Other than Apolar Residues													
A	8	+		+									
	12					+		+					
	15	+		+						+			
B	10	+	+										
	14		+			+							
CD	1					+							
	4					+		+					
D	5	+	+	+	+								
E	4	+	+	+		+							
	8								+	+			
	11	+		+									
	15	+	+	+		+							
	18		+	+					+	+			
G	5		+			+							
	8		+	+									
	12		+	+		+							
H	15	+	+	+		+							
	19		+	+	+								
	23				+		+						
(b) Sites Accommodating also Polar Residues													
A	11	+	+	+		+			+			Ser, Thr	*Chironomus* III; α dog
B	6								+	+		Pro, Gln	*Chironomus* I, III
	9	+		+					+	+		Thr	β *Propithecus*, lemur, γ man
	13	+	+		+	+						Thr	α dog
C	4			+					+	+		Thr	α, β man; Mb sperm whale

TABLE III (Continued)

Site		Val	Leu	Ile	Met	Phe	Tyr	Trp	Ala	Gly	Cys	Others	Organisms in which the other residues occur
(b) Sites Accommodating also Polar Residues (continued)													
E	12	+	+	+					+			Ser	α rabbit
	19	+	+	+	+					+		Thr	α rabbit
F	1	+	+		+	+	+					Asn, Ser	*Chironomus* III, α rabbit
FG	5	+		+								Asx	*Chironomus* III
G	11	+	+	+					+		+	His	α carp
	16	+	+	+	+							Ser	γ *Lemur fulvus*
H	8	+	+		+	+	+	+				Thr	β kangaroo
	11	+	+		+	+			+			His	β frog
	12	+	+			+			+			Thr	α dog
												Ser	lamprey

Thus, it appears that when a site has a strictly apolar function in the interior, not only glycine, as expected, but also alanine is unlikely to fill this function. *The presence of alanine at a site probably is the general sign that a number of functionally different amino acids can be accommodated at that site.*

Most of the internal sites not open to alanine or glycine will accommodate most of the other apolar amino acids.

The series of functionally typical apolar residues comprises valine, leucine, isoleucine, methionine, phenylalanine, and tyrosine. To these tryptophan may be added (Table III). Among the residues that occur at a site, when the two most distant from each other in the preceding series, sizewise, are considered, those in between will no doubt also be found in globin chains, if they are not already known. For instance, at H 12, where valine, leucine, and phenylalanine are found, isoleucine and methionine should also be found.

Tyrosine hardly occurs at strictly apolar residue sites, as the series of 20 such sites shows. With the exception of the special case of H 23, tyrosine has not been found at such sites. The inclusion of tyrosine among the "apolar" amino acids is warranted by its position on the polarity index scale. Yet, in fact, tyrosine is more frequently excluded from purely apolar than polar sites.

Tryptophan remains very rare; only five tryptophan sites are known in globins at this writing.

Internal sites at which polar amino acids occur are distinctly more variable than the set of sites at which no polar amino acids have been found yet. The mean number of different amino acids is 5.3 for the first group as against 2.5 for the second. Taking all internal sites together, there are two at which seven different residues have been found to date, five with six different residues, three with five different residues, and five with four different residues. About one half of the internal sites can thus accommodate as many different amino acid residues as, or more such residues than, the average external noncontact site (see Table VII below).

Thus, *at a number of interior sites, interiority does not imply a specially rigorous specification of tolerable amino acid residues.*

3.2. *Contact sites.* From a consideration of the data provided by Perutz, Muirhead, Cox, and Goaman [13] and by Bolton and Perutz [1] on contacts between chains in horse oxyhemoglobin and deoxyhemoglobin, it appears that thirteen out of 37 interchain contact sites, that is, one third of their number, are restricted to one type of chain.

Table IV gives a list of the amino acids found in different globin chains at sites characterized as contact sites for horse hemoglobin.

Table V shows the number of sites at which a given amino acid residue is actually involved in contacts (judging from the situation in horse hemoglobin) and the number of times the same residue occurs at different "contact sites" in globins existing as free protomeres. There is no significant shift in types of amino acids used at "contact sites" between chains that do and that do not associate to oligomeric molecules. *All kinds of amino acids are used at contact sites, apolar ones, polar hydrogen bond forming ones, charged ones.* The proportion of sites is rather similar for each of these three groups and it is not significantly different when, for each group, associating and nonassociating chains are compared.

Thus, there is, with one possible exception, nothing remarkable about the types of amino acids that are used for establishing interchain contact. Neither charged, nor hydrogen bond forming, nor even apolar residues are significantly favored for the formation of quaternary structure. This remark leaves open the question as to *which part* of a residue is actually used for making a contact. Perutz, Muirhead, Cox, and Goaman [13] have shown that the contacts are mostly made between the apolar sections of residues. For contacts between chains, it looks as though the protein moieties are "satisfied" as long as they find apolar groups of atoms that are sterically fit and don't "mind" what else there may be around on the residue provided there is no steric or no charge exclusion.

In the above comparison, proline, tryptophan, cysteine, and alanine have not been counted in the typical "apolar series," but listed separately, because each of these amino acids has in its way a peculiar behavior. It is striking that cysteine occurs at four contact sites in chains that actually make contact with others, whereas no cysteine residues occur at these sites in monohemic globins. Remarkably, three of the four cysteine sites are on helix G, in the region engaged in contacts $\alpha_1\beta_1$ (sites 11, 14, 18). Also, the fourth cysteine site, D 6, is listed for a $\alpha_1\beta_1$ contact. A fifth cysteine site, G 15, not listed as a contact site, is also in the region of helix G engaged in the $\alpha_1\beta_1$ contact. *It could be that this rare residue is sometimes specially selected for interchain contact function.* The total number of cysteine sites for all globins known at this writing is eight, with none for myoglobins, none for *Chironomus* globin, and one for the lamprey chain. The three cysteine sites not yet listed are E 16, F 9, and H 13. With the exception of G 11, none of the eight sites is classified as internal. The same site G 11 is, however, also given as a contact site. It can be concluded that cysteine apparently is not used for *internal* contacts. The only occurrence so far known of cysteine in monohemic molecules is in the lamprey chain, at site H 13 which is not a "contact site" in α or non-α chains of tetrahemic hemoglobins. The occurrence of cysteine at four if not five *interchain* contact sites thus seems to be quite significant.

Of 37 distinct contact sites, only nine appear as "invariant" in tetrahemic hemoglobins at the present writing. (Among these nine, site C 7 is also counted, although it has either phe or tyr in the non-α chains, since only tyr has been found in α chains and since only α chains are reported to be involved in contacts at that site).

However, if it is considered that the two types of chains may be engaged at a given contact site common to them in two different specific ways, contact sites common to both chains with residues that differ in the α and in the non-α chains, but are in either case unique, may also tentatively be listed among

TABLE IV

RESIDUES AT INTERCHAIN CONTACT SITES

Contact sites as defined for horse hemoglobin (see text). For globin chains from tetrahemic hemoglobins invariance is indicated by a dot when general, by a circle when only the α chains or the non-α chains have so far been found invariant, or when the invariant residue is not the same in both α and non-α chains. Under "contact in oxy- and deoxyhemoglobin," 1 stands for an $\alpha_1\beta_1$ contact, 2 for an $\alpha_1\beta_2$ contact ([12], [13], [1]). When only one of the two types of chains participates in interchain contact at a given site ([12], [13], [1]), the residues occurring at that site in the chain that does not participate in contact are between parentheses.

Site		Globin chain from tetrahemic hemoglobins: Invariance in α and/or non-α	Contact in hemoglobin: Oxy-	Contact in hemoglobin: Deoxy-	Residues in α chain	Residues in non-α chain	Monohemic globins: Myoglobins	Lamprey	*Chironomus*
B	11		1		Gly; Asp, Glu	(Gly, Ala)	Ile	Val	Tyr
	12	·	1		Arg	Arg	Arg	Lys	Ala
	15		1		Gly, Ala; Leu; Thr, Glm; His	Val, Leu, Ilu	Lys; Thr	Thr	Lys
	16	∘	1		Gly; Val; Ser	Val	Gly, Ser, Thr	Ser	Ala
C	1	∘	1		Phe, Tyr	Tyr	His	Thr	Asp
	2	·	2	2	Pro; Lys	Pro	Pro	Pro	Pro
	3	∘	2	2	Thr, Glm	Trp	Glu	Ala	Ser
	5	∘	2	2	Lys	Ser, Glm; Arg	Leu	Glm	Met
	6	∘ ∘	2	2	Thr	Arg	Glu	Glu	Ala
	7	·	2	2	Tyr	(Phe, Tyr)	Lys	Phe	Lys
CD	2			2	Pro, Ala	(Ser, Thr; Asp, Glu)	Asp	Pro	Pro, Thr
D	2		1		(Gly)	Pro, Ala	Glu	Ala	Leu
	6		1		(His)	Leu, Met; Cys	Lys	Lys	Lys
FG	3	·	2		Leu	(Leu)	His	Phe	Thr
	4	∘ ∘	2	2	Arg	His	Lys	Glm	His
	5	·	2	2	Val	Val	Ile, Val	Val	Asx
G	1	·	2	2	Asp	Asp	Pro	Asp	Glx
	2	·	2	2	Pro	Pro	Ile, Val	Pro	Leu
	3		2	2	Ala, Val	Ala, Glu	Lys	Glm	Asx
	4	·	2		(Asn)	Asn	Tyr, Phe	Tyr	Asx
G	10	∘	1		His, Asn	Asn	Glu, Asp	0	0
	11		1		His; Cys	(Val, Ilu)	Ala	0	Gly

TABLE IV (Continued)

Site		Invariance in α and/or non-α	Globin chain from tetrahemic hemoglobins: Contact in hemoglobin Oxy-	Deoxy-	Residues in α chain	Residues in non-α chain	Monohemic globins: Myoglobins	Lamprey	*Chironomus*
	13		1		Val, Leu	(Ala, Val, Ilu)	Ile	0	Val
	14		1		Val; Ser	Val, Glu; Thr; Cys	His, Glm	0	Ser
	17		1		(Ala, Met)	Gly, Ala, Ser	His, Glm	0	Lys
	18		1		Ala, Val, Phe; Ser, Asn; Cys	His, Arg, Glu	Ser, Ala	0	Ala
GH	2	◦	1		Gly, Pro	Gly, Glm	Pro, Ala	Ile	0
	5	•	1		Phe	Phe	Phe	Thr	Phe
H	1		1		(Thr; Pro)	Ser, Thr	Gly, Ala	Val	Gly
	2	◦	1		Pro	Pro; Ilu	Ala	Ala	Ala
	3		1		(Ala, Asp, Glu)	Ala, Val; Asp, Asx, Glu; Glm; Pro	Asp	Ala	Glu
	5	◦ ◦	1		His	Glm	Glm	Asp	0
	6		1		Ala, Asp; Met	Ala, Val; His	Gly, Ala	Ala	Ala
	9	◦	1		Asp	Glu; Glm	Asn, Thr, Ser, Lys	Glu	Gly
	10	◦	salt bridge		Lys	Ala, Lys	Lys	Lys	Thr
	23	•	2	2	Tyr	Tyr	Tyr	Tyr	Met
	24			2	(Arg)	His, Arg	Lys	0	0

TABLE V

NUMBER OF CONTACT SITES AT WHICH DIFFERENT AMINO ACID RESIDUES OCCUR

Compiled from data in Table IV. Only substitutions in the chain that is actually involved in a contact (in the case of horse hemoglobin) are recorded.

	α and non-α chains (Tetrahemic globins)		Monohemic globins	
	Number of sites	Proportion of total site count %	Number of sites	Proportion of total site count %
Gly	4		5	
Ala	8		11	
Val	10		4	
Leu	6		3	
Ile	2		6	
Met	2		2	
Phe	3		3	
Tyr	3		3	
Total "apolar series"	26	26.7	21	23.0
Pro	7		5	
Trp	1		0	
Cys	4		0	
Ser	6		5	
Thr	5		8	
Asn	3		1	
Glm	7		4	
Total "H bond forming residues"	21	21.7	18	19.8
Asp	4		6	
Glu	5		6	
Lys	2		9	
Arg	6		1	
His	8		5	
Total "charged residues"	25	25.7	27	29.6
Asx	1		3	
Glx	0		1	

"invariant" sites. They may indeed be invariant for one type of contact in the α chain, and invariant for another type of contact in the β chain. For instance, contact site H 5 with histidine in all α chains and glutamine in all non-α chains may be considered invariant. By virtue of this point of view, the number of invariant contact sites rises to 12. Thus, at the present writing, no more than one third of the contact sites, and perhaps less, are to be considered invariant.

Invariance is much larger for the $\alpha_1\beta_2$ than for the $\alpha_1\beta_1$ contacts, as already pointed out by Perutz, and Muirhead, Cox, and Goaman [13]. Of 15 sites involved in $\alpha_1\beta_2$ contacts (oxy- and deoxyhemoglobin), six are variable on a type of chain basis (Table IV). Of 21 sites involved in $\alpha_1\beta_1$ contacts, 18 are variable (salt linkage forming sites are not considered here). The more stable $\alpha_1\beta_2$ contacts are involved in the mechanism of oxygenation, during which "the contact $\alpha_1\beta_2$ undergoes drastic changes as the two subunits turn relative to each other by 13°" (Bolton and Perutz [1]). Thus, each contact residue on one chain may have to adapt to more than one group of atoms, or set of groups of atoms in the other chain, namely, to one set in the oxygenated and to another set in the deoxygenated state. It is plausible that such multiple stereochemical fitting reduces the evolutionary variability of the residues involved. The surprise, if any, is that it does not reduce it more drastically. On the other hand, Bolton and Perutz [1] report that the contacts $\alpha_1\beta_1$ undergo only slight changes upon deoxygenation. It is the $\alpha_1\beta_1$ dimer that is found in free solution upon spontaneous or electrolyte induced dissociation of tetrahemic hemoglobins [13]. The contact residues functioning between these two subunits, which have to adapt to essentially one situation and not to two sterically different ones, are free to vary.

Since the association of globin protomers to oligomers adds restrictions to the changeability of a certain number of molecular sites, one might expect globin chains, when engaged in tetrahemic hemoglobins, to "evolve" slightly more slowly than myoglobin chains. This is not verified by the only example that at present allows a precise checking of this point. In cattle and horse the α, β, and myoglobin chains have been analyzed for their amino acid sequences. The number of differences between cattle and horse are 24 for the α chains, about 31 for the β chains, and only 18 for the myoglobins.

The following conclusion can be drawn from the preceding discussion. *The formation of quaternary structure in hemoglobins has led to the effect of a relative invariance of the residues at contact sites, primarily at the $\alpha_1\beta_2$ contact sites, while not requiring a change in polarity of the residues involved in contact.*

Thereby, this set of contact sites should not fit well into any correlation between polarity and variability that might otherwise exist.

4. Frequency of amino acid substitutions during evolution

4.1. *Correlation between polarity and evolutionary variability.* We shall now consider variability in terms of the number of times amino acid substitution presumably has occurred at each molecular site during evolution ("evolutionary variability").

If all sites are considered for all chains, no significant correlation is found between polarity and variability in globins (Table VI).

Likewise (Table VI), no significant correlation between polarity and variability is found when different molecular sections, namely, the helical and the interhelical regions, are examined separately. There is no essential difference in the correlation between helical and nonhelical sections of the molecules. (The only mildly significant correlation of a section, that of region CD with

TABLE VI

THE CORRELATION BETWEEN POLARITY AND EVOLUTIONARY VARIABILITY
All globin chains (see legend to Table II) are considered. The correlation is calculated for the globin chain as a whole as well as for the individual sections of the chain.

Section	Sites	Length n	Correlation coefficient $r \pm$ standard deviation	Significance p (null hypothesis)
Total	1–165	165	0.09 ± 0.08	0.32
NA	1–12	12	-0.04 ± 0.35	
A	13–28	16	-0.11 ± 0.28	
B	30–45	16	0.17 ± 0.24	
C	46–52	7	$-0.14 \pm {0.42 \atop 0.48}$	
CD	53–60	8	$0.57 \pm {0.23 \atop 0.27}$	0.15
D	61–67	7	0.02 ± 0.46	
E	68–88	21	0.16 ± 0.24	
EF	89–97	9	$-0.19 \pm {0.29 \atop 0.25}$	
F	98–106	9	$0.18 \pm {0.25 \atop 0.29}$	
FG	107–111	5	$0.34 \pm {0.39 \atop 0.57}$	0.54
G	112–130	19	-0.03 ± 0.24	
GH	131–136	6	$0.27 \pm {0.48 \atop 0.56}$	
H	137–158	22	0.11 ± 0.22	0.62

$r = 0.57 \pm 0.27$ (significance level: 15 per cent) is to be expected as an extreme on a chance basis in a sample of 14 items).

However, a significant correlation between polarity and variability can be brought out if certain functional groups of sites are compared. This correlation is blurred out when the molecule is considered in its totality. The failure to find the correlation along linear sections of the molecule is no doubt due to the fact that the function with respect to which variability and polarity are both correlated, namely, exteriority, is not linearly distributed along the chain.

TABLE VII

MEAN SITE POLARITIES AND VARIABILITIES FOR ALL GLOBIN CHAINS
Two sites are common to the lists of internal and contact sites.

	Number of sites	Mean polarity per site	Mean variability per site
All sites	148	7.50 ± 0.1	4.2 ± 3.0
Internal sites	33	5.54 ± 0.81	3.1 ± 2.6
Contact sites	37	7.46 ± 1.78	3.4 ± 2.3
External noncontact sites	80	8.28 ± 1.88	5.0 ± 3.1

Table VII shows that the mean residue polarity and the mean variability both seem to increase by nearly the same factor as one goes from the set of internal sites to the set of external noncontact sites.

As expected (see the preceding section), contact sites do not participate in this numerical fit; their mean variability is not significantly higher than that of the internal sites, whereas their mean polarity is considerably higher. It must be pointed out that "contact sites" are considered here globally and defined as sites at which, in *some* chains, a contact with another chain is formed (on the basis of the situation in horse hemoglobin). There are of course chains that do not form these contacts. Therefore, the means in Table VII do not express the results of the contact function quantitatively, but only indicate a trend. If for every contact site, one considered only chains that are actually supposed to form a contact at that site, the mean variability for such a group should be lowered, and the contact sites should be further outside the polarity—variability correlation.

The same remark applies to the results obtained when groups of sites, defined by their exteriority, are investigated for individual helical sections (Table VIII).

TABLE VIII

MEAN SITE POLARITIES AND VARIABILITIES AND THEIR RELATIONSHIP FOR DIFFERENT HELICAL SECTIONS

Capital letters refer to helical sections. Data on helices B and C have been combined.

	Number of sites			Mean polarity per site			Mean variability per site		
	B + C	G	H	B + C	G	H	B + C	G	H
Internal sites	6	5	6	6.0	5.0	5.3	1.8	2.8	2.7
Contact sites	10	10	9	7.1	7.8	8.1	3.3	4.5	2.8
External non-contact sites	7	5	10	8.9	7.2	7.6	5.6	3.6	5.2

Mean polarity for this set of data: 7.0 ± 1.2
Mean variability for this set of data: 3.6 ± 1.2
Slope of regression line 0.68 ± 0.26. Intercept at -1.18.
Correlation coefficient $r = 0.71 \begin{smallmatrix} +0.15 \\ -0.26 \end{smallmatrix}$, $p = 0.007$.

Both polarity and variability at contact sites are always higher than at internal sites, but the relationship between contact sites and external noncontact sites is variable. The whole set of data shows a good correlation with a significance level at 0.007.

The mean variability at contact sites being much like that at internal sites, while the polarity at contact sites is much more like that found at external noncontact sites, it is seen again (Table VII) that contact sites must contribute to the blurring of a correlation between polarity and variability when all sites are considered together. It is to be noted that the contact sites do not have to be notably less polar, on the average, than the other external sites. Typical apolar residues, beginning with valine, have "polar requirement" values of 5.6 and less. The mean value for internal sites is 5.5. That for contact sites is 7.5, not far (taking into account the standard deviations) from the value of 8.3 that obtains for external noncontact sites.

4.2. *Correlation between polarity and exteriority and between variability and exteriority.* Because of the different possible orientations of residue side chains, the degree of exteriority of a residue can be established only by X-ray diffraction studies. But the position of residue sites along helical sections might allow one to define a degree of exteriority that is meaningful, if not in each individual case, at least on the average.

With this view in mind, exteriority $E(i)$ for site number i was characterized as follows. For each helix a site was chosen that seems to be exactly on the summit, or exactly in the valley, of a Perutz-wave ([11] Figure 1). Given this site and the number k, put $E(k) = \pm 1$, and define the $E(i)$ of the others by

$$E(i) = \cos \frac{2\pi}{3.6} (i - k) = \cos 100^\circ (i - k). \tag{1}$$

With P = polarity and V = evolutionary variability, the correlations $P(i)/E(i)$ and $V(i)/E(i)$ are then investigated with the help of a computer.

As Table IX shows, the correlation between exteriority thus defined and polarity is highly significant, as also are the differences between the correlation coefficients that are characteristic for certain groups of globins. Thus, the polarity-exteriority correlation coefficient seems to be a tool of some value for classifying globin chains. One is led to assume that the degree to which a globin protomer conforms to the correlation between polarity and exteriority is linked to function and, thus, determined by natural selection. Is a lowering of the correlation coefficient with respect to the ideal polarity-exteriority relationship due to the appearance of polar residues in the interior or due to apolar residues on the outside?

A comparison of the data of Tables II and IX shows that the higher the mean polarity of a type of globin chain, the better the correlation coefficient for polarity *versus* exteriority. Since myoglobins have a considerably larger number of charged groups on their surface than individual hemoglobin chains and since in spite of the occurrence of a number of polar groups in the interior of globin chains the polarity inside any one particular chain cannot be enhanced considerably, it is clear that the result of the above comparison of data implies that the lowering of the average polarity in the individual hemoglobin chains is brought about by placing apolar groups at the outside.

The relatively low polarity values for the two monohemic hemoglobin chains that have thus far been "sequenced," namely, the lamprey and *Chironomus* chains, are compatible with a hypothesis according to which low mean polarity is the more primitive condition in globins. If so, the passage from monohemic to tetrahemic hemoglobin did not require the lowering of surface polarity and

TABLE IX

Correlation Between Polarity and Exteriority for all Sites Belonging to Helical Sections of the Molecule

The correlation coefficient is

$$r = (\sum PiEi - \sum Pi \sum Ei/N)/[(\sum Pi^2 - (\sum Pi)^2/N)(\sum Ei^2 - (\sum Ei)^2/N)]^{1/2}$$

with P_i polarity of site i; E_i, defined for helices only, exteriority of site i; and N the number of sites. Thirteen deduced ancestral chains as given by Dayhoff [4] corresponding to six "α nodes" and seven "β nodes" were added to the contemporary chains. The standard deviations given after the individual values correspond to the approximately normal distribution of $\tanh^{-1}\rho$ (ρ: population correlation coefficient). The standard deviations of the group means (underlined) refer only to the averaging procedure of the individual values.

α chains					
Human	0.28 ± 0.09	Rabbit	0.31 ± 0.09	Sheep A	0.28 ± 0.09
Rhesus	0.25 ± 0.09	Horse	$0.23\,^{+0.09}_{-0.10}$	Llama	0.27 ± 0.09
Dog	0.29 ± 0.09	Pig	0.31 ± 0.09	Kangaroo	0.30 ± 0.09
Mouse	0.32 ± 0.09	Bovine	0.27 ± 0.09	Carp	0.28 ± 0.09
β chains					
Human	0.34 ± 0.09	Pig	0.26 ± 0.09	Goat A	0.33 ± 0.09
Rhesus	0.35 ± 0.09	Llama	0.35 ± 0.09		
Lemur	0.34 ± 0.09	Bovine	0.37 ± 0.09	Barbary Sheep	0.31 ± 0.09
Dog	0.35 ± 0.09	Sheep B	0.35 ± 0.09	Kangaroo	0.33 ± 0.09
Rabbit	0.33 ± 0.09	Sheep C	0.34 ± 0.09	Frog	$0.38\,^{+0.08}_{-0.09}$
Horse	0.29 ± 0.09	Sheep A	0.34 ± 0.09		
γ, δ, Fetal chains					
Human γ	0.35 ± 0.09	Sheep	0.36 ± 0.09		
Human δ	0.31 ± 0.09	Bovine	0.37 ± 0.09		
Monohemic globins					
Lamprey	0.31 ± 0.09	Mb Horse	$0.47\,^{+0.07}_{-0.08}$	Mb Kangaroo	0.46 ± 0.08
Mb sperm whale	$0.43\,^{+0.08}_{-0.09}$	Mb bovine	$0.50\,^{+0.07}_{-0.08}$	*Chironomus*	$0.20\,^{+0.09}_{-0.10}$

"Nodes"

α		β	
1	0.27 ± 0.09	7	0.35 ± 0.09
2	0.27 ± 0.09	8	0.34 ± 0.09
3	0.26 ± 0.09	9	0.33 ± 0.09
4	0.25 ± 0.09	10	$0.48\,^{+0.07}_{-0.08}$
5	0.27 ± 0.09	11	$0.38\,^{+0.08}_{-0.09}$
6	0.30 ± 0.09	12	$0.38\,^{+0.08}_{-0.09}$
		13	0.31 ± 0.09

Mean values

α	β	γ, δ, fetal	Non-α	Monohemic
0.28 ± 0.03	0.34 ± 0.03	0.35 ± 0.05	0.34 ± 0.12	0.46 ± 0.05

one would be led to surmise that quaternary structures arise as a proper "lock and key" interaction between residues, irrespective not only of the polarity of the residues involved, but of the overall polarity at the surface.

If, on the other hand, high polarity was the primitive condition at the surface of the common ancestor of myoglobin and hemoglobin chains, then the tendency to form tetrahemic hemoglobins has implied a general lowering of average polarity at the surface, the lowering being slightly more pronounced for α chains than for non-α chains.

The only reason why a monohemic globin like that of lamprey is called hemoglobin and not myoglobin is that it is found in the blood. This is neither a structural nor an evolutionary reason. It is noteworthy that from the polarity-exteriority correlation the lamprey chain behaves like a real hemoglobin, with a value intermediary between that of the average α chain and of the average non-α chain, and not like a myoglobin.

The frog β chain has the highest value of the β chains, though it is almost identical with that of the bovine β chain. The carp α chain has a value identical with that of man and sheep. Evidently, at least one of the primitive vertebrates carp and frog has not preserved the correlation coefficient characteristic of the common ancestor of the α and β chains. The difference in the correlation coefficient that developed along the two lines presumably corresponds to some functional requirement linked to the interaction between the two types of chains. This functional requirement has been satisfied in lower as well as in higher vertebrates. It may have been satisfied relatively fast at the beginning of the evolution of tetrahemic hemoglobins.

The values for the "nodes" (see legend to Table IX), that is, for the deduced ancestral chains, are close to the average value for α chains for α chain "nodes," and close to the average value for β chains for β chain "nodes." This consistency encourages one to think that the deduced ancestral chains are not too far from reality. A severe drop in the correlation coefficient for the ancestral chains would have made one suspicious in this respect.

Finally, the correlation between exteriority and evolutionary variability (Table X) is also significantly positive ($p = 0.02$); that is, the greater the

TABLE X

OVERALL CORRELATION BETWEEN POLARITY, EXTERIORITY, AND EVOLUTIONARY VARIABILITY OF GLOBIN SITES

The computations are done on all sites and all chains, including the "nodes." In the case of "exteriority," the sites considered are, however, restricted to the helical sections.

	r	p
Polarity/exteriority	0.33 ± 0.08	0.005
Polarity/variability	0.09 ± 0.08	0.32
Exteriority/variability	$0.21 {+0.10 \atop -0.09}$	0.02

exteriority, the higher the variability. We thus note that polarity and exteriority, as well as exteriority and variability are significantly correlated, whereas, as mentioned before and as again shown in Table X, polarity and variability are not significantly correlated when all sites (exterior and interior) are taken into account simultaneously. Such a discrepancy is possible if the figures that fit badly in the two other correlations are combined in the third one.

Thus, all the expected correlations have been found, including that between polarity and variability, provided the latter is established for sets of sites grouped according to exteriority. At the same time, the present results verify that by looking at the whole population of molecular sites it was indeed correct to conclude (Zuckerkandl, Derancourt, and Vogel [18]) that the evolutionary variability of polar and nonpolar amino acids does not differ significantly.

In summary, we find (1) that a certain polarity modulation along the molecule is characteristic of a type of globin protomer and probably linked to function; (2) that a great constancy in mean polarity per residue as observed for total chains is compatible with great variability of polarity at individual sites, even at internal sites, and along various sections of the molecule; (3) that many internal sites do not exclude polar residues at one time or another; (4) that polarity of residues on the other hand is not significantly lowered at interchain contact sites; (5) that all types of amino acids—apolar, polar uncharged, and charged—are found at such sites; (6) that cysteine is a candidate for specializing in the formation of interchain contact; (7) that the formation of quaternary structure leads to a decreased rate of evolutionary change at contact sites, but in general neither to absolute invariance nor to decreased polarity at these sites; (8) that the correlation between polarity and evolutionary variability is upset by this behavior of contact sites; and (9) that, however, polarity and exteriority as well as exteriority and variability are significantly correlated.

The fact that interchain contacts between globin protomers do not lead to absolute invariance at contact sites suggests that absolute invariance, when found in other proteins and brought in relation to molecular contacts, is due not to an intrinsic necessity of "freezing" residues engaged in intermolecular contact, but to an absolute invariance of the partner chain based in turn on functional reasons *other* than those of establishing intermolecular contact.

REFERENCES

[1] W. Bolton and M. F. Perutz, "Three dimensional Fourier synthesis of horse deoxyhaemoglobin at 2.8 Å resolution," *Nature*, Vol. 228 (1970), pp. 551–552.

[2] G. Buse, S. Braig, and G. Braunitzer, "The constitution of the hemoglobin (erythrocruorin) of an insect (*Chironomus thummi thummi, diptera*)," *Hoppe-Seyler's Z. Physiol. Chem.*, Vol. 350 (1969), pp. 1686–1691.

[3] J. Chauvet and R. Acher, "Sequence of frog hemoglobin β," *FEBS-Letters*, Vol. 10, (1970), pp. 136–140.

[4] M. O. Dayhoff, *Atlas of protein sequence and structure*, Vol. 4, Silver Spring, Md., National Biomedical Research Foundation, 1969.

[5] J. DERANCOURT, A. S. LEBOR, and E. ZUCKERKANDL, "Séquence des acides aminés, séquence des nucléotides et évolution," *Bull. Soc. Chim. Biol.*, Vol. 49 (1967), pp. 577–607.
[6] C. J. EPSTEIN, "Relation of protein evolution to tertiary structure," *Nature*, Vol. 203 (1964), pp. 1350–1352.
[7] ———, "Non-randomness of amino acid changes in the evolution of homologous proteins," *Nature*, Vol. 215 (1967), pp. 355–359.
[8] R. T. JONES, Personal communication, 1970.
[9] W. KAUZMANN, in *The Mechanism of Enzyme Action* (edited by W. McElroy and B. Glass), Baltimore, Johns Hopkins Press, 1954.
[10] J. C. KENDREW, "Side-chain interactions in myoglobin," *Brookhaven Symp. Biol.*, Vol. 15 (1962), pp. 216–228.
[11] M. F. PERUTZ, J. C. KENDREW and H. C. WATSON, "Structure and function of haemoglobin, II," *J. Mol. Biol.*, Vol. 13 (1965), pp. 669–678.
[12] M. F. PERUTZ and H. LEHMANN, "Molecular pathology of human haemoglobin," *Nature*, Vol. 219 (1968), pp. 902–909.
[13] M. F. PERUTZ, M. MUIRHEAD, J. M. COX and L. C. G. GOAMAN, "Three dimensional Fourier synthesis of horse oxyhaemoglobin at 2.8 Å resolution: the atomic model," *Nature*, Vol. 219 (1968), pp. 131–139.
[14] H. A. SOBER (editor) and R. A. HARTE (compiler), *Handbook of Biochemistry*, Cleveland, Chemical Rubber Co., 1968.
[15] H. VOGEL, "Two dimensional analysis of polarity changes in globin and cytochrome *c*," *Proceedings of the Sixth Berkeley Symposium on Mathematical Statistics and Probability*, Berkeley and Los Angeles, University of California Press, 1972, Vol. 5, pp. 177–191.
[16] C. R. WOESE, in *The Genetic Code*, New York, Harper and Row, 1967, p. 172.
[17] E. ZUCKERKANDL, "Hemoglobins, Haeckel's "biogenetic law," and molecular aspects of development," *Structural Chemistry and Molecular Biology* (edited by A. Rich and N. Davidson), San Francisco, W. H. Freeman and Co., 1968.
[18] E. ZUCKERKANDL, J. DERANCOURT and H. VOGEL, "Mutational trends and random processes in the evolution of informational macromolecules," *J. Mol. Biol.*, Vol. 59 (1971), pp. 473–490.
[19] E. ZUCKERKANDL and L. PAULING, "Evolutionary divergence and convergence in proteins," *Evolving Genes and Proteins* (edited by V. Bryson and H. J. Vogel), New York, Academic Press, 1965.

TWO DIMENSIONAL ANALYSIS OF POLARITY CHANGES IN GLOBIN AND CYTOCHROME *C*

HELMUT VOGEL
CENTRE NATIONAL DE LA RECHERCHE SCIENTIFIQUE, MONTPELLIER

1. Introduction

The classification of amino acid substitutions between protein chains has led to considerable success especially in the construction of phylogenetic trees that correctly, and in complete independence of the paleontological record or morphological facts, retrace many aspects of evolution (for example, M. O. Dayhoff [3]). The changes of physical properties that accompany those substitutions have not been as thoroughly investigated, at least not in close statistical conjunction with the substitutions themselves. The following attempt may open some new perspectives in this direction.

2. Procedure

By "class of proteins" we mean a set of homologous chains, generally functionally defined, that have been completely sequenced and that can include or exclude the reconstructed nodes in a phylogenetic tree (examples: all globins; all cytochromes *c*).

By "group of proteins" we mean a certain subset of a class of proteins which may or may not correspond to functional or taxonomic characteristics (examples: the α globins; all monohemic globins, the cytochromes *c* of all birds).

A "combination of chains" is a nonordered pair of chains.

The "combination of two groups" is the set of all combinations of chains, one member of the pair stemming from one of the groups, the other member from the other group. If the two groups are identical, we speak of an "in-group combination," otherwise of an "out-group combination."

In the present context, every combination is characterized by a point with the coordinates: N the number of sites where the two chains have different amino acids; P the difference of total polarities of the two chains. The polarities p_i of the amino acids are adapted from Woese [5] (Table I). Every combination is, according to the definitions above, listed only once, the sign of P being determined by the arbitrary order in which the chains are numbered. This has been done for 39 globin chains (see Table II) and for 25 cytochrome *c* chains (Table

TABLE I

POLARITIES OF AMINO ACIDS

Woese [5].

Cys	4.8	Pro	6.6	His	8.4
Leu	4.9	Thr	6.6	Gln	8.6
Ile	4.9	Ala	7.0	Arg	9.1
Phe	4.9	Ser	7.5	Asn	10.0
Try	5.2	Gly	7.9	Lys	10.1
Met	5.3			Glu	12.5
Tyr	5.4			Asp	13.0
Val	5.6				

Mean $\bar{p} = 7.43$

Standard deviation $(\overline{p^2} - \bar{p}^2)^{1/2} = 2.45$

II). Figures 1 and 2 show the distribution of the points (N, P). For every combination of groups, the set of points is sliced into an appropriate number of horizontal layers, such that the variation of N within any layer does not seem too great and, on the other hand, the layer still contains a reasonable number of points (at least 6). For every layer, the moments of the distribution of P up to sixth order are computed. The first two moments are fitted by determining the parameters p and ρ of a Bernoulli distribution (see Model 1) that has the same mean and variance (Table III). For this Bernoulli distribution, the higher moments are also computed and compared to the observed ones. If the Bernoulli distribution fits the observed one sufficiently well (as witnessed by a satisfactory correspondence of the higher moments or derived statistics like skewness, kurtosis, and so forth, which holds rather generally), the parameters of the former, p and ρ, are compared to the values of p and ρ that are predicted by totally random polarity changes only subject to the restrictions that the structure of the genetic code provides for one step mutations (see the end of Section 3).

3. Results

Quite roughly speaking, all points (N, P) together fill the inside of a parabola $N = AP^2$. For globins, due to their higher variability, the sector of this parabola traced by the points is much higher than for cytochromes. Nevertheless, for both classes the width is about the same, $A \approx 0.03$. Such a behavior would result from a random polarity change; if an amino acid substitution were connected with an average polarity change p, and if this change could with equal probabilities be positive or negative, within N substitutions the expectancy for the total polarity change would be 0, and its expected standard deviation about $\frac{1}{2}pN^{1/2}$. Thus, there should be a parabola, outside of which at any height some 20 per cent of the points can be found. As a first rough approximation, one thus obtains $p \approx 3.5$.

TABLE II

List of Investigated Chains

The parallel numbering systems for cytochrome c correspond to the arrangements used in Figures 2 and 4, respectively.
The printing arrangement within each group is in decreasing order of polarity.
Sources: [1], [2], [4] as stated in table, and [3] if unstated.

Globins: Alpha	Globins: Beta	Globins: Others	Cytochrome c		
1 Human	14 Human	30 Human γ	2 1 Horse	10 8 Pigeon	13 17 Turtle
2 *Rhesus*	15 *Rhesus*	31 Human δ	3 2 Sheep	8 9 Duck	16 16 Tuna
3 Dog [4]	16 Lemur	32 Sheep fetal	6 3 Rabbit	9 10 Penguin	17 15 Lamprey
4 Mouse	17 Dog [4]	33 Bovine fetal	7 4 Kangaroo	11 11 Chicken	14 14 Bullfrog
5 Rabbit	18 Rabbit	34 Lamprey	5 5 Grey Whale		15 13 Dogfish
6 Horse	19 Horse	35 Myogl. Whale	4 6 Dog		12 12 Rattlesnake
7 Pig	20 Pig	36 Myogl. Kangaroo	1 7 Human		
8 Bovine	21 Llama	37 Myogl. Horse	21 21 Horn Worm Moth		22 22 *Neurospora*
9 Sheep A	22 Bovine	38 *Chironomus* [1]	20 20 Silk Worm Moth		23 23 Yeast
10 Goat A	23 Sheep B	39 Myogl. Bovine	19 19 Screw Worm Fly		24 24 *Candida*
11 Llama	24 Sheep C		18 18 *Drosophila*		25 25 Wheat
12 Kangaroo	25 Sheep A				
13 Carp	26 Goat A				
	27 Barbary Sheep				
	28 Kangaroo				
	29 Frog [2]				

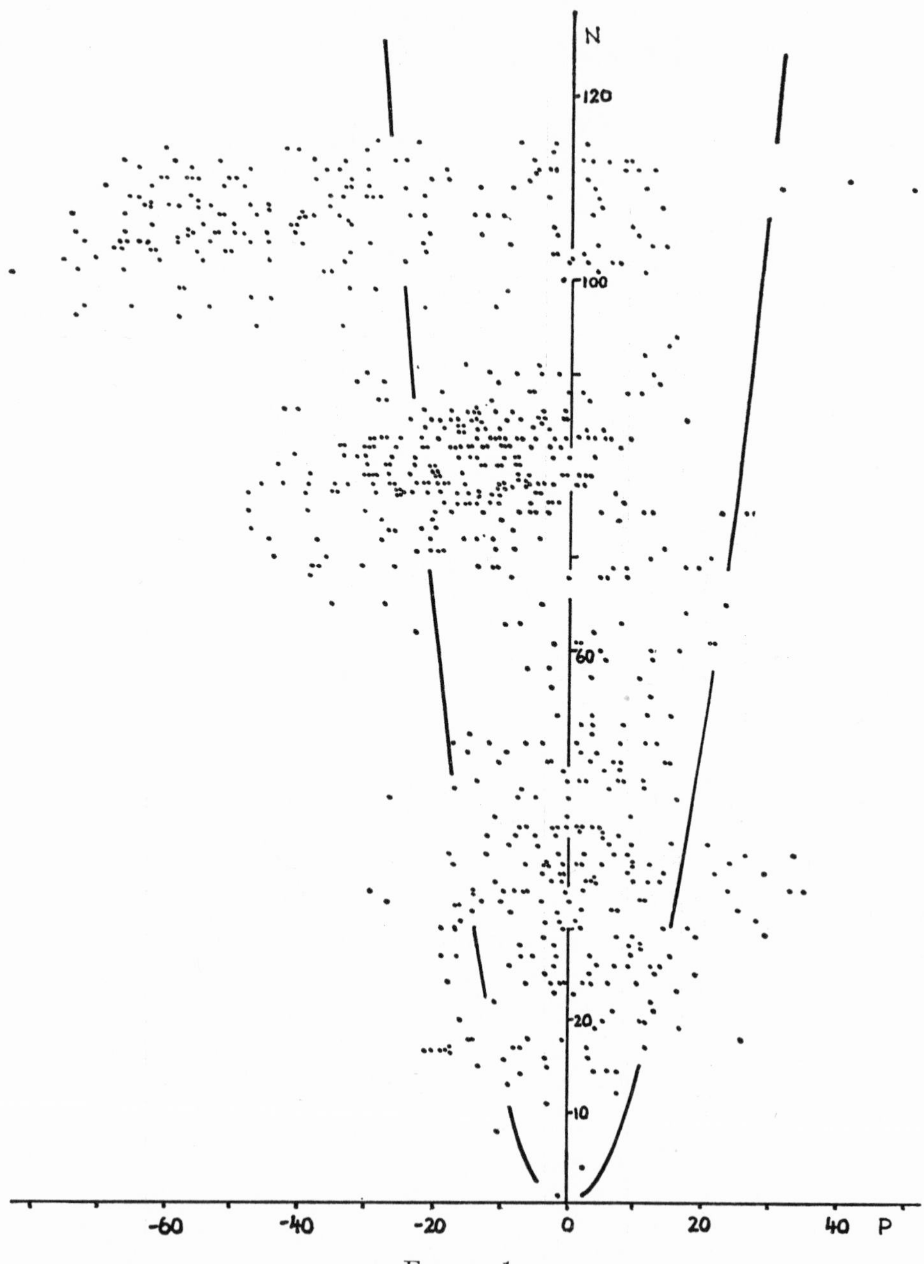

FIGURE 1

Number of substitutions N *versus* polarity difference P for globins.

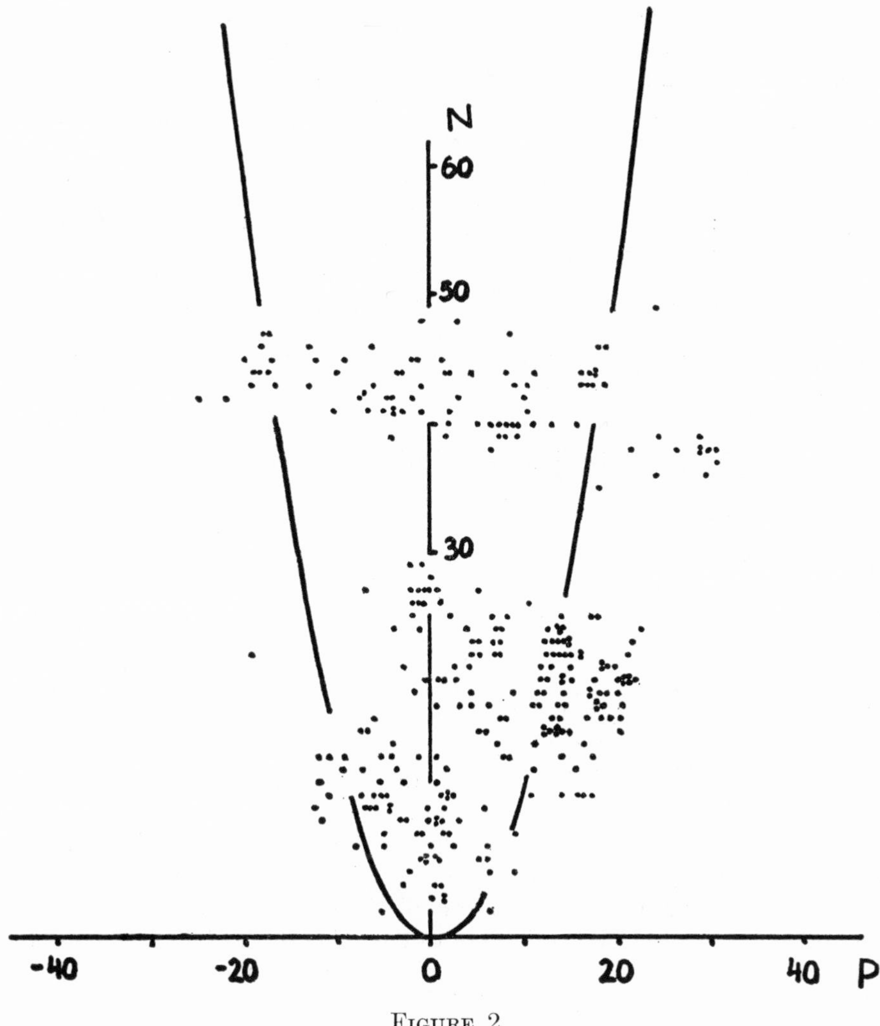

FIGURE 2

Number of substitutions N *versus* polarity change P for cytochrome *c*.

For globins and cytochromes alike, the distribution of the points is uneven: discrete islands are sometimes separated by almost empty areas. These islands are not always centered around $P = 0$. Each of them, generally speaking, corresponds to a certain combination of (taxonomical or functional) groups. Their eccentricity is often particularly patent if the two groups combined are taxonomically or functionally different (for example, birds/reptilia, amphibia, fish, displaced to the right, centered around $P \approx 16$; human/other mammals, birds displaced to the left, centered around $P \approx -10$). This evidently expresses a corresponding tendency in the polarity differences; reptilia (with the exception of the turtle), amphibia and fishes have a lower polarity than birds, man has a lower one than the rest of the warm blooded animals.

But also within the same group (that is, for an in-group combination), displacements of the center of the point cloud from $P = 0$ may occur. Such is the case, for example, for the insect cytochromes or, in a lesser degree, for the α globins. This obviously means that the order of numbering the proteins within that group, being generally at least somewhat systematic, just corresponds to the order of rising or falling polarities (see below).

Each of these islands representing in-group or out-group combinations displays a much smaller dispersion of the P values than all the points taken together. On the other hand, the parabolic shape of those individual islands is less pronounced (better still for β globins), mostly due to the generally very limited extension in N direction for every group. If one nevertheless, on the testimony of the total distribution and of the fortunate cases like β globins, accepts the parabolic shape and the corresponding Bernoulli model, one finds for every individual combination of groups a much smaller mean polarity change p than for the total distribution. In some instances, generally for in-group and out-group combinations of relatively uniform taxonomy, p goes down to about 0.3.

Nearly all combinations of groups (out-group and in-group) show a p that is significantly smaller than $p_{\text{rand}} = 2.6$. Exceptions are all the α globins combined with each other ($p = 2.45 \pm 0.66$) and all the monohemic globins combined with each other ($p = 3.68 \pm 1.06$). If one picks out of the latter group the only really related subset of chains, namely, the myoglobins, one finds again $p = 2.80 \pm 0.10$, that is, practically the random value. Some combinations have p values down to ⅙ or ⅑ of the random value (birds/birds, insects/insects, human/insects). For the globins, the combination β/β, the one most numerous in points, presents the most perfect example for an absolutely symmetric, nonskew distribution with a kurtosis close to the value three of a normal distribution. Here, p (1.33 ± 0.31) is about half of the random value. This is true although the fetal sheep and cattle chains as well as human γ and δ are included. On the other hand, the α/α distribution is almost as wide as randomness prescribes.

"Right" and "left" with respect to the ordinate axis are evidently matters of the arrangement of the chains: since polarity establishes an ordering relation, there exists an arrangement such that all points lie to the right, for example. In Figure 4, this principle has been followed within each taxonomical group. Permuting the groups themselves would not quite succeed in putting all points to the right; the best solution would be birds—mammals—reptilia—amphibia—fish—insects, but there is an overlap between each pair of this series, also expressed by the fact that out-group combination point sets of the respective two groups cross the ordinate axis.

The most extended in-group point set is understandably that of the com-

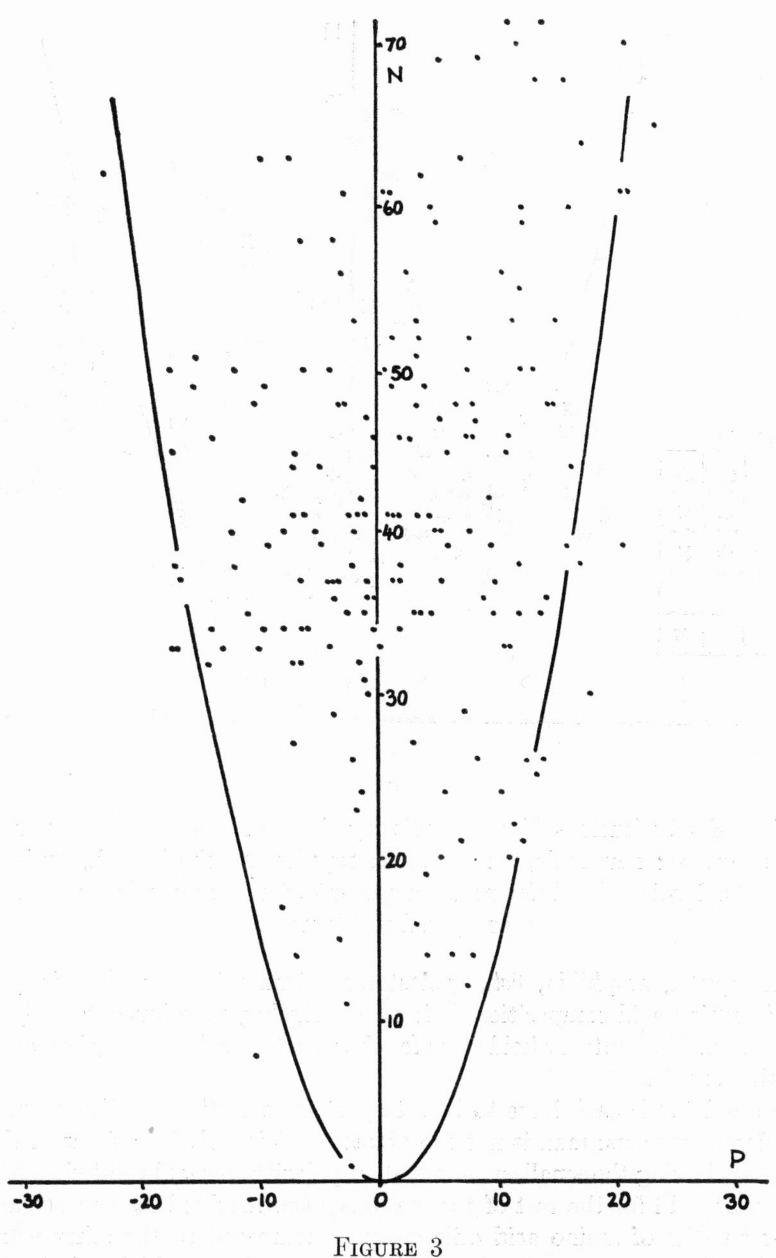

FIGURE 3

Number of substitutions N *versus* polarity change P for β globins (including human γ, δ, and sheep, bovine fetal).

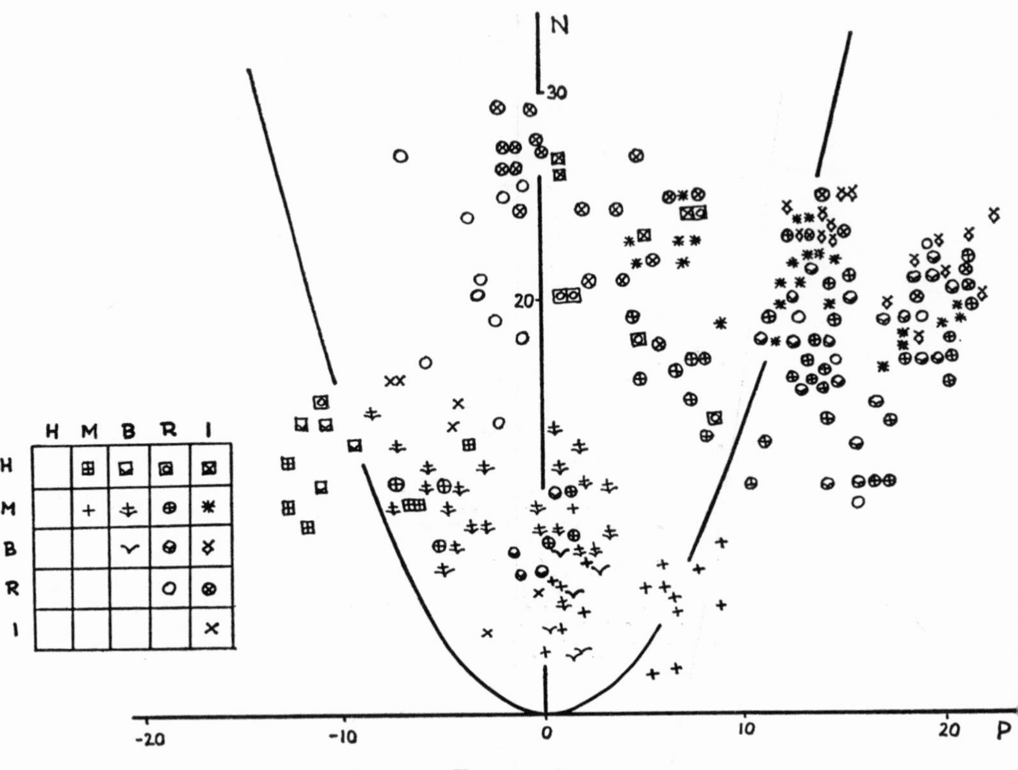

FIGURE 4

Number of substitutions N *versus* polarity change P for animal cytochrome c. H: human, M: mammals, B: birds, R: reptilia, amphibia, fish, cyclostoma, I: insects. Notice the different arrangement of the chains in each group as compared to Figure 2.

pound reptilia, amphibia, fish, cyclostoma. Mammals vary more in polarity than in amino acid composition, being a relatively young but extremely diversified group. The contrary holds true for the insects. Birds are very homogeneous in both respects.

Man and turtle each have to be set apart from their respective groups, but for different reasons: man is not too extravagant in polarity (if put at the end after dog, having the smallest mammalian polarity, it would not change the p and ρ that hold for the rest of the mammals too much); but man has a much higher number of amino acid differences N compared to the other mammals than these have among each other. Conversely, turtle has a high polarity totally outside of the rest of its group, rattlesnake included, but is very conservative in its N. In this sense, man is something between bird and reptile; turtle is a sort of bird.

4. Random polarity changes as dictated by the genetic code

Consider a protein in which amino acid i has an abundance a_i and a polarity p_i. If all possible base replacements occur with the same probability, the frequency of the amino acid substitution $i \rightarrow k$ will be given by a weight factor ν_{ik} that expresses the number of possible base replacements leading from i to k. This is done under the assumption that the different codons for a certain amino acid occur with equal frequency.

A substitution between amino acids represented by codon quartets, if possible in one step, will get a weight 1 (for example, Ala-Val). For the substitution of a quartet coded acid by a duet coded one, like Thr-Asn, the weight will be $\frac{1}{2}$, which may be interpreted by cutting the Thr abundance in two, since only the two codons that permit a transition to Asn are in question. Likewise, for Leu-Ile, one gets $\frac{2}{3}$, conversely for Ile-leu, $\frac{4}{3}$.

Quite generally, one obtains the weight ν_{ik} for the substitution $i \rightarrow k$ by counting all possible transitions leading from any codon of i to any codon of k and dividing by the number of codons for i. The weights lie between $\frac{1}{6}$ (for example, Ser-Try) and 3 (for example, Phe-Leu).

With a polarity change $\delta_{ik} = p_k - p_i$ for the substitution $i \rightarrow k$, the expectation for the *mean* polarity change per substitution and its standard deviation in a randomly selected substitution are plainly

$$\text{(1)} \qquad \bar{\delta} = \frac{\sum_{i,k} \delta_{ik}\nu_{ik}a_i}{\sum_{i,k} \nu_{ik}a_i}, \qquad \sigma = (\overline{\delta^2} - \bar{\delta}^2)^{\frac{1}{2}}.$$

Unless the above summation is arbitrarily restricted, $\bar{\delta}$ is very close to 0, some possible departure from 0 being only due to a preponderance in abundance either of the polar or the unpolar amino acids. Anyhow, the "equilibrium protein" has $\bar{\delta} = 0$, whether it is defined as a protein with five per cent abundance for every amino acid, or as one in which the abundances are proportional to the number of codons available for the different amino acids. The actual abundances in globin and cytochrome *c* yield very small $\bar{\delta}$. The appropriate measure for average polarity change per substitution, as far as only its absolute size is concerned, is evidently the standard deviation σ. This will be used under the name of p_{rand} for comparison with the observed polarity changes according to Model 1.

For concrete cases, one obtains the values shown in Table III for $\sigma = p_{\text{rand}}$. By its very construction, polar and apolar amino acids each group residing together in their special "quarters," the code keeps the mean polarity change smaller than it would be for really random changes. If any amino acid could be freely substituted for any other one, the average absolute size of the polarity change would be

$$\text{(2)} \qquad p_{\text{free}} = 1.45\,(\overline{p_i^2} - \bar{p}_i^2)^{\frac{1}{2}},$$

TABLE III

AVERAGE POLARITY CHANGE PER SUBSTITUTION

Abundance	p_{rand}	p_{free}
5% protein	2.68	3.55
Globin	2.98	4.10
Cytochrome *c*	2.60	3.58
Codon equilibrium protein	2.58	3.53

p_i meaning the polarity of amino acid i (Table I). The completely random p values thus obtained and listed in the second column of Table III are generally about 40 per cent bigger than those taking account of the restrictions dictated by the code.

5. Model 1

Every amino acid substitution is connected with a polarity change of equal absolute value p; this change can be positive or negative, with probabilities ρ and $1 - \rho$, respectively.

If there is a total of N substitutions between two chains, the probability that among these exactly μ correspond to a polarity increase (and $N - \mu$ to a decrease) is

$$W(\mu) = \binom{N}{\mu} \rho^{\mu}(1 - \rho)^{N-\mu}. \tag{3}$$

The resulting Bernoulli distribution of the total polarity change P, which is expressed in terms of μ by

$$P = p(2\mu - N), \tag{4}$$

has the following moments (in a form convenient for recursive computation):

$$\overline{P} = pN(2\rho - 1), \tag{5}$$

$$\overline{P^2} = 4p^2N\rho(1 - \rho) + \overline{P}^2, \qquad \sigma_P = 4p^2N\rho(1 - \rho), \tag{6}$$

$$\overline{P^3} = \overline{P}\sigma_P^2(3 - 2/N) + \overline{P}^3, \tag{7}$$

$$\overline{P^4} = 3\sigma_P^4(1 - 2/N) + 4p^2\sigma_P^2 + 2\overline{P}^2\sigma_P^2(3 - 4/N) + \overline{P}^4, \tag{8}$$

$$\overline{P^5} = 10\overline{P}^3\sigma_P^2(1 - 2/N) + \overline{P}\sigma_P^4(15 - 50/N + 24/N^2) + 4\overline{P}p^2\sigma_P^2(5 - 2/N) + \overline{P}^5. \tag{9}$$

From the observed values for P and ρ, according to (5) and (6), the parameters of the model can be computed:

$$p = \frac{1}{N}(N\sigma_P^2 + \overline{P}^2)^{1/2}, \tag{10}$$

$$\rho = \tfrac{1}{2}[1 + \overline{P}/(N\sigma_P^2 + \overline{P}^2)^{1/2}]. \tag{11}$$

TABLE IV

PARAMETERS OF MODEL 1

M: mammals; B: birds; RAF: reptilia, amphibia, fish, *Cyclostoma;* I: insects; P: plants; Hum: human; Tur: turtle; α: α globins; β: non-α globins (β, γ, δ, fetal); Myo: myoglobins; Mono: monohemic chains (myoglobin, lamprey *Chironomus*); Chir: *Chironomus.*

Combination	No. of points evaluated	Mean no. of substitutions	Parameters of model 1	
			Step length	Probability for right step
Cytochrome *c*				
Hum/MB	11	11.9	1.17	0.15
Hum/RAF	6	18.5	0.74	0.66
Hum/I	4	25.2	0.63	0.62
Hum/P	4	40.3	1.97	0.50
M/M	15	5.4	1.54	0.78
M/B	24	10.6	1.14	0.42
M/RAF	30	16.6	1.49	0.76
M/I	24	21.0	1.20	0.74
M/P	24	41.3	2.04	0.55
B/B	6	5.2	0.40	0.81
B/RAF	20	17.7	1.11	0.90
B/I	16	22.3	1.02	0.87
B/P	16	41.7	1.80	0.56
RAF/RAF	10	21.1	0.47	0.35
RAF/I	20	24.2	1.34	0.56
RAF/P	20	42.8	1.74	0.48
I/I	6	11.8	0.80	0.27
I/P	16	43.9	1.49	0.43
P/P	6	40.8	1.04	0.66
M/Tur	6	11.2	1.15	0.40
B/Tur	4	9.0	0.30	0.46
Tur/RAF	5	17.4	1.01	0.96
Tur/I	4	22.0	1.01	0.88
Tur/P	4	42.8	1.82	0.58
Globin				
α/α	63	26.1	2.45 ± 0.66	0.52 ± 0.09
α/α carp	12	68.2	1.32	0.47
α/β	74	74.1	1.50 ± 0.65	0.39 ± 0.08
α/Mono	78	104.0	1.31	0.49
α/Myo	52	106.6	2.06	0.42
α/Chir	13	101.2	1.09	0.19
β/β	218	45.0	1.33 ± 0.31	0.50 ± 0.05
β/Mono	68	111.6	2.04 ± 0.24	0.45 ± 0.01
β/Chir	19	105.3	0.78	0.13
Mono/Mono	15	71.3	3.68 ± 1.06	0.48 ± 0.04
Myo/Myo	6	22.5	2.80	0.49

Using these in (7), (8), (9), one can determine the higher moments of the Bernoulli distribution and compare them to observed ones, either directly or after transforming them into skewness and kurtosis:

$$\beta_1 = \overline{P^3}/\sigma_P^3 \qquad \beta_2 = \overline{P^4}/\sigma_P^4. \tag{12}$$

6. Model 2

The polarity change connected with a substitution is itself a random variable p with a probability distribution $f(p)$, that does not depend on the site at which the substitution occurs nor on time. If

$$\varphi(\zeta) = \int_{-\infty}^{\infty} e^{\zeta p} f(p)\, dp \tag{13}$$

is the generating function of $f(p)$, that is, essentially its Laplace transform, and if N subsequent independent substitutions are considered, the total polarity change P by those N substitutions has a distribution with the generating function

$$\Phi(\zeta) = \varphi(\zeta)^N. \tag{14}$$

The moments of that distribution are, consequently,

$$\begin{aligned}
\overline{P} &= N\overline{p},\\
\overline{P^2} &= N\overline{p^2} + N(N-1)\overline{p}^2, \qquad \sigma_P^2 = N\sigma_P^2,\\
\overline{P^3} &= N\overline{p^3} + 3N(N-1)\overline{p}\overline{p^2} + N(N-1)(N-2)\overline{p}^3,\\
\overline{P^4} &= N\overline{p^4} + 3N(N-1)(\overline{p}\overline{p^3} + \overline{p}^2)\\
&\quad + 6N(N-1)(N-2)\overline{p}^2\overline{p^2} + N(N-1)(N-2)(N-3)\overline{p}^4,
\end{aligned} \tag{15}$$

and so forth.

Thus, given the observed distribution of the P, one could get an idea of the underlying distribution $f(p)$, for example, from its moments, the coefficients of the MacLaurin expansion of its Laplace transform:

$$\begin{aligned}
\overline{p} &= \overline{P}/N\\
\overline{p^2} &= \overline{P^2}/N - (N-1)\overline{P}^2/N^2\\
\overline{p^3} &= \overline{P^3}/N - 3(N-1)\overline{p}\overline{p^2} - (N-1)(N-2)\overline{p}^3,
\end{aligned} \tag{16}$$

and so forth.

Heuristically, this procedure suffers from hyperparametritis, because *any* distribution of the P could be exactly fitted this way. However, one could decide whether the resulting $f(p)$ looks anyway reasonable, for example, like the random distribution predicted by the genetic code.

At the moment, the available set of data hardly permits doing this in a meaningful manner. Therefore, throughout this paper only Model 1 is referred to.

7. Conclusions

7.1. Model 1 satisfactorily describes all the combinations of groups investigated, as is established by the fact that once the model parameters p and ρ have been determined from the first two observed moments, the higher moments are predicted by the model with the accuracy to be expected considering the size of the sample of points. There is no need at present to recur to the more complete distribution of Model 2, which in Model 1 is simplified to two δ shaped peaks at $\pm p$.

7.2. In the case of cytochrome *c*, every combination of taxonomical groups, even for a group as wide as "plants," yields a point set that is much narrower than the genetic code predicts. During the evolution of these groups as well as for larger portions of the phylogenetic tree that include some of the corresponding branchings, a mechanism has been acting that has kept polarity at a desired level, evidently by suppressing substitutions connected with too high a polarity change. In cases of extreme constancy of polarity, as within the birds or the insects, practically only the "central tower" of Figure 5 can have been used, that corresponds to substitutions as harmless as, for example, Phe-Val or Ala-Ser. Even Ser-Thr, for example, must already have been too radical.

7.3. Just within this framework, the adaptive character of the relatively large changes of polarity from group to group is particularly accentuated. Since the distribution of *all* the cytochrome *c* points is even somewhat wider than the parabola predicted by the code, one has to admit that during the whole not only of animal, but even of vertebrate, evolution not only the reins that have curbed polarity changes have been let loose, but that even to a certain degree substitutions with higher than average polarity effect have been encouraged.

7.4. For the globins, the taxonomical grouping has not yet been done. Within the functional groups (α, β, monohemic), the above conclusions also hold true, although in a lesser degree.

The analysis described above may be generally applied to any property of a protein chain other than polarity that depends additively on the corresponding property of the component amino acids. By its two dimensional character, the method disposes of the objection that can sometimes be raised against similar discussions, namely, that a lack or smallness of differences in a given property be just due to the fact that there are so few substitutions between the considered chains. More detailed investigations are under way, taking into account also polarity variations between sections of chains.

REFERENCES

[1] G. Buse, S. Braig, and G. Braunitzer, "The constitution of the hemoglobin (erythrocruorin) of an insect (*Chironomus thummi thummi, Diptera*)," *Hoppe-Seyler's Z. Physiol. Chem.*, Vol. 350 (1969), pp. 1686–1691.

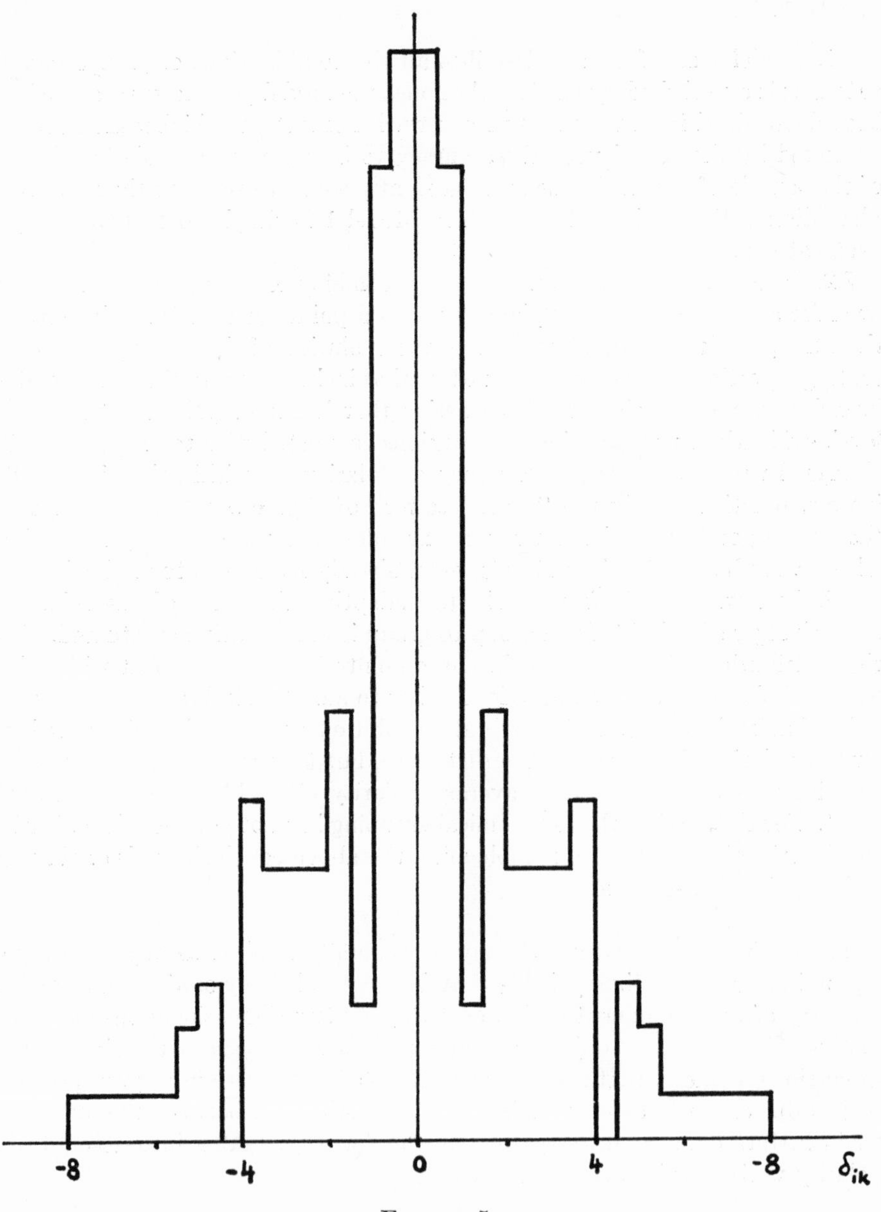

FIGURE 5

Frequency of polarity changes δ_{ik} in one step mutations on random movement within the genetic code; abundances of amino acids for codon equilibrium protein.

[2] J. Chauvet and R. Acher, "Sequence of frog hemoglobin β," *FEBS-Letters*, Vol. 10 (1970), pp. 136–140.

[3] M. O. Dayhoff, *Atlas of Protein Sequence and Structure*, Vol. 4, Silver Spring, Md., National Biomedical Research Foundation, 1969.

[4] R. T. Jones, Personal communication.

[5] C. W. Woese, in *The genetic code*, New York, Harper and Row, 1967, p. 172.

EVOLUTION OF MAMMALIAN DNA

D. E. KOHNE
CARNEGIE INSTITUTION OF WASHINGTON
J. A. CHISCON
PURDUE UNIVERSITY
and
B. H. HOYER
CARNEGIE INSTITUTION OF WASHINGTON

1. Introduction and rationale behind experiments

This communication describes the measurement of DNA nucleotide sequence changes which have occurred since the divergence of various primates. These measurements, coupled with paleontological evidence for times of divergence of primates, allow an estimation of the rate of nucleotide sequence change during different periods of primate evolution. These data strongly suggest the possibility that the rate of nucleotide sequence change may have been a function of the generation times of the species involved. The rate appears to be faster in species with short generation times. This suggests that the rate of molecular evolution has not been constant through evolutionary time.

A great deal of information about evolutionary events and processes has been inferred from careful studies of fossil records. Early DNA studies [1], [4], [11] generally corroborated classical evolutionary findings, and at the same time provided some new understanding of molecular processes in evolution. For the most part, the early work with DNA simply measured the quantities of DNA which would reassociate when mixtures of DNA from various species were allowed to react. Repeated sequences had not been recognized and their significance was not appreciated [2]. At present it is possible to separate the repeated sequences from the nonrepeated sequences and to measure relationships between species using either fraction.

DNA reassociation is the primary tool used for exploring the evolution of DNA. Since a wide variety of scientists are interested in evolution, an attempt has been made to provide enough background material to be helpful to those with little knowledge of DNA. Readers well versed in DNA lore could well ignore this material and skip to another section.

1.1. *Basic characteristics of* DNA. DNA is a long linear polymer constructed of four distinct chemical subunits called nucleotides. These nucleotides differ in their bases, containing either Adenine (A), Guanine (G), Cytosine (C), or Thymine (T). The biological information of the DNA molecule is stored in the ordered sequence of its bases. In its natural state DNA exists as a double strand molecule with each strand winding around the other in a helical fashion. The two strands are held together by specific interactions between the individual nucleotides in each strand. A always interacts or pairs with T while G and C always pair together. A-T and G-C are called complementary base pairs. In DNA whose strands have never been separated (native DNA) each base in one strand is believed to be paired with its complementary base in the other. Perfect base pair matching is then presumed to be present between the strands. The complementary strands of a double strand DNA molecule are readily separated or dissociated into two single DNA strands. Under the proper conditions, these complementary DNA strands can reassociate and reform a stable double strand DNA molecule. This "sequence recognition" property of complementary DNA strands is extremely useful for studying the evolution of DNA, and is utilized to compare the genetic material of different organisms [4], [11].

Reassociation of DNA can be measured in several ways, all of which depend upon some physical chemical differences between dissociated (single strand DNA) and reassociated (double strand DNA). A particularly useful technique for measuring DNA reassociation utilizes the properties of hydroxyapatite crystals. Under the proper conditions hydroxyapatite will absorb double strand or reassociated DNA while single strand DNA does not absorb to the hydroxyapatite and can be washed away [2], [6].

1.2. *Thermal stability of* DNA. The thermal stability of double strand DNA can be used as an indicator of the degree of perfection of base pair matching between the strands of a double strand DNA molecule. Double strand DNA heated in solution will dissociate into single strands at a specific temperature, depending on the salt concentration and DNA composition. The thermal stability of reassociated DNA can be compared to that of native DNA which has perfect, or nearly perfect, matching between the bases of its component strands. Reassociated DNA can have perfection of base pair matching similar to that of native DNA. It is possible to alter nucleic acid strands so that a small fraction of the nucleotides can no longer interact with a complementary nucleotide. These partially complementary strands can then reassociate to form stable double strand molecules whose thermal stability is less than that of reassociated, unaltered molecules. Lack of complementarity in 1.5 per cent of the nucleotides will lower the thermal stability of the double strand molecule by about 1°C [10].

1.3. *Rationale of experiments.* To determine the rate of nucleotide sequence change since the divergence of various species, one must know (a) the time since the divergence of the species in question, and (b) the number of nucleotide changes which have occurred during that time. Paleontology provides us with estimates of (a), and the number of nucleotide changes can be experimentally determined. Complementary single DNA strands from different animal species can interact to form a "hybrid" double strand molecule. One strand of this "hybrid" molecule is radioactive and is from one species while the other is nonradioactive and from a different species. If the two strands are only partially complementary in the reassociated region, the "hybrid" reassociated DNA will have a lower thermal stability than perfectly base pair matched DNA. The difference in thermal stability between the "homologous" and "hybrid" DNA is a measure of the extent of nucleotide changes which have occurred since divergence of the two species in question. A reasonable estimate of the actual percentage of nucleotide changes can be obtained by using the experimentally determined observation that 1.5 per cent nucleotide pair mismatches lowers the thermal stability 1°C. Figure 1 presents the rationale used for determining the extent of nucleotide substitutions which have occurred since the divergence of the two species.

The value for per cent base changes which have occurred since divergence is an average divergence value for the many different nucleotide sequences present in the DNA used for the experiment. Furthermore, multiple changes have undoubtedly occurred at many sites in extensively diverged DNA's (for example, man *versus* galago). However, only those nucleotide changes existing today can be detected, so the extent of change seen between distantly related species are minimum values. Values obtained for closely related species should be very nearly correct since the probability of multiple changes occurring at any one site will be low.

A primary requirement for comparing the DNA's of different species is that one species of DNA forms predominately "hybrid" double strand DNA molecules during the reassociation period. This requirement can be met by properly adjusting the radioactive and nonradioactive DNA concentrations to control the rates of reassociation. The radioactive DNA used must be kept at a low enough concentration so that the collision of two radioactive complementary strands is very improbable. The nonradioactive DNA concentration must be high enough so that virtually all of the nonradioactive DNA sequences will find complementary sequences and reassociate during the time of incubation. Thus, a radioactive DNA strand will form a stable double strand "hybrid" molecule *if and only if there is a complementary sequence for it in the nonradioactive DNA*.

Not all of the DNA of higher organisms can be used for these experiments. All higher organisms examined thus far have contained large fractions or families of repeated DNA sequences (Table I) [2]. A family of repeated nucleotide sequences is composed of many member sequences, each of which can reassociate with any other member of that family. There are at least several families in each cell and the member sequences of each family are generally not identical to one another in sequence, but are similar enough to reassociate together. It is known that related species contain at least some of the same families of repeated DNA sequences, since the repeated DNA of one species will reassociate with that of another species [1], [4]. If two related species contain similar repeated DNA, it seems probable that they both inherited this repeated DNA from their most recent common ancestor. Since it is not known whether these particular family DNA sequences were identical or only similar at the time of divergence, it is not possible to determine how long it took to accumulate the

I. Original DNA sequence in the most recent common ancestor of species X and Z.

The asterisk indicates nucleotide substitution.

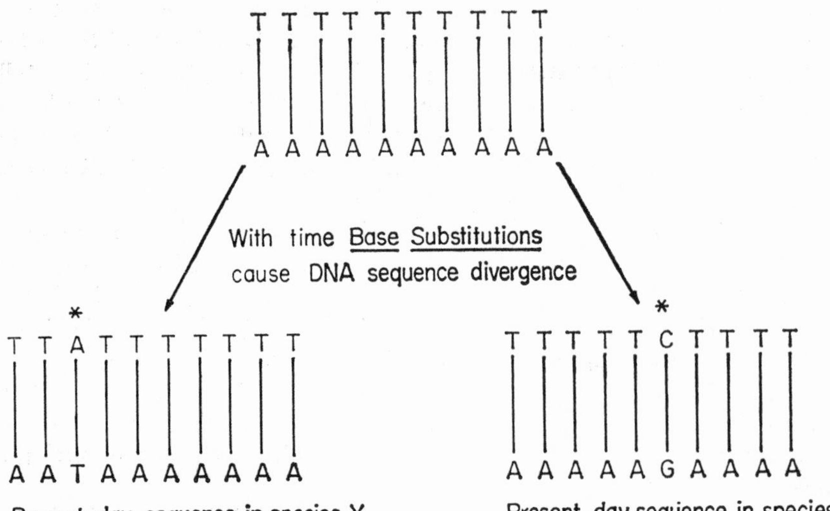

II. The DNA SEQUENCES from species X and Z can be compared in a test tube by forming a *HYBRID* DOUBLE STRAND DNA molecule. Single strand DNA sequences do not have to be perfectly complementary in order to reassociate to form a stable double strand hybrid molecule.

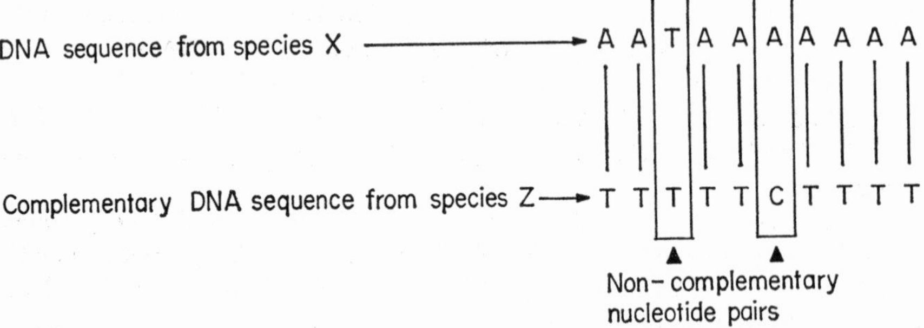

III. Double strand DNA is most *Thermal Stable* when all of the nucleotides in one strand are properly paired with their complementary nucleotide.

When all of the nucleotides in the double strand region are *not* paired their complementary nucleotide, the thermal stability is *lowered.*

1.5 per cent noncomplementary nucleotide pairs = 1°C lowering of thermal stability.

IV. The per cent of noncomplementary nucleotide pairs seen for a particular hybrid represents the extent of nucloeitde sequence change which has occurred since the time of divergence of the species involved.

FIGURE 1

Determination of the extent of nucleotide substitutions which have occurred since the divergence of two species.

TABLE I

CHARACTERISTICS OF MAMMALIAN DNA

A. Repeated DNA
   (1) Repeated DNA is composed of groups or families of DNA sequences. *Each* member of a family can interact and reassociate with *any other* member of that family.
   (2) There are at least several families in each cell. The number of members per family is about 10^5 for mammals.
   (3) Family members are generally similar but not identical to each other. There are exceptions to this.
   (4) Repeated DNA reassociates *very* rapidly because of the high concentration of each sequence in each cell. It is easily purified.

B. Nonrepeated DNA
   (1) Each nonrepeated sequence is present one time per haploid cell.
   (2) Nonrepeated DNA sequences reassociate *very* slowly because of their low concentration per cell. They are readily purified.
   (3) *Most* of the potential genetic information is contained in this fraction.

number of differences observed in the "hybrid" reassociated repeated DNA. Thus, the comparison of DNA in its repeated form will not yield the true rate of nucleotide sequence change when correlated with the time since divergence.

2. Extents of nucleotide change since the divergence of various mammals

The basic experiments involve reassociating radioactive nonrepeated DNA from various animals with total nonradioactive DNA prepared from the same or from other organisms (Table II). Radioactive nonrepeated DNA's was obtained by reassociating the radioactive DNA (human, green monkey, mouse) for a sufficient time to reassociate the repeated DNA sequences. The DNA was then fractionated on hydroxyapatite and the nonrepeated fraction isolated [2], [6].

TABLE II

PROCEDURE FOR COMPARING MAN DNA SEQUENCES WITH CHIMP DNA SEQUENCES

I. Isolate man radioactive nonrepeated DNA sequences. There are about 5×10^9 nucleotides per cell in this fraction. All mammals have about the same amount of DNA per cell. DNA used in these experiments has been sheared to *400 nucleotide long pieces.*

II. A. Mix a small amount of human *radioactive nonrepeated* DNA with a large amount of *nonradioactive* DNA from the Chimp. Denature and incubate.
    B. For the control, a small amount of *human radioactive nonrepeated* DNA is mixed with a large amount of *human nonradioactive* DNA. Denature and incubate. In this case the radioactive DNA will form perfectly nucleotide pair matched duplexes.

III. Isolate both the hybrid [Man-Chimp (A)] and control [Man-Man (B)] double strand molecules and determine their thermal stabilities (T.S.):

$$\Delta \text{ T.S.} = [\text{T.S. of Man-Man duplex}] - [\text{T.S. of Man-Chimp hybrid}].$$

IV. Per cent noncomplementary nucleotide pairs (or per cent nucleotide substitutions since the divergence of Man and Chimp)

$$= [\Delta \text{ T.S. } (^\circ\text{C})]\left(\frac{1.5\% \text{ noncomplementary nucleotide pairs}}{1^\circ\text{C lowering of T.S.}}\right).$$

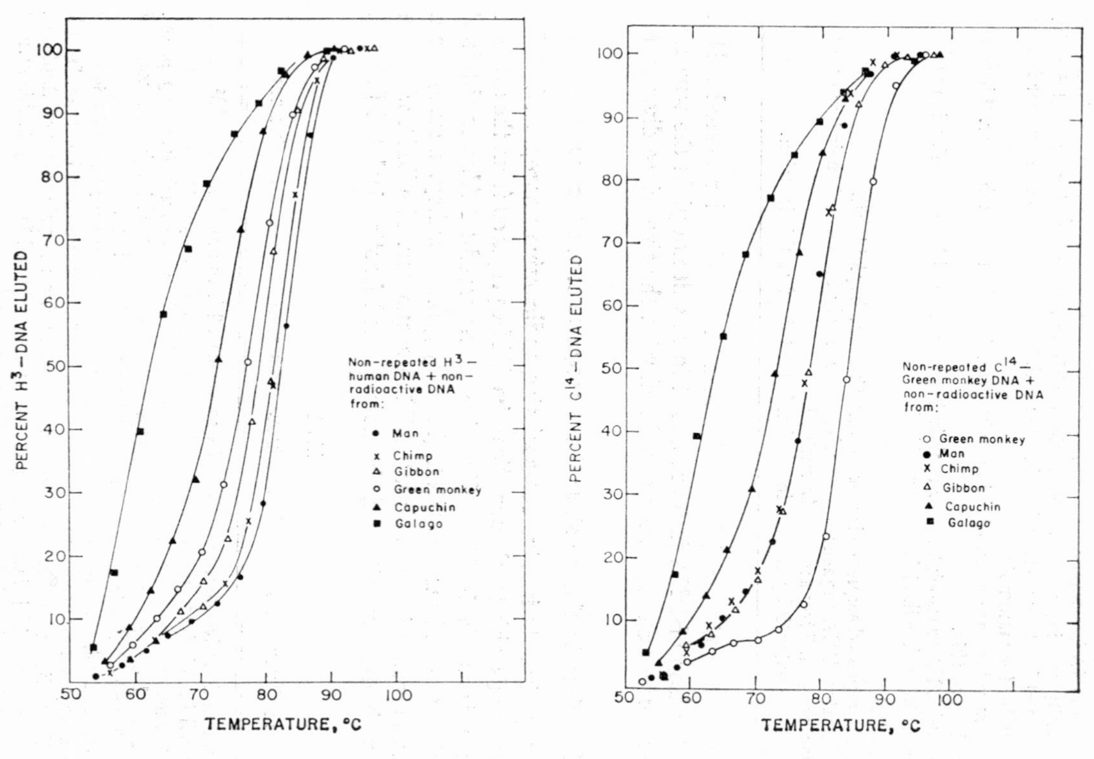

FIGURE 2

Thermal stability profiles.

On the left is shown per cent H^3-DNA eluted plotted against temperature, and on the right per cent C^{14}-DNA eluted plotted against temperature. The basic experimental procedure followed for these studies is given in the text.

TABLE III

EXTENT OF NUCLEOTIDE SEQUENCE DIFFERENCE BETWEEN VARIOUS SPECIES

DNA compared	Δ T.S. (°C) (Control T.S. − hybrid T.S.)	Per cent nucleotide substitutions observed since the time of divergence of the two species compared
Man–Chimp	1.7	2.5
Man–Gibbon	3.4	5.1
Man–Green monkey	6.1	9
Man–*Rhesus*	5.5	8.3
Man–Capuchin	10.5	15.8
Man–Galago	∼28	42
Mouse–Rat	20	30
Cow–Sheep	7.5	11.2
Green monkey–Man	6	9
Green monkey–Chimp	6.2	9.3
Green monkey–Gibbon	6	9
Green monkey–Capuchin	11.5	17.3
Green monkey–Galago	∼28	∼42

Figure 2 presents thermal stability profiles obtained in these experiments, while Table III summarizes the data. The stability profiles of the hybrids show several qualitative features of significance. Almost all of the human DNA reacted with most of the species. It appears that no large class of DNA has diverged so rapidly that hybrids cannot be formed at this temperature of incubation. At the other extreme, almost all of the DNA has diverged appreciably, so there is very little of a highly conserved class of DNA sequences (for example, the man-galago comparison).

The basic experimental procedure followed for these studies was: mix a large amount of nonradioactive DNA (2 to 4 milligrams in less than one milliliter) with a small amount of nonrepeated radioactive DNA (1 to 2 micrograms). Add EDTA (pH = 8.0) to a final concentration of 2.5 millimolar and dissociate the DNA by placing the vial in a boiling water bath for one minute. EDTA helps prevent degradation of the DNA during incubation. Care should be taken to heat the DNA for as short a time as possible when the DNA is in a solution of a low salt concentration. After heating, make the DNA to 0.8 M PB (1.2 M Na^+) and incubate the mixture at 63°C. The DNA was incubated to a DNA C_0t (the product of the molar concentration of DNA monomers and time in seconds) equivalent to 10,000 in 0.14 M PB 51°C (at 2 mg/ml of nonradioactive DNA in 0.8 M PB, where PB is a sodium phosphate buffer solution. This involves incubating the DNA for about 90 hours at 63°C). The final volume was one milliliter. At the end of the incubation period, freeze the samples for later analysis. For analysis take an aliquot of the sample (about 400 micrograms of DNA) and adjust the salt concentration to 0.14 M PB and pass this sample over a water jacketed hydroxyapatite (two grams of Bio Rad HTP) column equilibrated to 51°C and 0.14 M PB. Wash the unreassociated DNA off the column. The thermal stability of these samples was measured by raising the temperature of the column automatically (0.9°C rise in 2.5 minutes) and collecting, with a fraction collector, the effluent (0.14 M PB) which pumped through the column at about five milliliters per minute. Each point on Figure 2 represents 40 milliliters of effluent (0.14 M PB) and a 1.8°C temperature rise. After collection, precipitate each fraction, collect on a filter, and assay for radioactivity with a scintillation spectrometer. In order to gain maximum reproducibility, the thermal elution procedure was completely automated. For the sake of efficiency, some of the incubation mixtures contained nonrepeated DNA from two different species (H^3-man DNA and C^{14}-green monkey DNA). These DNA's were at such low concentration that they were unable to react significantly with themselves (that is, H^3-DNA with H^3-DNA).

In comparing the divergence of different pairs of species, the same classes of DNA's must be compared. For the purposes of the calculations and considerations to follow, we have compared the average divergence of *all* the nonrepeated DNA. The measure of the average divergence is the temperature at which 50 per cent of the *total* radioactive DNA is in a hybrid form. This temperature is referred to as the thermal stability of the hybrid molecules. Two parameters are used to calculate the thermal stability of a hybrid pair: the extent of reaction of the radioactive DNA with the nonradioactive DNA, and the temperature elution profile of the hybrids formed (Figure 2).

2.1. *Interpretation of base substitution data.* Both the extent of reassociation between the various primate DNA's and the thermal stability measurements agree with the paleontological view of the relationships between the primates tested [6]. As expected, for example, chimp DNA is most closely related to man DNA, and *Rhesus* DNA to green monkey DNA. The nucleotide substitutions studied here are those which have been "fixed" during the evolution of these species. Further, the substitution values for the more distantly related species are likely to be underestimates due to multiple changes at the same site, as was discussed earlier.

Figure 3 depicts the present day view of the phylogenetic relationships among the primates studied here. Chimp and man, according to paleontological estimates, diverged from their most recent common ancestors about 10 to 20 million years ago. The extent of divergence of the man and chimp DNA's is, then, the sum of changes occurring since divergence in the chimp line plus those occurring in the human line. The difference seen between chimp and human DNA's today (2.5 per cent) is equal to ΔA (the per cent nucleotide changes occurring in the human line) plus ΔF (the per cent changes occurring in the chimp line). Using the data listed on Figure 3, it is possible to determine how many changes occurred in each line. The rationale for this is as follows. Form hybrids between radio-

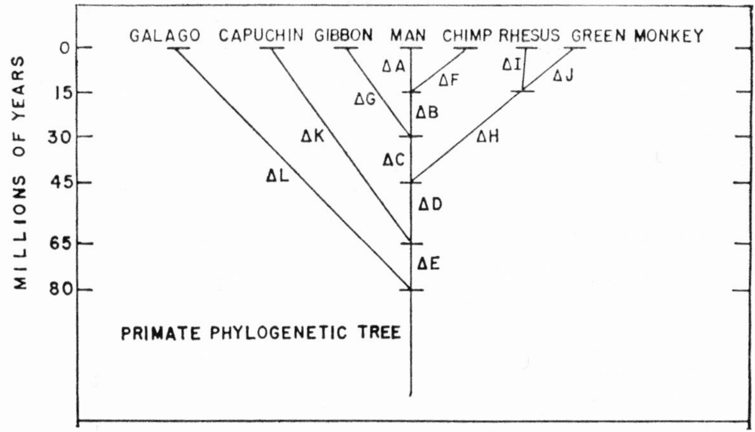

DNA's Compared	*Per Cent Changes Since Divergence*
Man*–Chimp	$\Delta A + \Delta F = 2.5$
Man*–Gibbon	$\Delta A + \Delta B + \Delta G = 5.1$
Man*–Green monkey	$\Delta A + \Delta B + \Delta C + \Delta H + \Delta J = 9$
Man*–*Rhesus*	$\Delta A + \Delta B + \Delta C + \Delta H + \Delta I = 8.3$
Man*–Capuchin	$\Delta A + \Delta B + \Delta C + \Delta D + \Delta K = 15.8$
Man*–Galago	$\Delta A + \Delta B + \Delta C + \Delta D + \Delta E + \Delta L \approx 42$
GM*–*Rhesus*	$\Delta J + \Delta I = 2.9$
GM*–Man	$\Delta J + \Delta H + \Delta C + \Delta B + \Delta A = 9$
GM*–Chimp	$\Delta J + \Delta H + \Delta C + \Delta B + \Delta F = 9.3$
GM*–Gibbon	$\Delta J + \Delta H + \Delta C + \Delta G = 9$
GM*–Capuchin	$\Delta J + \Delta H + \Delta D + \Delta K = 17.3$
GM*–Galago	$\Delta J + \Delta H + \Delta D + \Delta E + \Delta L \approx 42$

FIGURE 3

Present day view of the phylogenetic relationships among several primates.

active green monkey nonrepeated DNA and man and chimp DNA's and then determine their thermal stabilities. The extent of difference between chimp and man DNA's has been measured to equal $\Delta A + \Delta F = 2.5$ per cent. If all of the changes had occurred in the chimp line, $\Delta F = 2.5$ per cent, while $\Delta A = 0$. In this case the man-green monkey hybrid difference would be $\Delta A + \Delta B + \Delta C + \Delta H + \Delta J = X + 2.5$. If the same number of changes had occurred in both the chimp ΔF and human ΔA, then $\Delta A + \Delta B + \Delta C + \Delta H + \Delta J = \Delta F + \Delta B + \Delta C + \Delta H + \Delta J$. Analysis of this type allows the assignment of rough values of extent of change for the various periods during the primate evolution. Table IV presents these values. The data used in these analyses is not yet good enough to precisely determine the values for each period. However, it is good enough so that large differences between values are meaningful. The

TABLE IV

RATES OF CHANGE DURING DIFFERENT EVOLUTIONARY ERAS
ΔD, ΔE, ΔK, and ΔL are estimated.

Per cent nucleotide change during period		Length of period $\times 10^6$ years	Per cent change per 10^6 years
ΔA	1.1	15	0.074
ΔB	1.45	15	0.097
ΔC	1.20	15	0.08
ΔD	$\sim$ 4.45	20	0.22
ΔE	12.8	15	0.85
ΔF	1.4	15	0.093
ΔG	2.55	30	0.085
ΔH	3.45	30	0.12
ΔI	1.1	15	0.073
ΔJ	1.8	15	0.12
ΔK	$\sim$ 8.2	65	0.13
ΔL	$\sim$21	80	0.26

values of ΔA, ΔB, ΔC, ΔF, ΔG, ΔH, ΔI, ΔJ can be directly determined, while the values for ΔD, ΔE, ΔK, ΔL can be estimated by assuming that after divergence the number of changes in each line was roughly the same. This, clearly, is the case in those periods where definite data is available. For example, after the divergence of human and green monkey the extent of change in the human line and the green monkey line is roughly the same. A similar situation exists for the human-chimp and human-gibbon lines. The estimated value for ΔE is undoubtedly low due to multiple changes which have occurred at many of the nucleotide sites.

2.2. *Rate of nucleotide sequence change.* The average rate of nucleotide sequence change since the time of divergence of two species can be calculated from the data of Table III and estimates of the time since divergence of two species. Divergence times have been inferred from the fossil record. There is, however, by no means unanimous agreement among paleontologists on the assignment of divergence times from this record. The divergence times used in Table V are generally considered to be realistic values in the light of present knowledge [12], [13] (p. 208), [14], [15].

Table V presents the calculated average rates of change since the divergence of the various primate species. Table IV presents rates of change calculated for the different time periods during the development of the primate line. The absolute values for the rates calculated in Table IV are probably not very accurate. Much better data is needed to generate precisely accurate rate values by the method of analysis used. The data appear sufficiently good, however, to support the changing rate trend seen in Table IV. It appears that the rate of change was much higher during the early part of the primate line than during recent primate evolution.

Table V also presents rate of change data for rodent (rat *versus* mouse)

TABLE V

RATES OF NUCLEOTIDE FIXATION DURING EVOLUTION

Estimates of time since divergence are from paleontological studies [12], [13], [14], [15]. The last column gives average rates of nucleotide substitution.

DNA's compared	Per cent nucleotide substitutions observed since the time of divergence	Million years since divergence × 2	Rate of nucleotide fixation since divergence (per cent/year) × 10^6
Man–Chimp	2.5	30	0.08
Man–Gibbon	5.1	60	0.08
Man–Green monkey	9.0	90	0.1
Man–*Rhesus*	8.3	90	0.09
Man–Capuchin	15.8	130	0.12
Man–Galago	42	160	0.26
Mouse–Rat	30	20	1.5
Cow–Sheep	11.2	50	0.22

comparisons. Calculated on an absolute time basis, the rate of change of rodent nucleotide sequences is about 10 times faster than that for the primates or bovids. These data plus the rate data of Table IV suggest that the rate of nucleotide sequence change has not been the same for all species over evolutionary time. It must again be emphasized that the large difference in the rates seen for primates and rodents depend upon the divergence times used for these calculations. Wilson and Sarich [16] have recently questioned the paleontologically determined times of divergence for man and chimp. They contend from amino acid sequence and immunological evidence that the actual times of divergence of these species has been much more recent than the bulk of the fossil evidence indicates. The contention basic to their interpretation is that the *rate of molecular evolution has been the same for all species and has also been constant over absolute time.* Obviously, if this can be shown to be true, the degree of molecular change can be used as an evolutionary clock and times of divergence can be calculated with good accuracy from biochemical data, if just one fairly accurate time of divergence is known. Wilson and Sarich used their data and a time of divergence for old world monkeys (for example, *Rhesus* and green monkey) and man of 30 million years to calculate that the man-chimp divergence time was about five million years ago. Table VI shows the nucleotide sequence change data treated in a similar manner. The divergence time estimate of mouse-rat derived in this way is far larger than that estimated from the fossil record.

It must again be emphasized that the observed large difference in rates of nucleotide sequence divergence for primates and rodents depends on the divergence times used for these calculations.

TABLE VI

CALCULATIONS OF DIVERGENCE TIMES PREDICTED FROM ASSUMPTIONS

The assumptions are that rate of DNA molecular evolution is constant with absolute time and the same in all species, and further, that man and green monkey diverged as early as 30 million years ago. (The fossil dates of divergence can't easily be reconciled with those calculated from a constant rate assumption.)

DNA's compared	Per cent nucleotide substitutions since divergence	Million years since divergence	Divergence time estimates based on fossil evidence (million years)
Man–Green monkey	9	30	30–45
Man–Chimp	2.5	$\frac{2.5}{9} \times 30 = 8$	10–25
Mouse–Rat	30	$\frac{30}{9} \times 30 = 100$	5–20

2.3. *Generation time correlation of rate of nucleotide change.* Table VII shows the values obtained when the rate of nucleotide sequence change is calculated in terms of generation time. The generation time used for the calculation was in each case the shortest generation time of the two animals compared. The rates calculated on this basis are surprisingly similar. Table VII also includes data for the rat-mouse nonrepeated DNA comparisons and also data on cow-sheep comparisons done by Laird, McConaughy, and McCarthy [10]. The rates of divergence (corrected for generation time) seen in these lines are very similar to the primate values, even though wide differences in generation times exist. The data of Table VII give a strong suggestion that the mutational events seen in evolution may be generation time dependent [9]. It seems highly improbable

TABLE VII

GENERATION TIME BASED RATE OF NUCLEOTIDE SUBSTITUTION

DNA's compared	*A* Per cent nucleotides fixed per year since divergence	*B* Estimated generation time using shortest generation time of pair	Per cent nucleotide substitutions per generation ($A \times B$)
Man–Chimp	0.08×10^{-6}	~10 years	0.8×10^{-6}
Man–Gibbon	0.08×10^{-6}	~10	0.8×10^{-6}
Man–Green monkey	0.1×10^{-6}	2–4	0.3×10^{-6}
Man–*Rhesus*	0.09×10^{-6}	2–4	0.27×10^{-6}
Man–Capuchin	0.1×10^{-6}	2–4	0.3×10^{-6}
Man–Galago	0.26×10^{-6}	1–2	0.39×10^{-6}
Mouse–Rat	1.5×10^{-6}	0.33 year	0.5×10^{-6}
Cow–Sheep	0.22	1–2 years	0.33×10^{-6}

that fluctuating mutational events in each of the three groups could fortuitously give essentially the same result. In calculating the values for Table VII, it is assumed that the present day generation times provide a reasonable estimate of the generation time history of the various species' lines.

The changes in the absolute time based rates of nucleotide change seen in Table IV could also be explained by taking into account the probable generation time history of the species' lines. If the generation times have increased with the development of the present day primates, the slowing of the absolute time based rate of change would be due to a decrease in the number of generations per million years as the primate lines evolved.

It appears from the fossil records that generation times may have been lengthening during the development of the primate line. The "precursor" primate was apparently a small rodent-like creature which is presumed to have had a short generation time. There is today a general, but not absolute, correlation between animal size and generation time. In general the larger the animal the longer its generation time.

The data on nucleotide substitution rates suggests that the rate of molecular evolution may have been different for different species. If true, this would be a troublesome observation. An assumption basic to evolutionary studies has been that morphological and molecular similarity of two species is an index of the temporal relationship between those species. In this view, species which have diverged most recently would be more similar molecularly and morphologically. This situation is certainly true if the rate of molecular evolution has been the same for all species at any time during evolution. If, however, the rate of molecular and morphological change can be different for different species' lines during the same evolutionary period, the situation becomes difficult. In this case it is possible (particularly in the case of closely related species) that the similarity between species will give an erroneous view of the relative times of divergences of the species in question. Table VIII illustrates this point. It is

TABLE VIII

EFFECT OF RATE MODEL ON INTERPRETATION OF SIMILARITY DATA

I. Data. Of the three species X, Y, and Z, X and Y are most similar to one another. Species Z is equally dissimilar to species X and Y.

II. Assumption. The rate of morphological and molecular change has been the same for all species during any evolutionary period.
   A. From this assumption, the pattern of relatedness among the three species is that species X and Y had diverged more recently than X and Z or Y and Z.

III. Assumption. The rate of morphological and molecular change has been different for different species during any evolutionary period.
   A. From this assumption, the pattern of relatedness could be as is described in IIA. If the rate change in different species is variable enough, species X and Z could have diverged more recently than species X and Y or Z and Y.

important, then, to know something about the rate at which change has occurred during the evolution of the various species' lines, since different rate models give different interpretations for the similarities seen between species.

3. The nature of the nucleotide substitutions

A wide range of effects of nucleotide substitutions on an organism is possible [5]. Nucleotide substitutions in DNA sequences which have *no function* will not affect the organism. It is not known how much of the total cell DNA is actually needed. There may be a class of DNA which can tolerate no nucleotide changes. Such a class of DNA will suffer nucleotide substitutions but these changes will be selected against and not passed along. It is known that very little of the DNA falls in this class. Another class of DNA may determine the templates for proteins which require an unchanged sequence of amino acids. Roughly, one fourth of the total nucleotide substitutions in this class will have no effect on the amino acid sequence, since in many triplets the third base can be changed to any other base without changing the amino acid inserted into the protein (for example, UCU, UCC, UCA, UCG all code for the amino acid serine). The remainder of the DNA (not in one of the above three classes) can be considered in a class intermediate between these defined extremes.

As is apparent from the above paragraph, little is known about the effect of the nucleotide substitutions which have occurred during evolution. The above discussion implies that many base changes which have occurred could have little if any effect on the phenotype of the organism, and are not likely to have much selection pressure on them. There is a chance, then, that many, if not most, of the nucleotide substitutions measured in the experiments reported here may be neutral or close to neutral in their effect on the organism. If this is so, an enormous number of nucleotide substitutions can accumulate in the DNA with little effect on the organism.

Perhaps the best available indication that many changes can accumulate in a genome lies in the behavior of the Treffers mutator strain of *E. coli* [3]. This bacterial strain has a mutation rate hundreds of times higher than the normal *E. coli*. It has been shown that an average of seven nucleotide substitutions per bacterium per cell division occurs in the mutation strain of bacteria [3]. These bacteria grow virtually as well and as rapidly on minimal media as do wild type bacteria, even after many thousands of cell divisions. This directly indicates that many nucleotide substitutions can occur in a living organism without appreciably affecting its short term viability. Over the period during which the mutator strain has been studied, it would not be unreasonable to say that most of the base changes which have occurred are "neutral." Perhaps it would be better to say that the great majority of nucleotide substitutions have had little selective value on them. What the effect of these changes will be over a longer time is not known.

Many base changes can accumulate in the Treffers bacteria without affecting the viability of the bacteria. This indicates that the fraction of the total base changes which produce a lethal effect is relatively small. A rough estimate of the ratio of lethal base changes to the total base changes can be made. Each bacterium contains on the average seven new base substitutions after every cell division. If the ratio of lethal to total base changes is one seventh, the bacterial population would not increase in number. It is known that the mutator strain grows virtually as well as the wild type. The available data indicate that in minimal media cultures roughly 10 per cent of the mutator bacteria are auxotrophs (they have mutated so that they will no longer grow on minimal media). This indicates that roughly one in 25 nucleotide changes is lethal. It seems reasonable to say further that the majority of the other 24 changes has little effect on the viability of the bacteria, since they grow so well. The 1/25 ratio is a rough one and the ratio may well be much smaller.

It does seem, then, that in a relatively simple organism such as a bacterium a large number of base changes can occur and accumulate, apparently with little effect on the bacterial viability. This raises the very real possibility that a similar situation exists in higher organisms. If so, most of the nucleotide sequence changes studied here may well have had little effect on the evolution of the organism. Moreover the rates of nucleotide fixation observed may be a reasonable reflection of the actual base substitution rate during evolutionary time.

3.1. *Divergence of DNA sequences expressed in a specific tissue.* Recent application of DNA:RNA hybridization technique have enabled the isolation from specific tissues of DNA sequences which have been expressed as RNA [8]. These isolated expressed DNA sequences (E-DNA) have been utilized to ask whether E-DNA sequences diverge at the same average rate as do the bulk of the DNA sequences [8]. This approach was tried because we do not know the function of the nonrepeated DNA sequences examined in the earlier paragraphs. The one function which some of these nonrepeated DNA sequences have is that they are expressed as RNA. Unfortunately, we do not know what the function of the RNA is, but at least the E-DNA sequences have been shown to be active in present day creatures.

The experiments to determine the rate of divergence of the E-DNA relative to the bulk DNA are similar to those already described earlier for the bulk DNA sequences. Cow radioactive nonrepeated E-DNA (from brain or liver) was reacted with sheep and pig nonradioactive DNA's. The thermal stabilities of the hybrid molecules were then determined and compared to the thermal stability of hybrids formed by reacting the total radioactive nonrepeated cow DNA with sheep and pig nonradioactive DNA.

About six per cent of the nonrepeated DNA was isolated as being expressed in the liver while about three per cent was detected as being expressed in the brain (these values are minimum indications of the actual extent of expression). Both liver and brain E-DNA diverged at the same average rate as the total nonrepeated DNA. Expression, then, does not appear to confer any great selective advantage to a DNA sequence.

4. Conclusion

The current view of the role of DNA indicates that evolutionary changes at the organismal level must have been preceded by some quantitative or qualitative change in the DNA. Detectable in the DNA are repeated DNA sequences, nonrepeated DNA sequences, base changes, and translocations. These are the present day "residues" of historical events which occurred during the evolutionary history of the DNA. Correlating molecular changes with organismal changes is always a difficult task. Comparison of the pattern and character of these "DNA fossils" with the patterns derived from the classical fossil record should, however, provide new insight into the nature of evolutionary forces. The evidence available thus far allows a very tentative correlation to be made. During the period where the classical fossil record indicates that extensive speciation occurred in the primates it appears that (a) DNA was added to the presumptive human genome at a much faster rate than during other periods for which data is available [7], and (b) the rate of nucleotide sequence divergence was also much higher relative to other periods.

It is not known whether these increased rates are the result or the cause of the extensive speciation seen during this period. The correlation of nucleotide sequence change with generation time suggests that the generation time of the early primate ancestors of man which lived during this period was relatively short. The rate of mixing of genetic material to provide new genetic combinations (for example, translocations, inversions, and so forth) should also be generation time dependent. It does not seem unreasonable to speculate that the rapid addition, divergence, and mixing of the DNA may have played a role in the burst of speciation seen during this period. Much more extensive work needs to be done, however, before any strong statements can be made.

REFERENCES

[1] E. T. Bolton, R. J. Britten, T. J. Byers, D. B. Cowie, B. H. Hoyer, Y. Kato, B. J. McCarthy, M. Miranda, and R. B. Roberts, "Biophysics section report," *Carnegie Institution of Washington Year Book*, Vol. 63 (1964), pp. 366–398.

[2] R. J. Britten and D. E. Kohne, "Repeated sequences in DNA," *Science*, Vol. 161 (1968), p. 529.

[3] E. C. Cox and C. Yanofsky, "Altered base ratios in the DNA of an *Escherichia coli* mutator strain," *Proc. Nat. Acad. Sci. U.S.A.*, Vol. 58 (1967), pp. 1895–1902.

[4] B. H. Hoyer and R. B. Roberts, "Studies of DNA homology by the DNA-agar technique," *Molecular Genetics*, New York, Academic Press, 1967, pp. 425–479.

[5] J. L. King and T. H. Jukes, "Non-Darwinian evolution: random fixation of selectively neutral mutations," *Science*, Vol. 164 (1969), pp. 788–798.

[6] D. E. Kohne, "Isolation and characterization of bacterial ribosomal RNA genes," *Biophys. J.*, Vol. 8 (1968), pp. 1104–1110.

[7] ———, "Evolution of higher organism DNA," *Quart. Rev. Biophys.*, Vol. 3 (1970), pp. 317–375.

[8] D. E. Kohne and M. J. Byers, "Studies on the evolutionary divergence of expressed DNA sequences," in preparation.

[9] D. E. Kohne, J. A. Chiscon, and B. H. Hoyer, "Evolution of primate DNA," in preparation.
[10] C. D. Laird, B. L. McConaughy, and B. J. McCarthy," Rate of fixation of nucleotide substitutions in evolution," *Nature*, Vol. 224 (1969), pp. 149–154.
[11] J. Marmur, R. Rownd, and C. L. Scnildkraut, "Denaturation and renaturation of DNA," *Progr. Nucl. Acid Res. Mol. Biol.*, Vol. 1 (1963), pp. 232–300.
[12] M. McKenna, Personal communication.
[13] A. S. Romer, *Vertebrate Paleontology*, Chicago, University of Chicago Press, 1966.
[14] E. Simons, "The early relatives of man," *Sci. Amer.*, Vol. 211 (1964), pp. 50–62.
[15] G. G. Simpson, "The nature and origin of supraspecific taxa," *Cold Spring Harbor Symp. Quant. Biol.*, Vol. 24 (1959), pp. 255–272.
[16] A. C. Wilson and V. M. Sarich, "A molecular time scale for human evolution," *Proc. Nat. Acad. Sci., U.S.A.*, Vol. 63 (1969), pp. 1088–1093.

DARWINIAN *VERSUS* NON-DARWINIAN EVOLUTION IN NATURAL POPULATIONS OF *DROSOPHILA*

FRANCISCO J. AYALA
UNIVERSITY OF CALIFORNIA, DAVIS

1. Scientific hypotheses: natural selection

The goal of science is to discover patterns of relations among recorded phenomena, so that a few principles can explain a large number of propositions concerning these phenomena [2]. The scientific value of a theory depends on its explanatory power, that is, on its ability to encompass many subsidiary hypotheses into a single comprehensive set of mutually consistent principles. But in order to be accepted in science the applicability of a theory needs to be demonstrated.

Demonstration, or proof, of a hypothesis or theory concerning the empirical world is a gradual process which is never irrevocably completed. The process of demonstration requires, first, to show that the hypothesis or theory is consistent with the relevant facts. Moreover, the hypothesis or theory needs to be confirmed by empirical tests. Empirical tests are experiments or observations which may conceivably have diverse outcomes only some of which are compatible with the hypothesis tested while most of them would lead to its rejection. If the tests are of such a nature that any conceivable outcome or state of affairs be compatible with the hypothesis tested they contribute nothing to the scientific verification of the hypothesis. The value of an empirical test is measured by the *a priori* likelihood of its outcome being incompatible with the hypothesis.

The synthetic theory of evolution, or the theory of evolution by natural selection has a considerable explanatory power. The central concept of the theory is the principle of natural selection—the differential reproduction of genetic variants. Natural selection is the main process directing the evolution of organisms by promoting their adaptation to their environments. The principle of natural selection, together with some subsidiary and generally well authenticated hypotheses, can explain a large number of facts concerning the living world; like the diversity of organisms, their gradual change through historical time, and their remarkable adaptations to the environments where they live. The synthetic theory of evolution by natural selection is, indeed, the single most encompassing biological theory.

The theory of evolution by natural selection appears to be consistent with all or most facts known in every field of biology. The principle of natural selection can account for different patterns, rates and outcomes of evolutionary processes. Phylogenetic radiations as well as lack of phyletic diversification, rapid and slow rates of evolutionary change, abundant and limited genetic variation within a population—these and many other alternative occurrences can all be explained by postulating the existence of appropriate environmental challenges. This is, of course, as it should be. If evolutionary processes are to be explained by a single major directional process, this process must be able to explain the multifarious ways in which evolution in fact occurs. However, it is often difficult or impossible to obtain direct evidence demonstrating the role of natural selection in a given evolutionary situation. This is in part due to the historical character of biological evolution; it is difficult to get information concerning events which occurred in a more or less remote past. But there is also a methodological handicap. Obtaining direct evidence is hampered precisely by the broad explanatory power of the theory of evolution by natural selection. It is difficult to plan observations or to design experiments most of whose conceivable outcomes are incompatible with natural selection. The scientific validity of the theory of evolution by natural selection rests largely on its consistency with the entire range of available facts. Empirical evidence for the theory is, more often than not, indirect—it supports the theory of natural selection because it is incompatible with alternative hypotheses.

The hypothesis has recently been advanced that much evolutionary change may be the result of random processes [11], [15]. This has been labeled "non-Darwinian evolution" [15], or more appropriately, "evolution by random walk" [8]. According to this hypothesis most genetic variants, particularly variants observed at the molecular level, are adaptively neutral, and therefore not subject to natural selection. The proponents of the theory agree that a large fraction of all new mutations is harmful. These mutants are eliminated or kept at very low frequencies by natural selection. It is, however, argued that a substantial fraction of all occurring mutations is adaptively neutral. Carriers of alternative genotypes do not differ in their adaptedness to the environment. The frequencies in populations of these adaptively neutral variants is, accordingly, not subject to natural selection. Since no natural population consists of an infinite number of individuals, random sampling affects the frequencies of adaptively neutral variants and leads ultimately to the elimination of some variants and the fixation of others. It would then be possible that many genetic differences between species—especially differences observed at the molecular level—have no adaptive or evolutionary significance. They are evolutionary "noise" rather than "signal."

The suggestion that random processes play a major role in evolutionary change is an interesting addition to modern development of the theory of evolution. Evolution by random walk may provide a mechanistic explanation, alternative to natural selection, of evolutionary change. The theory of evolution by random walk has the advantage of leading to some rather precise predictions concerning the pattern of genetic variation. It, therefore, lends itself to observational verification or refutation. Experiments and observations to test the predictions of random walk evolution test also, at least indirectly, the validity of the theory of natural selection.

Evolution by natural selection and evolution by random walk are not incompatible. As noted above, proponents of random walk evolution admit that some genetic variants have deleterious effects and are, therefore, subject to natural selection. Similarly, the theory of natural selection admits that gene frequencies are affected by other processes besides natural selection, like mutation, migration, and random sampling. The issue between the two theories is whether most of the genetic variation occurring in natural populations, and most of the genetic differences between species, are the result of random processes or are the result of natural selection. The rest of this paper gives a partial summary of a series of observations and experiments made to decide that question.

2. Genetic variation in *Drosophila*

These studies deal with a group of neotropical species related to *Drosophila willistoni.* The group includes at least twelve species, whose overall geographic distribution extends from Mexico and southern Florida, through Central America and the West Indies, to tropical South America, as far down as southern Brazil and northern Argentina. Six species are siblings, nearly indistinguishable from each other by external morphology. Four of the siblings, *D. willistoni, D. tropicalis, D. equinoxialis,* and *D. paulistorum* have wide, largely overlapping geographic distributions [29].

Genetic variation in the *D. willistoni* group has been studied with standard techniques of starch gel electrophoresis [3], [4], [5], [24]. A summary of the amount of genetic variation in *D. willistoni* is given in Table I. For each of 27 gene loci the table gives the total number of wild genes sampled. This is simply the number of wild individuals studied multiplied by two, except for sex linked genes which are carried by males in single dose. The number of populations sampled is the number of different local populations studied. The amount of genetic variation at each locus is measured in two ways. First, the proportion of populations in which the gene is polymorphic. A gene is taken as polymorphic when the most common allele in the population has a frequency no higher than 0.95. The second, more precise, measure of variation is the mean frequency of individuals which are heterozygous at a given locus. The estimates of frequency of heterozygotes are made assuming that the locus is at Hardy-Weinberg equilibrium.

The information in Table I is summarized, first, by the proportion of polymorphic loci per population. This is simply the average over all loci of the proportion of polymorphic loci per population. This statistic is 56.3 ± 6.2 per cent for the genes sampled in the table. The data are summarized also by the

TABLE I

GENETIC VARIATION AT 27 GENE LOCI OF *Drosophila willistoni*

Gene	Sample size	Number of populations sampled	Frequency of polymorphic populations	Frequency of heterozygous individuals
Lap-5	10348	80	1.00	.629 ± .007
Est-2	7012	61	.668	.117 ± .008
Est-3	3914	42	.727	.108 ± .009
Est-4	9692	78	.885	.270 ± .012
Est-5	10432	81	.333	.089 ± .006
Est-6	2418	48	1.00	.285 ± .030
Est-7	6819	67	1.00	.601 ± .009
Aph-1	2041	10	.800	.136 ± .020
Acph-1	3066	24	.588	.102 ± .018
Acph-2	874	10	.900	.151 ± .023
Adh	5916	56	.515	.103 ± .015
Mdh-2	6680	58	.175	.040 ± .008
α-Gpdh	7032	57	.044	.021 ± .005
Idh	3168	63	.370	.086 ± .023
G3pdh	190	27	.500	.148 ± .053
Odh-1	1088	30	.765	.180 ± .019
Odh-2	40	4	.333	.060 ± .024
Me-1	2882	61	.275	.079 ± .013
Me-2	1368	15	.733	.235 ± .044
To	4949	22	.333	.131 ± .029
Tpi-2	2478	59	.105	.041 ± .010
Pgm-1	2636	28	.815	.186 ± .025
Adk-1	2150	40	1.00	.527 ± .019
Adk-2	2580	60	.722	.171 ± .023
Hk-1	1620	44	.333	.099 ± .016
Hk-2	2228	49	.556	.138 ± .032
Hk-3	2060	47	.071	.039 ± .009

Mean frequency of polymorphic loci per population 0.563 ± .062
Mean frequency of heterozygous loci per individual 0.177 ± .031

proportion of genes at which an individual is heterozygous. This is obtained by averaging over all loci the frequency of heterozygotes at each locus. On the average, an individual of *D. willistoni* is heterozygous at 17.7 ± 3.1 per cent of its genes. A similar degree of genetic variation has been found in the sibling species, *D. tropicalis* (18.6 ± 3.3) and *D. equinoxialis* (21.3 ± 3.0).

Genetic variation of the same order of magnitude has also been observed in other species of *Drosophila*. The following are estimates of the per cent of the genome heterozygous per individual: *D. simulans*, 8.0 [16], [20]; *D. athabasca*, 8.7 [16]; *D. persimilis*, 10.5 [21]; *D. ananassae*, 10.8 [9]; *D. pseudoobscura*, 12.3 [23]; *D. melanogaster*, 17.8 [16], [20]; *D. affinis*, 25.3 [16]. Similarly, the proportion of heterozygous loci per individual is estimated to be about five or six per cent in organisms as different as man [10], horseshoe crab [28], and old-field mouse [26]. The proportion of heterozygous loci per individual is 8 to 11 per cent in the house mouse [25], [27].

The genes examined in *D. willistoni*, as well as in all the studies just quoted, were chosen because simple assay techniques were available. They can be considered a random sample with respect to variation of loci coding for soluble proteins, since they were selected independently of how variable they were. However, they all belong to a single class of genes, namely, those coding for soluble proteins. Two other major classes of genes exist, regulatory genes and those coding for structural proteins. We do not know how variable regulatory genes are. Our ignorance about variation in genes coding for structural proteins is also nearly complete (see [19], however, for variation in fibrinopeptides among species of artiodactyls). It must also be pointed out that even for the genes coding for soluble proteins the estimates of heterozygosity are only gross approximations. The biochemical techniques used, as well as the sampling procedures are potentially subject to various biases which may affect the estimates in an undetermined direction and by an undetermined amount. It is, however, likely that they underestimate the amount of variation [3], [17]. The remarkable fact is that different investigators using different techniques to study different genes in different organisms have obtained estimates of genetic variation falling within a narrow range of values. It may be concluded that in organisms as different as *Drosophila* flies, horseshoe crabs, mice, and men, about 50 per cent of the genes coding for soluble proteins are polymorphic in a given population, and an individual is heterozygous at about ten per cent of its genes. These estimates are likely to be correct as to the order of magnitude.

To estimate the absolute number of polymorphic loci in a given species we need to know the total number of functional gene loci in the species. Estimates obtained by dividing the number of nucleotide pairs per haploid DNA (between 10^9 and 10^{10} for insects and mammals) by the average number of nucleotide pairs (about 500) per gene (cistron) run in the millions. These estimates give the upper limit of the number of functional loci per individual. Considerations based on classical genetic analyses give a lower limit of 10,000 loci per individual in organisms like man and *Drosophila*. It is likely that the number of loci in these organisms is of the order of 10^5. The total number of genes which are polymorphic in a given population must, then, be at least in the thousands. This is, indeed, a considerable amount of genetic variation, and considerably more than what many geneticists were willing to accept as recently as five years ago.

The question which concerns us now is: what is the evolutionary significance of all this genetic variation? Are most genetic polymorphisms adaptively neutral or, on the contrary, are they maintained by natural selection? To answer this question we must look at the *pattern* of the genetic variation.

3. Genetic variation between populations

The amount of genetic variation varies considerably from locus to locus of *Drosophila willistoni*. The proportion of heterozygous individuals is greater than 50 per cent at the *Lap-5*, *Est-7*, and *Adk-1* loci, but four per cent or less at

TABLE II

GENETIC VARIATION AT THE *Lap-5* LOCUS OF *Drosophila willistoni*

Locality	Sample size	Alleles .96	.98	1.00	1.03	1.05	Other	Frequency of heterozygous individuals
Jaque	194	.03	.18	.34	.37	.06	.02	.709
Teresita	260	.03	.14	.32	.36	.14	.02	.728
P. Lopez	402	.002	.08	.30	.57	.04	.002	.578
Betoyes	180	.01	.13	.26	.54	.06	.01	.618
Mitu	160	.03	.09	.23	.41	.24	.01	.710
Caracas	316	.01	.11	.23	.52	.12	—	.646
Macapa	56	—	.11	.23	.63	.04	—	.543
Belem	74	—	.10	.22	.47	.22	—	.674
Santarem	492	.02	.14	.39	.43	.02	.01	.649
Tefe	172	—	.11	.34	.42	.13	—	.678
Mirassol	1806	.01	.07	.25	.57	.09	.002	.618

Mdh-2, *α-Gpdh* and *Hk-3* (Table I). In sharp contrast, the amount and the pattern of the variation remain remarkably constant from locality to locality [4], [5]. To illustrate this difference I have selected six loci, two very polymorphic (*Lap-5* and *Est-7*), two moderately polymorphic (*Est-5* and *Pgm-1*), and two with little polymorphism (*α-Gpdh* and *Mdh-2*). Tables II through VII give for each of the six loci the number of genes sampled, the allelic frequencies, and the proportion of heterozygous individuals expected on the assumption of Hardy-Weinberg equilibrium. At each locus one allele, usually the most common, has been arbitrarily designated 1.00. Others are named with reference to that standard. For instance an allele .98 codes for a protein that in the gels migrates towards the anode 2 mm less than the standard. Only a few representative populations are included in the tables. Their geographic position is

TABLE III

GENETIC VARIATION AT THE *Est-7* LOCUS OF *Drosophila willistoni*

Locality	Sample size	Alleles .96	.98	1.00	1.02	1.05	Other	Frequency of heterozygous individuals
Jaque	174	.03	.27	.46	.20	.03	—	.673
Teresita	216	.02	.20	.45	.30	.03	—	.669
P. Lopez	294	.02	.15	.58	.20	.05	.003	.560
Betoyes	111	.06	.14	.53	.22	.05	—	.646
Mitu	224	.06	.25	.50	.14	.05	.004	.658
Macapa	41	—	.12	.39	.29	.20	—	.709
Belem	61	—	.10	.64	.16	.10	—	.545
Santarem	406	.01	.15	.55	.21	.07	—	.620
Tefe	147	.02	.12	.45	.31	.11	—	.679
Mirassol	920	.01	.13	.57	.26	.01	.02	.586

TABLE IV

GENETIC VARIATION AT THE *Est-5* LOCUS OF *Drosophila willistoni*

Locality	Sample size	Alleles .95	1.00	1.05	Other	Frequency of heterozygous individuals
Jaque	188	.01	.98	.01	—	.032
Teresita	238	.02	.97	.01	—	.058
P. Lopez	400	.04	.95	.01	.002	.101
Betoyes	178	.02	.96	.02	—	.076
Mitu	224	.04	.94	.02	—	.115
Caracas	212	.02	.96	.02	—	.078
Macapa	48	.02	.96	.02	—	.081
Belem	58	.02	.98	—	—	.034
Santarem	444	.03	.96	.01	—	.079
Tefe	174	.03	.96	.01	.01	.078
Mirassol	1976	.03	.96	.01	—	.086

TABLE V

GENETIC VARIATION AT THE *Pgm-1* LOCUS OF *Drosophila willistoni*

Locality	Sample size	Alleles .96	1.00	1.04	Other	Frequency of heterozygous individuals
Jaque	42	—	.95	.05	—	.091
Teresita	36	—	.92	.08	—	.153
P. Lopez	188	.03	.88	.08	.01	.222
Betoyes	134	.02	.98	.01	—	.044
Macapa	12	—	.83	.17	—	.320
Santarem	20	.05	.85	.10	—	.265
Tefe	20	.05	.85	.10	—	.265
Mirassol	40	.05	.88	.08	—	.226

TABLE VI

GENETIC VARIATION AT THE *α-Gpdh* LOCUS OF *Drosophila willistoni*

Locality	Sample size	Alleles .94	1.00	1.06	Other	Frequency of heterozygous individuals
Jaque	48	—	1.00	—	—	.000
Teresita	48	—	1.00	—	—	.000
P. Lopez	476	.002	.99	.002	.002	.013
Betoyes	256	—	.99	.004	.004	.016
Caracas	222	.01	.95	.05	—	.105
Macapa	54	—	1.00	—	—	.000
Belem	74	.01	.99	—	—	.027
Santarem	498	.002	.996	—	.002	.008
Tefe	174	—	1.00	—	—	.000
Mirassol	1102	.004	.98	.02	.004	.046

TABLE VII

GENETIC VARIATION AT THE *Mdh-2* LOCUS OF *Drosophila willistoni*

Locality	Sample size	.86	.94	Alleles 1.00	1.06	Other	Frequency of heterozygous individuals
Jaque	50	—	—	.98	.02	—	.039
Teresita	48	—	—	1.00	—	—	.000
P. Lopez	476	.004	.004	.99	—	—	.017
Betoyes	256	—	.004	.996	—	—	.008
Caracas	220	—	.02	.93	.05	.005	.168
Macapa	54	—	—	1.00	—	—	.000
Belem	74	.03	.03	.93	.01	—	.129
Santarem	486	.004	.002	.99	—	.002	.016
Tefe	178	—	.03	.93	.02	.02	.095
Mirassol	1024	.001	.03	.95	.01	.004	.090

indicated in Figure 1. The localities in Tables II through VII are all in continental South America. They embrace an enormous territory extending from Panama (Jaque) to the Amazon delta (Belem), and from Caracas to south Brazil (Mirassol). For additional details concerning these populations see [5], [29].

The observed pattern of genetic variation in *D. willistoni* suggests the following generalizations: (1) there is considerable variation from locus to locus in the amount of polymorphism; (2) at a given locus there is great similarity from population to population as to the amount and pattern of the genetic variation; (3) nevertheless, differences between localities occur, with some allelic frequencies being characteristic of certain regions or localities. For instance, as illustrated in Table II, allele 1.03 is the most common in continental populations of *D. willistoni*, allele 1.00 is the second most common, and alleles .98, 1.05 and .96 occur at lower frequencies. These five alleles have been found in every adequately sampled population. Yet allele 1.05 reaches frequencies of 0.24 and 0.22 in Mitu and Belem, respectively, but only 0.02 in Santarem, 0.04 in P. Lopez and Macapa. Any explanation of the genetic variation must account for the observed pattern of the variation.

The ultimate source of gene variability is mutation. In a local population gene variants may also be introduced or removed by migration between neighboring populations. The theory of random walk evolution argues that most of the observed genetic variation is adaptively neutral and, therefore, not subject to natural selection. Gene frequencies are the result of random sampling which occurs every generation. Then, the effective number of neutral alleles which can be maintained in a population is

$$n = 4Nu + 1, \tag{1}$$

where N is the effective size of the population and u is the rate of mutation to neutral alleles [12]. The use of (1) is handicapped by lack of accurate estimates

FIGURE 1

Geographic origin of the samples used in this study: 6, Jaque, Panama; 14, Teresita, Colombia; 18, Mitu, Colombia; 22, southeast of Caracas, Venezuela; 32, Macapa, Brazil; 33, Belem, Brazil; 34, Santarem, Brazil; 38, Tefe, Brazil; 45, Mirassol, Brazil; 50, P. Lopez, Colombia; 52, Betoyes, Colombia; 53, Martinique; 54, St. Lucia; 55, St. Vincent; 56, Bequia; 57, Carriacou; 58, Grenada.

concerning N and u. The rate of mutation to neutral alleles is unlikely to be much higher than 10^{-6}, and is likely to be of the order of 10^{-7} per gene per generation. We do not know the effective size of a breeding population of *D. willistoni*. If it is approximately one tenth of the reciprocal of the mutation rate, the average effective number of alleles would be 1.4 per locus; and the average heterozygosity per locus would be about 29 per cent. This is approximately what is found in populations of *D. willistoni*.

There are, however, some difficulties. First, assuming, as is reasonable, that the mutation rate remains constant from population to population, differences in average population size should result, unless they are all effectively infinite, in differences in the effective number alleles. Consider two isolated patches of tropical forest, one with an effective population size of *D. willistoni* four times as large as the other. Four times as many alleles per locus should be found in one as in the other population. We find that, on the contrary, the number of alleles at a given locus remains fairly constant from population to population. A conceivable escape from this difficulty consists in arguing that independently of the total number of flies in a given locality, the population is subdivided in breeding colonies of approximately equal size. These breeding colonies should all be of a size approximately one tenth of the reciprocal of the mutation rate, or about 10^6 individuals. It seems difficult to understand why such remarkable constancy of effective population size would in fact occur.

A more serious difficulty is that, according to the *theory of random evolution*, different sets of alleles should occur in different populations. Whenever the same alleles are found in different populations, their frequencies should be uncorrelated. This prediction clearly follows from the claim that allelic frequencies are the result of random sampling. This prediction is completely negated by the observations. At every locus the same alleles are found again and again in different populations, at frequencies which are highly correlated.

4. The role of migration

The great degree of similarity between populations in the pattern of genetic polymorphisms is incompatible with the hypothesis that these polymorphisms are the result of unrelated, random processes. It has been recently suggested, however, that if a certain amount of migration occurs between neighboring populations, the species may effectively approximate a single panmictic population [13]. This may seem a possible hypothesis because it could explain, even with selective neutrality, the similarity of allelic frequencies throughout the distribution of the species. Different localities would effectively represent samples of one single interbreeding population. The similarity of allelic frequencies between populations can be accounted for if

$$Nm > 4, \tag{2}$$

where N is the effective size of the *local* population, and m is the rate of migration per generation between neighboring populations. The observed levels of heterozygosity per individual and of polymorphic loci per population (10 and 50 per cent, respectively) would obtain [13] if

$$4N_e u \approx 0.1, \tag{3}$$

where N_e is the effective size of the *species*, and u is the mutation rate as in (1).

It is difficult to estimate even approximately the effective population size of *D. willistoni*. The geographic distribution of this species extends over several million square kilometers. Throughout this enormous territory *D. willistoni* is often the most abundant drosophilid. An experienced collector typically obtains several hundred to several thousand individuals in two or three hours within a few hundred square meters. There is no indication that the removal of these individuals affects substantially the size of the population. Collections made in consecutive days in the same site yield approximately equal numbers of flies. The total number of *D. willistoni* flies living at a given time is doubtless much larger than 10^9. Taking 10^9 as a lower estimate of the effective size of the species and assuming that the mutation rate is of the order of 10^{-7} [13], we obtain $4N_e u + 1 = 401$. If the species *D. willistoni* forms effectively a single panmictic population, we should observe at each locus a large number of alleles each at very low frequencies. We find instead a small number of alleles at high and intermediate frequencies.

The theory of evolution by random walk does not claim that selective differences between genotypes are in fact zero in all places and at all times. This is, on theoretical grounds, extremely unlikely. The theory claims rather that allelic frequencies will be regulated by random processes rather than by selection whenever the differences in selective value between genotypes is sufficiently small. If N_e is the effective size of the population and s measures the selective advantage or disadvantage of the alleles, random sampling will have an important effect in loci satisfying the condition ([7], [14])

$$|N_e s| \leqq 1. \tag{4}$$

If the whole species approximates a single panmictic population, N_e will be greater than 10^9 in *D. willistoni*. Even extremely small differences in selective value, say of the order of 10^{-7} or 10^{-8}, would be the major factor determining gene frequencies.

Within the framework of population genetics theory the most serious difficulty with the maintenance of large amounts of variation by natural selection is that it imposes a "genetic load" on the population. This matter cannot be fully discussed here. In outline the difficulty is as follows. If genetic variation is maintained at a given locus by natural selection, some genotype or genotypes will be, by definition, adaptively inferior to others. The average fitness of the population will, then, be lower than a population monomorphic for the most fit genotype. If there are many polymorphic loci maintained by selection, the average fitness of the population will be substantially lower than that of the optimal genotype. For instance, assume that the average fitness of the population is reduced by one per cent at each polymorphic locus. If genes affect fitness independently of each other the mean fitness of a population segregating at 1000 loci is $(0.99)^{1000} = 4 \times 10^{-5}$. It seems impossible that a population would survive with a genetic load that reduced its fitness to 0.00004 of its potential reproductive capacity.

This difficulty has been answered in many ways. The proponents of random walk evolution believe that the objection is nevertheless valid and that, therefore, most genetic polymorphisms must be selectively neutral. As shown above if the species constitutes effectively a single breeding population, very small selective differences would be sufficient to maintain polymorphisms. The difficulty from genetic load theory consequently vanishes. For instance, if the mean fitness of the population is on the average, decreased at each polymorphic locus by 10^{-5}, the average fitness of a population polymorphic at 1000 independent loci would be $(0.99999)^{1000} = 0.99$. With 10,000 polymorphic loci the mean fitness of the population would be 0.90. A reduction of ten per cent from its optimal reproductive potential can doubtless be sustained by *D. willistoni*. A *Drosophila* female produces several hundred mature eggs throughout its lifetime. Even a human female possesses several hundred oocytes capable to develop into mature eggs. To maintain a constant population size only two need, on the average, to develop into reproducing adults.

5. Genetic variation in geographically isolated populations

The hypothesis that, with selective neutrality, migration explains the constancy of pattern of the genetic polymorphisms not only encounters serious theoretical difficulties, but is incompatible with empirical evidence. First, this hypothesis fails to account for the occurrence of local and regional differentiations. Why should two populations, such as P. Lopez and Betoyes, have very similar allelic frequencies at some loci, for example, *Est-5*, *α-Gpdh* and *Mdh-2*, but quite different at *Pgm-1?* Examples of local and regional differences in allelic frequencies have been found at essentially every locus of *D. willistoni* studied [4], [5]. Some examples are given below.

To ascertain whether migration may account for the similarity of gene frequencies we conducted a study of six oceanic islands: Martinique, St. Lucia, St. Vincent, Bequia, Carriacou, and Grenada. They belong to the Windward Group of the Lesser Antilles (see Figure 1). Although relatively small islands, they range in size from about ten square miles (Bequia and Carriacou) to four hundred (Martinique). These islands were not connected with each other or with continental South America in geological history. Their *Drosophila* populations must derive from small numbers of founders which reached them by accidental transport. Substantial differences in the chromosomal polymorphisms between the islands and the continent, and between different islands, attest to their geographic isolation [4]. Allelic frequencies at the six loci listed for the continental populations are given in Tables VIII through XIII. If the poly-

TABLE VIII

GENETIC VARIATION AT THE *Lap-5* LOCUS IN SIX ISLAND POPULATIONS OF *Drosophila willistoni*

Locality	Sample size	Alleles .96	.98	1.00	1.03	1.05	Frequency of heterozygous individuals
Martinique	264	—	.06	.53	.40	.01	.555
St. Lucia	280	—	.03	.56	.39	.01	.530
St. Vincent	258	.004	.03	.59	.37	.004	.509
Bequia	354	—	.05	.59	.35	.02	.532
Carriacou	306	.003	.05	.49	.42	.03	.573
Grenada	266	.01	.04	.49	.42	.04	.584

TABLE IX

GENETIC VARIATION AT THE *Est-7* LOCUS IN SIX ISLAND POPULATIONS OF *Drosophila willistoni*

Locality	Sample size	Alleles .96	.98	1.00	1.02	1.05	Other	Frequency of heterozygous individuals
Martinique	153	.05	.14	.63	.16	.01	.01	.552
St. Lucia	156	.01	.06	.74	.16	.03	.01	.417
St. Vincent	230	.01	.10	.70	.17	.03	—	.478
Bequia	258	.01	.09	.67	.21	.02	—	.493
Carriacou	262	.01	.10	.65	.21	.03	—	.521
Grenada	233	.02	.12	.65	.17	.03	.004	.528

TABLE X

GENETIC VARIATION AT THE *Est-5* LOCUS IN SIX ISLAND POPULATIONS OF *Drosophila willistoni*

Locality	Sample size	Alleles .95	1.00	1.05	Other	Frequency of heterozygous individuals
Martinique	264	.02	.97	.02	—	.059
St. Lucia	278	.01	.91	.08	.004	.166
St. Vincent	258	.02	.97	.01	.004	.061
Bequia	352	.01	.96	.02	—	.072
Carriacou	306	.04	.90	.06	.003	.193
Grenada	262	.07	.93	.004	—	.135

TABLE XI

GENETIC VARIATION AT THE *Pgm-1* LOCUS IN SIX ISLAND POPULATIONS OF *Drosophila willistoni*

Locality	Sample size	Alleles .96	1.00	1.04	Frequency of heterozygous individuals
Martinique	286	—	.77	.23	.359
St. Lucia	234	—	.70	.30	.423
St. Vincent	174	.02	.79	.20	.342
Bequia	384	—	.91	.09	.166
Carriacou	218	.01	.98	.01	.036
Grenada	198	.02	.97	.02	.059

TABLE XII

GENETIC VARIATION AT THE *α-Gpdh* LOCUS IN SIX ISLAND POPULATIONS OF *Drosophila willistoni*

Locality	Sample size	Alleles .94	1.00	1.06	Frequency of heterozygous individuals
Martinique	292	—	1.00	—	.000
St. Lucia	280	.004	.99	.01	.021
St. Vincent	236	—	.99	.01	.025
Bequia	294	—	1.00	—	.000
Carriacou	220	—	1.00	—	.000
Grenada	194	.01	.99	—	.010

TABLE XIII

GENETIC VARIATION AT THE *Mdh-2* LOCUS IN SIX ISLAND POPULATIONS OF *Drosophila willistoni*

Locality	Sample size	Alleles .86	.94	1.00	1.06	Other	Frequency of heterozygous individuals
Martinique	292	.003	.01	.99	—	—	.027
St. Lucia	280	—	—	1.00	—	—	.000
St. Vincent	236	—	—	1.00	—	—	.000
Bequia	294	—	—	.997	—	.003	.007
Carriacou	220	—	.005	.995	—	—	.009
Grenada	194	—	—	.99	.01	—	.010

morphisms were adaptively neutral, allelic frequencies should be uncorrelated between the islands and the continental populations, and between different islands. In fact they are very similar.

The overall similarity of allelic frequencies throughout the distribution of *D. willistoni*, including the oceanic islands, could conceivably be explained by postulating that allelic variants are selectively neutral but different alleles mutate at different rates. The frequencies of alleles would simply reflect the rates at which they arise by mutation. A most serious difficulty with this hypothesis is its *ad hoc* character. It is advanced *post facto* to account for an observed state of affairs. There is no evidence whatsoever to support the claim that alleles found in natural populations at high frequencies arise by mutation at higher rates than alleles occurring at low frequencies. Moreover, this hypothesis fails to account for the existence of local and regional differences in gene frequencies. Enough has been said above about local differentiation. Here I will point out two major instances of regional differentiation. Allele 1.03 of the *Lap-5* locus is the most common in the eleven continental populations of Table II, and in fact through most of the distribution area of *D. willistoni* [5]. In the six islands of Table VIII, allele 1.00 becomes the most common. At the *Est-7* locus, the frequency of allele 1.00 increases from about 0.52 in continental populations (see Table III and [5]) to about 0.67 in the six islands of the lesser Antilles (Table IX). Local differentiation and regional clines in gene frequencies have been observed in allozyme polymorphisms studied in other organisms [1], [23], [24].

6. Genetic variation between species

My colleagues and I have studied for several years allozyme variation in several species of the *D. willistoni* group, especially *D. willistoni* and its siblings, *D. tropicalis, D. equinoxialis,* and *D. paulistorum.* These species have largely overlapping geographic distributions. Often they can all be found in the same collection sites. The comparative study of genetic variation in these species provides important clues as to the evolutionary significance of the polymorphisms. Only a brief summary of the relevant facts can be presented here.

The first generalization concerns the amount of genetic variation. Similar degrees of overall variation are found in the sibling species. As stated above, the average proportion of polymorphic loci per individual is 17.7 per cent in *D. willistoni,* 18.6 per cent in *D. tropicalis,* and 21.3 per cent in *D. equinoxialis.* Moreover, at a given locus the amount of genetic variation is positively correlated between the species. If a locus is very polymorphic in one species, it is in most cases also highly polymorphic in the other sibling species. If a given locus is nearly monomorphic in one species, it is generally so in the other species (Table XIV). The correlation coefficients in the amount of heterozygosity per locus are: *D. willistoni-D. tropicalis,* 0.83; *D. willistoni-D. equinoxialis,* 0.60; *D. tropicalis-D. equinoxialis,* 0.70.

Let us now turn to the *pattern* of the variation. First, I shall consider loci with high levels of polymorphism. Tables XV and XVI give the allelic frequencies of *Lap-5* and *Est-7* in *D. tropicalis.* Comparison of these two tables with Tables II and III reveal that the constellation of allelic frequencies is remarkably similar in the two species. At *Lap-5,* 1.03 is the most common allele in both species,

TABLE XIV

GENETIC VARIATION AT 27 LOCI IN THREE SPECIES OF *Drosophila*

Gene	Frequency of heterozygous individuals *D. willistoni*	*D. tropicalis*	*D. equinoxialis*
Lap-5	.629	.545	.439
Est-2	.117	.137	.126
Est-3	.108	.321	.566
Est-4	.270	.321	.373
Est-5	.089	—	.090
Est-6	.285	.247	.256
Est-7	.601	.528	—
Aph-1	.136	.183	.133
Acph-1	.102	.124	.308
Acph-2	.151	—	.296
Adh	.103	.335	.232
Mdh-2	.040	.009	.010
α-Gpdh	.021	.016	.027
Idh	.086	.078	.088
G3pdh	.148	.099	.196
Odh-1	.180	.125	.293
Odh-2	.060	—	.165
Me-1	.079	.120	.025
Me-2	.235	.197	.308
To	.131	.101	.131
Tpi-2	.041	.041	.037
Pgm-1	.186	.038	.444
Adk-1	.527	.525	.474
Adk-2	.171	.137	.096
Hk-1	.099	.094	.151
Hk-2	.138	.112	.146
Hk-3	.039	.049	.122

alleles 1.00 and 1.05 occur at intermediate, and allele .98 at somewhat lower, frequencies. At *Est-7*, allele 1.00 is the most frequent in both species, .98 and 1.02 occur at intermediate frequencies, and alleles .96 and 1.05 at still lower frequencies. *D. willistoni* and *D. tropicalis* are reproductively completely isolated from each other. No gene exchange whatsoever occurs between them (see [29] and references therein). The gene pools of *D. willistoni* and *D. tropicalis* have evolved independently for probably many millions of generations. It is incredible that independent random sampling of these two gene pools would have produced after such a long process so similar distributions of allelic frequencies. We must conclude that at these two loci some directional process like natural selection is maintaining a similar pattern of genetic variation in the two species. We can generalize this finding. At every highly polymorphic locus that we have studied the pattern of allelic frequencies found in one of these sibling species is also found in some other sibling species.

We turn now to those loci which have intermediate or low levels of poly-

TABLE XV

GENETIC VARIATION AT THE *Lap-5* LOCUS OF *Drosophila tropicalis*

Locality	Sample size	Alleles .96	.98	1.00	1.03	1.05	Other	Frequency of heterozygous individuals
Jaque	560	—	.04	.23	.51	.20	.01	.640
Teresita	490	—	.02	.20	.51	.25	.02	.635
P. Lopez	128	—	.02	.34	.64	.01	—	.475
Betoyes	348	—	.01	.14	.84	.02	—	.282
Mitu	50	—	.04	.20	.56	.16	.04	.599
Caracas	96	—	—	.28	.68	.04	—	.454
Macapa	180	—	—	.09	.74	.12	.04	.402
Belem	226	—	—	.06	.74	.19	.01	.414
Santarem	70	.01	.07	.23	.43	.16	.11	.724
Tefe	24	—	.04	.25	.58	.13	—	.580
Mirassol	12	—	—	.25	.75	—	—	.375

morphism. I will include in this category those loci in which the most frequent allele has an average frequency of 0.90 or higher. About half of the 30 loci that we have studied fall in this category. These loci can be divided into two classes, which I shall call rigid and flexible loci.

Rigid loci are those with a predominant allele which is identical (as judged by the electrophoretic mobility of the allozyme for which it codes) in all the sibling species (and in all other species of the group which have been studied so far). An example of a rigid locus is the *α-Gpdh* gene. Allele 1.00 has an average frequency of about 0.99 in each of seven species of the group studied in our laboratory. The allelic frequencies of *α-Gpdh* in several localities of *D. tropicalis*

TABLE XVI

GENETIC VARIATION AT THE *Est-7* LOCUS OF *Drosophila tropicalis*

Locality	Sample size	Alleles .96	.98	1.00	1.02	1.05	Other	Frequency of heterozygous individuals
Jaque	484	.03	.13	.52	.29	.03	.004	.625
Teresita	510	.03	.13	.58	.23	.03	.002	.593
P. Lopez	120	.02	.14	.73	.08	.03	—	.436
Betoyes	236	.03	.18	.64	.15	.01	—	.535
Mitu	62	.02	.11	.68	.18	.02	—	.459
Macapa	160	.03	.09	.53	.33	.02	—	.594
Belem	162	.02	.09	.44	.43	.02	—	.609
Santarem	66	—	.02	.52	.41	.06	—	.476
Tefe	17	—	—	.76	.18	.06	—	.381

TABLE XVII

GENETIC VARIATION AT THE *α-Gpdh* LOCUS OF *Drosophila tropicalis*

Locality	Sample size	Alleles .94	1.00	1.06	Other	Frequency of heterozygous individuals
Jaque	56	—	1.00	—	—	.000
Teresita	48	—	1.00	—	—	.000
P. Lopez	142	—	1.00	—	—	.000
Betoyes	1222	.001	.998	.001	.001	.005
Caracas	66	.02	.97	.02	—	.059
Macapa	54	—	1.00	—	—	.000
Belem	222	.005	.99	.005	—	.018
Santarem	68	.01	.99	—	—	.029
Tefe	24	—	1.00	—	—	.000

and *D. equinoxialis* are given in Tables XVII and XVIII. About half of the loci with intermediate or low levels of polymorphism are rigid. Of the genes listed in Table I, *α-Gpdh, Idh, Me-2, Hk-2,* and *Hk-3* are rigid loci.

The most likely causal explanation of these rigid polymorphisms is that one and the same allele is favored by natural selection in all the species of the group. Within the physiological background of these species one of the allozymes is more effective than the others. It might be that the physicochemical properties of the enzyme are such that only one of its multiple configurations can perform all its essential functions.

Flexible loci are those in which the predominant allele is not the same in all species. An example of a flexible locus is *Mdh-2*. Tables XIX and XX give the allelic frequencies in *D. tropicalis* and *D. equinoxialis*. In *D. willistoni* (Tables VII and XIII), allele 1.00 is the most frequent in every locality. The most common allele in *D. tropicalis* is .86, and .94 in *D. equinoxialis*. Allele 1.00 is absent or has frequencies no higher than 0.01 in every population of *D. equi-*

TABLE XVIII

GENETIC VARIATION AT THE *α-Gpdh* LOCUS OF *Drosophila equinoxialis*

Locality	Sample size	Alleles .94	1.00	1.06	Other	Frequency of heterozygous individuals
Jaque	42	.02	.96	.02	—	.092
Teresita	56	—	1.00	—	—	.000
P. Lopez	362	.01	.99	.003	—	.023
Betoyes	598	.01	.98	.005	.002	.033
Caracas	44	.02	.95	—	.02	.088
Macapa	22	—	1.00	—	—	.000
Belem	46	.02	.98	—	—	.043
Tefe	502	.01	.99	.002	.002	.028

TABLE XIX

GENETIC VARIATION AT THE *Mdh-2* LOCUS OF *Drosophila tropicalis*

Locality	Sample size	Alleles .82	.86	.94	Other	Frequency of heterozygous individuals
Jaque	44	—	1.00	—	—	.000
Teresita	48	—	1.00	—	—	.000
P. Lopez	142	—	1.00	—	—	.000
Betoyes	1206	.001	.997	.003	—	.007
Caracas	66	—	1.00	—	—	.000
Macapa	54	—	1.00	—	—	.000
Belem	224	—	.996	—	.004	.009
Santarem	68	—	.99	—	.01	.029
Tefe	24	—	1.00	—	—	.000

noxialis and *D. tropicalis.* Allele .86 is absent or very rare in every population of *D. willistoni* and *D. equinoxialis.* Allele .94 is rarely found in *D. willistoni* or *D. tropicalis.* Other flexible loci in the *D. willistoni* group are *Est-5, Mdh-2, G3pdh, Odh-2, Me-1, To,* and *Tpi-2.*

What is the causal explanation of the situation encountered in these flexible loci? Clearly, it is not that only one of the allozymes has the physiological properties necessary to be fully functional. Different allozymes occur with high frequencies in different species. A conceivable explanation is that allelic variation at these loci is selectively neutral. It would then be a historical accident—the result of random sampling—that one allele has high frequency in one species but low frequency in some other species. However, if the alleles were adaptively identical in a given species their frequencies should vary from one local population to another. Yet we find that within a given species the allelic frequencies are highly correlated among the populations, including those of geographically isolated islands. (For *Mdh-2* see Tables VII, XIII, XIX, and XX; in particular,

TABLE XX

GENETIC VARIATION AT THE *Mdh-2* LOCUS OF *Drosophila equinoxialis*

Locality	Sample size	Alleles .86	.94	1.00	Frequency of heterozygous individuals
Jaque	40	—	1.00	—	.000
Teresita	24	—	1.00	—	.000
P. Lopez	362	.006	.99	.003	.019
Betoyes	598	—	.997	.003	.007
Caracas	44	.02	.98	—	.044
Macapa	24	—	1.00	—	.000
Belem	46	—	1.00	—	.000
Tefe	494	.002	.99	.01	.016

compare Table VII with Table XIII.) Considering the constancy of allelic frequencies throughout a given species, a more likely explanation is that different alleles are selectively favored in different species. This can be due to the well known phenomenon of genetic coadaptation [8]. Organisms are well integrated systems. The selective value of a given genetic variant is unlikely to be the same regardless of the remainder of the genotype. Rather, it depends on the genetic background in which it exists. Genetic coadaptation affecting an allozyme locus has been demonstrated in *D. pseudoobscura* [22]. Experiments in our laboratory (see below) have shown that allelic variants at the *Mdh-2* are not selectively neutral.

7. Evidence of natural selection

The study of allelic variation in natural populations of the *D. willistoni* group provides conclusive evidence against the theory of random walk, or non-Darwinian, evolution. The pattern of the variation within and between species is inconsistent with the hypothesis that most of the variation is adaptively neutral. Laboratory experiments described below provide direct evidence of the operation of natural selection maintaining several allozyme polymorphisms.

The rationale of these experiments is simple. For the study of each locus, two or more experimental populations are started in laboratory cages. The environmental treatment (amount of food and space, temperature, and so forth) as well as the genetic background is made identical. The only variable is the initial allelic frequency. The populations are allowed to run for a number of generations. Predictions as to the course of events are derived from the hypotheses of natural selection and of random walk. If natural selection is operative the allelic frequencies will change in a determinate fashion as to direction and rate of change. In a typical situation allelic frequencies in the various cages will converge towards equilibrium values, which are independent of the initial frequencies. The rate of change is a function of the genotypic selective values and permits the estimation of these values.

If the genetic variation is selectively neutral, changes in gene frequency will be mostly the result of accidents of sampling through the generations. These changes will be erratic as to direction. Their magnitude will be an inverse function of population size. If population size is in the thousands, changes in gene frequencies from one to another generation will be very small. If the alleles are adaptively identical but mutate at different rates, a directional process is superimposed over the random process of change. Changes in gene frequency due to differential mutation rate are, however, very slow. The rate of change in the frequency of a given allele cannot be greater than the mutation rate to that allele.

Using experimental populations, we have studied five gene loci in the *D. willistoni* group. Two loci, *Lap-5* and *Est-7*, are highly polymorphic in natural populations; one locus, *Est-5*, is moderately polymorphic; and two loci, *α-Gpdh* and *Mdh-2*, are nearly monomorphic. The *α-Gpdh* is a *rigid* locus; *Mdh-2* is a *flexible* locus. These studies are still in progress. Only preliminary and brief accounts can be given here. Yet the significance of the results is already clear.

The studies at the *Lap-5*, *Est-7*, and *Est-5* are being conducted by Mr. Jeffrey Powell, a graduate student in our laboratory. Several hundred inseminated females of *D. willistoni* collected in March 1970 in Mirassol, São Paulo, Brazil, were brought to the laboratory. Each female was placed in a separate culture giving raise to a "strain." The genetic constitution of the strain was ascertained. Flies from about 50 different strains were used to start each experimental population. By advisedly combining the number of flies of each strain introduced in a given cage, the initial allelic frequencies can be manipulated as desired. Each allele introduced in a cage was present in at least 30 of the strains used to establish that cage. The allele was thus introduced in many different genetic combinations, which must be a fair sample of the genetic combinations in which it occurs in the natural population. The results are given in Figures 2, 3, and 4, where the frequency of a given allele (ordinate) has been plotted through time (abscissa). Each line gives the allelic frequency in a separate cage. Average population size is about 5,000 individuals per cage; one full generation takes approximately 30 days. The cages are kept at 25°C. Similar experiments are being conducted at 19°C with results qualitatively similar to those given in the figures.

The results at the *Lap-5* and *Est-5* loci (Figures 2 and 3) are unambiguous. In each case the populations converge towards an equilibrium frequency. This

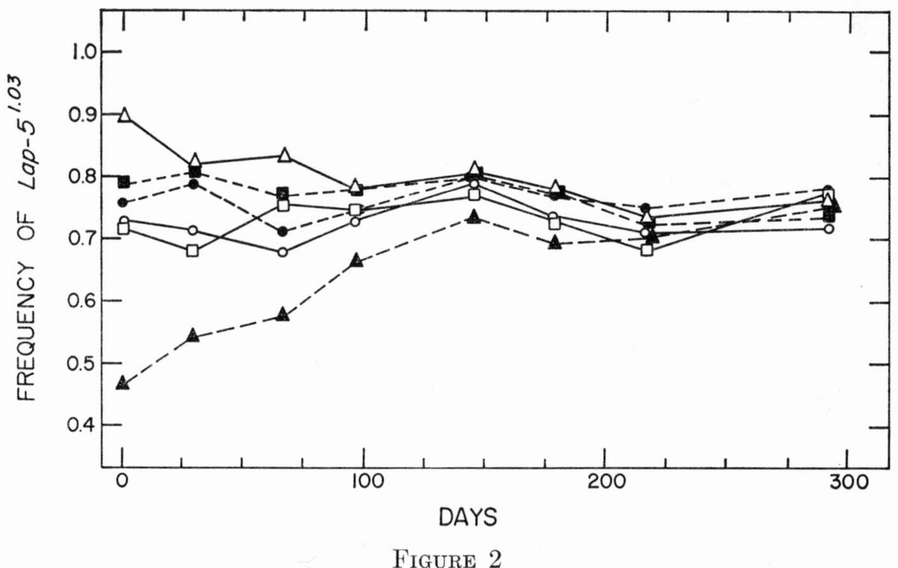

FIGURE 2

Changes in the frequency of allele 1.03 at the *Lap-5* locus in six experimental populations of *Drosophila willistoni*. The populations are kept at 25°C.

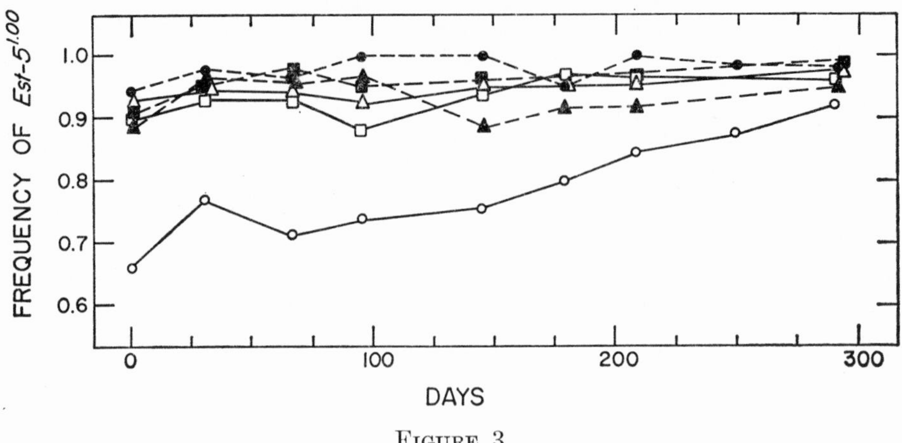

FIGURE 3

Changes in the frequency of allele 1.00 at the *Est-5* locus in six experimental populations of *Drosophila willistoni*. The populations are kept at 25°C.

is the outcome predicted by the hypothesis of natural selection. The results are incompatible with the hypothesis of adaptive neutrality. This latter hypothesis predicts that only erratic changes will occur in gene frequencies. There have not been enough generations for directional mutation to cause any appreciable change in allelic frequencies.

The results at the *Est-7* locus (Figure 4) are different. They show no large selective differences among the genotypes. Whether weak natural selection is occurring, or no selection at all is not clear. More generations are required to decide this question. The results are compatible with the hypothesis that the allelic variants in the experimental populations are selectively neutral. Experimental studies at the *Est-5* locus of *D. pseudoobscura* (which, like *Est-7* in *D. willistoni* is sex linked) have shown no evidence of natural selection either [30]. Ambiguous results have been obtained with a sex linked esterase locus in *D. melanogaster* [18]. We should recall, however, that in natural populations variation at the *Est-7* locus is not adaptively neutral.

The studies of the *α-Gpdh* and *Mdh-2* loci were made with the descendants of flies collected in the Llanos of Colombia. Two population cages with different initial frequencies were started for each species. Each population was started with about 30 different strains established as above. The results show that natural selection is operating at the *α-Gpdh* as well as at the *Mdh-2* locus. The initial frequencies of allele 1.00 at *α-Gpdh* were, in *D. equinoxialis*, 0.702 and 0.807. After three generations allele 1.00 has increased in both cages to frequencies 0.771 and 0.826, respectively. (The frequency of allele 1.00 in the natural population is 0.99.) The results in the *D. willistoni* and *D. tropicalis* cages are similar.

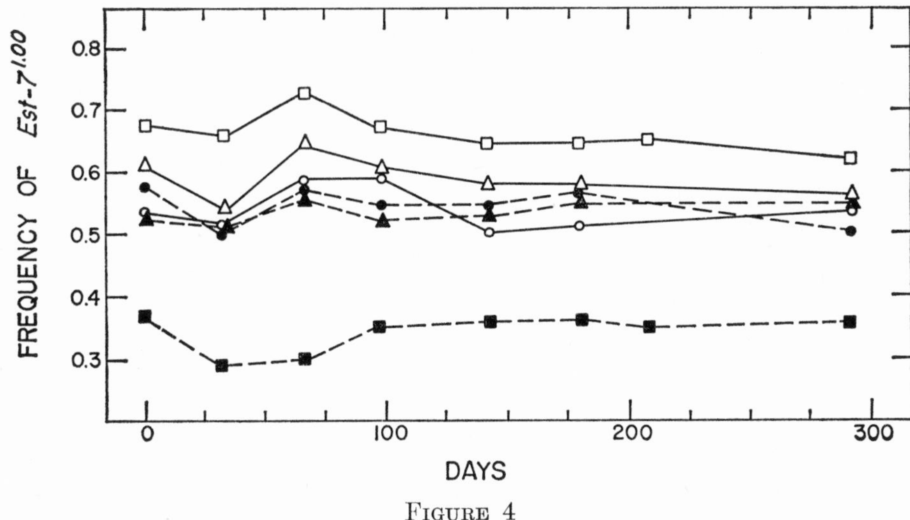

FIGURE 4

Changes in the frequency of allele 1.00 at the *Est-7* locus in six experimental populations of *Drosophila willistoni*. The populations are kept at 25°C.

Mdh-2 is a flexible locus in the *D. willistoni* group. In natural populations of *D. equinoxialis*, allele .94 has a frequency of about 0.99, while allele .86 occurs at frequencies usually lower than 0.01. In *D. tropicalis*, allele .94 occurs at frequencies lower than 0.01, while allele .86 reaches a frequency of 0.99 or higher. The results of this experiment are summarized in Table XXI. Although the environmental treatment is the same for all cages, the allelic frequencies follow opposite trends in the two species. Allele .86 decreases, while allele .94 increases in *D. equinoxialis*. In *D. tropicalis* allele .86 increases in frequency, while allele .94 decreases. There is evidence of natural selection at the *Mdh-2* in both species. Which allele is selectively favored depends on the genetic background in which it occurs. In the experimental populations of *D. equinoxialis*, as in natural

TABLE XXI

FREQUENCY OF ALLELE .86 AT THE *Mdh-2* LOCUS IN TWO EXPERIMENTAL POPULATIONS OF *Drosophila equinoxialis* AND TWO OF *D. tropicalis*

The frequency of allele .94 is one minus the frequency of allele .86.

Species	Cage	Initial frequency	Frequency after three generations
D. equinoxialis	I	0.161 ± 0.018	0.049 ± 0.010
D. equinoxialis	II	0.025 ± 0.002	0.009 ± 0.007
D. tropicalis	I	0.927 ± 0.013	0.977 ± 0.006
D. tropicalis	II	0.918 ± 0.014	0.970 ± 0.007

populations, allele .94 is at a selective advantage. In *D. tropicalis* allele .94 is selected against, while allele .86 is selectively favored.

8. Summary and conclusions

The theory of "non-Darwinian evolution" or "evolution by random walk" proposes that most genetic variation, particularly that observed at the molecular level, is adaptively neutral and therefore not subject to natural selection. The consequence follows that most of the genetic differences between species, espepecially differences responsible for amino acid substitutions in proteins, might be evolutionary "noise," with no adaptive significance. This paper summarizes a study of genetic polymorphisms in several species of the *Drosophila willistoni* group. The evidence indicates that most of the genic variation observed in these species is *not* adaptively neutral.

Drosophila willistoni, D. equinoxialis, and *D. tropicalis* are three sibling species whose geographic distributions overlap through much of the American tropics. These species possess large amounts of genetic variation. In *D. willistoni,* 56.3 ± 6.2 per cent loci are polymorphic in a given population. A locus is considered polymorphic when the frequency of the most common allele is no greater than 0.95. An individual *D. willistoni* is heterozygous at 17.7 ± 3.1 per cent of its loci. These figures are based on the study of 27 randomly selected genes coding for enzymes. Similar amounts of genetic variation exist in *D. equinoxialis* and *D. tropicalis.*

The amount of genetic variation varies considerably from locus to locus. However, at a given locus there is great similarity among different populations in the amount and in the pattern of genetic variation. Nevertheless, differences between populations occur, with some allelic frequencies being characteristic of certain regions or localities. The localities surveyed embrace an enormous territory extending from Panama to the Amazon delta and to the state of São Paulo in southern Brazil. They include six small oceanic islands. The similarity of the pattern of variation among continental populations covering such enormous territory, and between the continental and the isolated island populations is incompatible with the hypothesis that the polymorphisms are adaptively neutral. The similarity cannot be accounted for by migration between populations, since populations geographically isolated from each other have also similar genetic frequencies.

The amount of variation at a given locus is positively correlated among the sibling species. Comparisons between species as to the pattern of the variation are as follows. At very polymorphic loci similar distributions of allelic frequencies are found in different species. This result is incompatible with the hypothesis that allelic variation is selectively neutral at these highly variable loci.

Loci with intermediate and low levels of polymorphism can be classified as *rigid*—if one and the same allele is the most common one in all the species of the *willistoni* group, or *flexible*—when the most common allele varies from species to species. The evidence supports the hypothesis that allelic frequencies at loci with low levels of variation are for the most part regulated by natural selection. Which allele is selectively favored depends in the case of flexible loci on the genetic background of the species.

The predominant role of natural selection in determining the amount and kind of genetic variation is confirmed by the study of change in gene frequencies in laboratory populations.

REFERENCES

[1] R. W. Allard and A. L. Kahler, "Patterns of molecular variation in plant populations,', *Proceedings of the Sixth Berkeley Symposium on Mathematical Statistics and Probability'* Berkeley and Los Angeles, University of California Press, 1972, Vol. 5, pp. 237–254.

[2] F. J. Ayala, "Biology as an autonomous science," *Amer. Sci.*, Vol. 56 (1968), pp. 207–221.

[3] F. J. Ayala, C. A. Mourão, S. Pérez-Salas, R. Richmond, and T. Dobzhansky, "Enzyme variability in the *Drosophila willistoni* group. I. Genetic differentiation among sibling species," *Proc. Nat. Acad. Sci. U.S.A.*, Vol. 67 (1970), pp. 225–232.

[4] F. J. Ayala, J. R. Powell, and T. Dobzhansky, "Enzyme variability in the *Drosophila willistoni* group. II. Polymorphisms in continental and island populations of *Drosophila willistoni*," *Proc. Nat. Acad. Sci. U.S.A.*, Vol. 68 (1971), pp. 2480–2483.

[5] F. J. Ayala, J. R. Powell, M. L. Tracey, C. A. Mourão, and S. Pérez-Salas, "Enzyme variability in the *Drosophila willistoni* group. IV. Genic variation in natural populations of *Drosophila willistoni*," *Genetics*, Vol. 70 (1972), pp. 113–139.

[6] J. M. Burns and F. M. Johnson, "Esterase polymorphism in the butterfly *Hemiargus isola*. Stability in a variable environment," *Proc. Nat. Acad. Sci. U.S.A.*, Vol. 68 (1971), pp. 34–37.

[7] J. F. Crow, "Darwinian and non-Darwinian evolution," *Proceedings of the Sixth Berkeley Symposium on Mathematical Statistics and Probability*, Berkeley and Los Angeles, University of California Press, 1972, Vol. 5, pp. 1–22.

[8] T. Dobzhansky, *Genetics of the Evolutionary Process*, New York, Columbia University Press, 1970.

[9] J. H. Gillespie and K. Kojima, "The degree of polymorphism in enzymes involved in energy production compared to that in nonspecific enzymes in two *Drosophila ananassae* populations," *Proc. Nat. Acad. Sci. U.S.A.*, Vol. 61 (1968), pp. 582–585.

[10] H. Harris, *The Principles of Human Biochemical Genetics*, New York, Elsevier, 1970.

[11] M. Kimura, "Genetic variability maintained in a finite population due to mutational production of neutral and nearly neutral isoalleles," *Genet. Res.*, Vol. 11 (1968), pp. 247–269.

[12] M. Kimura and J. F. Crow, "The number of alleles that can be maintained in a finite population," *Genetics*, Vol. 49 (1964), pp. 725–738.

[13] M. Kimura and T. Ohta, "Protein polymorphism as a phase of molecular evolution," *Nature*, Vol. 229 (1971), pp. 467–469.

[14] ———, "Population genetics, molecular biometry and evolution," *Proceedings of the Sixth Berkeley Symposium on Mathematical Statistics and Probability*, Berkeley and Los Angeles, University of California Press, 1972, Vol. 5, pp. 43–68.

[15] J. L. King and T. H. Jukes, "Non-Darwinian evolution: Random fixation of selectively neutral mutations," *Science*, Vol. 164 (1969), pp. 788–798.

[16] K. Kojima, J. Gillespie, and Y. N. Tobari, "A profile of *Drosophila* species' enzymes assayed by electrophoresis. I. Number of alleles, heterozygosities, and linkage disequilib-

rium in glucose-metabolizing systems and some other enzymes," *Biochem. Genet.*, Vol. 4 (1970), pp. 627–637.

[17] R. C. LEWONTIN and J. L. HUBBY, "A molecular approach to the study of genic heterozygosity in natural populations. II. Amounts of variation and degree of heterozygosity in natural populations of *Drosophila psuedoobscura*," *Genetics*, Vol. 54 (1966), pp. 595–609.

[18] R. J. MACINTYRE and T. R. F. WRIGHT, "Response of esterase-6 alleles of *Drosophila melanogaster* and *Drosophila simulans* to selection in experimental populations," *Genetics*, Vol. 53 (1966), pp. 371–387.

[19] G. A. MROSS and R. F. DOOLITTLE, "Amino acid sequence studies on artiodactyl fibrinopeptides. II. Vicuna, elk, pronghorn, antelope and water buffalo," *Arch. Biochem. Biophys.*, Vol. 122 (1967), pp. 674–684.

[20] S. J. O'BRIEN and R. J. MACINTYRE, "An analysis of gene-enzyme variability in natural populations of *Drosophila melanogaster* and *D. simulans*," *Amer. Natur.*, Vol. 103 (1969), pp. 97–113.

[21] S. PRAKASH, "Genic variation in a natural population of *Drosophila persimilis*," *Proc. Nat. Acad. Sci. U.S.A.*, Vol. 62 (1969), pp. 778–784.

[22] S. PRAKASH and R. C. LEWONTIN, "A molecular approach to the study of genic heterozygosity in natural populations. III. Direct evidence of coadaptation in gene arrangements of *Drosophila*," *Proc. Nat. Acad. Sci. U.S.A.*, Vol. 59 (1967), pp. 398–405.

[23] S. PRAKASH, R. C. LEWONTIN, and J. L. HUBBY, "A molecular approach to the study of genic heterozygosity in natural populations. IV. Patterns of genic variation in central, marginal, and isolated populations of *Drosophila pseudoobscura*," *Genetics*, Vol. 61 (1969), pp. 841–858.

[24] R. C. RICHMOND, "Enzyme variability in the *Drosophila willistoni* group. III. Amounts of variability in the superspecies *D. paulistorum*," *Genetics*, Vol. 70 (1972), pp. 87–112.

[25] R. K. SELANDER, W. G. HUNT, and S. Y. YANG, "Protein polymorphism and genic heterozygosity in two European subspecies of the house mouse," *Evolution*, Vol. 23 (1969), pp. 379–390.

[26] R. K. SELANDER, M. H. SMITH, S. Y. YANG, W. E. JOHNSON, and J. B. GENTRY, "Biochemical polymorphism and systematics in the genus Peromyscus. I. Variation in the old-field mouse (*Peromyscus polyonotus*)," *Univ. Texas Publ.*, Vol. 7103 (1971), pp. 49–90.

[27] R. K. SELANDER and S. Y. YANG, "Protein polymorphism and genic heterozygosity in a wild population of the house mouse (*Mus musculus*)," *Genetics*, Vol. 63 (1969), pp. 653–667.

[28] R. K. SELANDER, S. Y. YANG, R. C. LEWONTIN, and W. E. JOHNSON, "Genetic variation in the horseshoe crab (*Limulus polyphemus*), a phylogenetic relic," *Evolution*, Vol. 24 (1970), pp. 402–414.

[29] B. SPASSKY, R. C. RICHMOND, S. PÉREZ-SALAS, O. PAVLOVSKY, C. A. MOURÃO, A. S. HUNTER, H. HOENIGSBERG, T. DOBZHANSKY, and F. J. AYALA, "Geography of the sibling species related to *Drosophila willistoni*, and of the semispecies of the *Drosophila paulistorum* complex," *Evolution*, Vol. 25 (1971), pp. 129–143.

[30] T. YAMAZAKI, "Measurement of fitness at the esterase-5 locus in *Drosophila pseudoobscura*," *Genetics*, Vol. 67 (1971), pp. 579–603.

PATTERNS OF MOLECULAR VARIATION IN PLANT POPULATIONS

R. W. ALLARD and A. L. KAHLER
UNIVERSITY OF CALIFORNIA, DAVIS

1. Introduction

It has recently been argued [8], [9] that rate of evolution at the molecular level is greater than can be accounted for by natural selection and hence that a large part of observed molecular changes must be selectively neutral. This discussion will be concerned with some experimental results, from studies of a number of different species of plants, that bear on this question. These results will be illustrated here in terms of several representative examples taken from studies of various general enzyme systems in two of the plant species under study, cultivated barley (*Hordeum vulgare*), and the Slender Wild Oat (*Avena barbata*).

The electrophoretic procedures followed are standard ones [6], [11] and consequently they need not be described here. In applying the electrophoretic techniques, we have adopted the procedure of working out the formal genetics of all bands that appear at different migrational distances for five or more enzyme systems in each species chosen for study. An example of such a full analysis of banding patterns is given in Figure 1. This figure shows, in schematic form, some banding patterns observed in a worldwide survey of esterases in cultivated barley and its wild ancestor. Bands appear in seven zones, designated A through G.

Formal genetic studies show that banding patterns for differences in migrational distances are governed by a single locus in each zone, or by seven loci in all. In the A and B zones homozygotes are single banded and heterozygotes double banded. In the C and D zones, homozygotes are double banded and heterozygotes are quadruple banded (triple banded when leading and trailing bands for two alleles are in juxtaposition). Null alleles (no band), which are recessive to alleles which produce bands, are found at the B, D, E, F and G loci. Loci A, B, and C are very tightly linked, as shown in Figure 2. Locus A is located between loci B and C, 0.0023 to the right of B and 0.0048 to the left of C. Inheritance of banding patterns for five other enzyme systems (phosphatase, leucine aminopeptidase, peroxidase, amylase, and malate dehydrogenase) which have been studied in barley also usually feature codominance.

Supported in part by grants from the National Institutes of Health (GM-10476) and from the National Science Foundation (GB-13213).

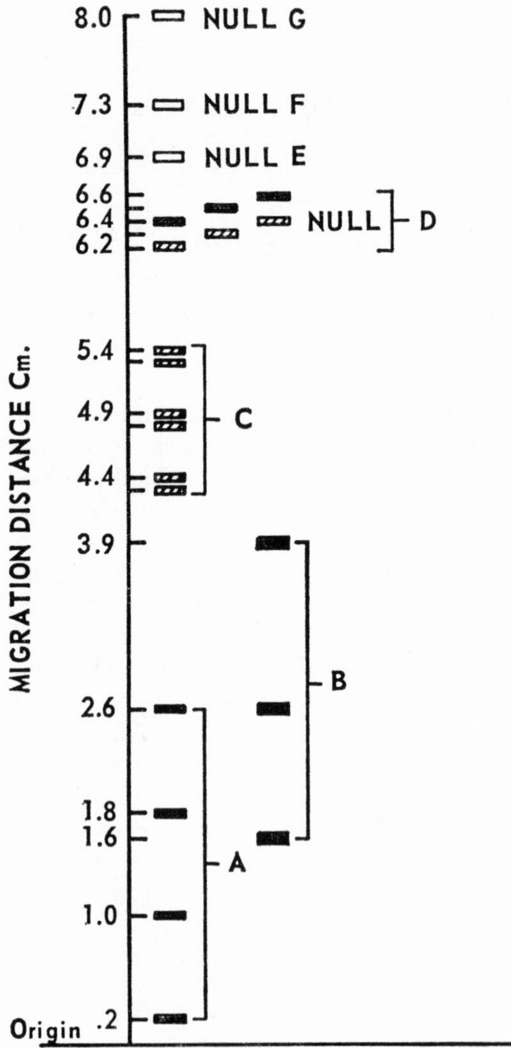

FIGURE 1

Schematic representation showing the migrational distances of some esterase electrophoretic variants observed in a worldwide survey of cultivated barley and its wild ancestor [6].

Formal genetic studies of all bands observed for these six enzyme systems establish that they are governed by 17 loci in total. This assumes that one invariant phosphatase band which has appeared uniformly in a worldwide sample of barley represents a single locus which is fixed for one allele. Since this band has appeared in all of the more than 400,000 barley plants which have been examined, either mutation rate is unusually low at this locus, or mutations which affect migrational distance are lethal, that is, selection is very strong at this locus. The remaining 16 loci have from two up to ten or more allelic forms. This particular sample of enzymes therefore indicates that about 93 per cent of loci are not only capable of mutating to allelic forms affecting migrational distance, but also that the mutant forms can and have become established in populations. In a mutation rate study of five polymorphic loci in barley, more than 68,000 individuals, representing more than 680,000 possible mutational events, have now been examined. Since no mutants have yet been found, it

B ← 0.0023 ± 0.0007 → A ← 0.0048 ± 0.0008 → C

← 0.0059 ± 0.0010 →

FIGURE 2

Linkage relationships among esterase loci *A*, *B*, and *C* in cultivated barley [6].

appears that the rate of mutation for migrationally detectable amino acid substitutions in these enzyme molecules is probably not higher than $1/10^5$ and hence that loci governing these variants are not unusually mutable.

2. Patterns of geographical variability in barley

A study of enzymatic variation in about 1500 entries in the world collection of barley maintained by the U.S. Department of Agriculture has shown that different populations from within any single local area often differ sharply in allelic frequencies. Allelic frequencies also differ from one ecological situation to another within a limited geographical area. Over longer distances clines are discernible, up to and including clines of continental proportions. Table I gives

TABLE I

RELATIVE ALLELIC FREQUENCIES (WITHIN THREE MAJOR CONTINENTAL AREAS) IN THE ESTERASE *B* LOCUS IN BARLEY [7]

Alleles	Relative allelic frequency		
	Europe	Middle South Asia	Orient
$B^{1.6}$	.06	.01	.00
$B^{2.0}$	.02	.03	.02
$B^{2.7}$	.86	.68	.37
$B^{3.9}$	.05	.10	.29
B^{N}	.01	.18	.32

an example of such a cline. This table shows the relative frequencies of five different alleles of the esterase *B* locus in Europe, Middle South Asia and the Orient. The slow migrating $B^{1.6}$ allele occurs in frequency of six per cent in

Europe but falls off in frequency to the east until it is very rare in the Orient. The $B^{2.7}$ allele, which is frequent in Europe, also falls off in frequency to the east. The $B^{3.9}$ and B null alleles, on the other hand, are rare in Europe and increase in frequency to the east. Extensive differentiation in allelic frequencies is thus the rule in barley on both macro- and microgeographical scales. It is equally the rule in various natural populations of plants we have studied, with one significant exception which will be discussed later. This extensive differentiation in allelic frequencies between different populations of plants is in accord with predictions based on the drift of neutral alleles [8]. However, it is also in accord with the proposition that the differentiation is adaptive and that it reflects the effects of selection operating in different ways in the different environments associated with different geographical areas.

3. Barley Composite Crosses II and V

3.1. *Description.* In discussing experiments which bear on the factors that are responsible for these observed patterns of variability in plants, we will focus attention on two experimental populations of barley, Composite Crosses II and V, and use them as a base line from which to make comparisons with various natural populations. Composite Cross V (CCV) was developed from intercrosses among 30 varieties of barley representing the major barley growing areas of the world [13]. In 1937 the 30 parents were crossed in pairs and during the next three years the F_1 hybrids of each cycle were again paircrossed to produce a single hybrid stock involving all 30 parents. This hybrid stock was then allowed to reproduce by its natural mating system, which in barley is one of about 99 per cent self fertilization and about 1 per cent outcrossing. The initial selfed generation, designated F_2, was grown in 1941 and the F_3 and all subsequent generations were grown from random samples of seeds taken from the harvest of the preceding generation. The plot was managed according to normal agricultural practice with no conscious selection practiced at any time. Viable seed has been maintained by keeping part of the harvest of each generation in storage and growing these reserve seeds at about ten generation intervals. The F_4 is the earliest generation for which viable seed of CCV is available at present.

Composite Cross II (CCII) is a substantially older experimental population than CCV and it also differs from CCV in parentage and method of synthesis. CCII was developed in 1929 by pooling equal amounts of F_1 hybrid seed from the 378 intercrosses among its 28 parents. The management of the two populations since their syntheses has, however, been the same. It should be noted that, in the 28 years that CCV and 42 years that CCII have been grown at Davis, California, temperature, rainfall, and many other factors of the environment have fluctuated sharply from year to year, and that they have also fluctuated in longer cycles.

3.2. *Parents.* Each of the parents of Composite Crosses II and V is an entry in the world barley collection maintained by U.S. Department of Agriculture.

The introduction of items into this collection is on the basis of a small sample of seeds. Thereafter each entry is maintained by growing a short row (from seed of on-type plants) whenever the seed supply is nearly exhausted. Consequently, entries in this collection are subject to severe founder effect at the time of their introduction into the collection, and also to recurring drastic restriction in population size (often to $N < 10$) thereafter. Furthermore, since barley is nearly completely self fertilized, this imposes an additional restriction on effective population size. Drift is therefore expected to be an extremely powerful force within each entry in the world collection. Consequently, it can be predicted that any polymorphism, original or new, involving *neutral alleles* will be transient within entries of the world collection and that fixation for one allele will occur at any polymorphic locus in a very few generations. In other words, the entries in the world collection are expected to be homogeneous and homozygous pure lines.

To determine whether this is the case, the genotype of each of the 28 entries in the world collection which were parents of CCII and 30 entries which were parents of CCV was determined by assaying 100 or more individuals within each entry. The sort of result obtained can be illustrated with the sample of data given

TABLE II

Allelic Variability Within the 28 Parents of Barley Composite Cross II [1]

Locus	Alleles	Monomorphic parents	Polymorphic parents 2 Alleles	3 Alleles	4 Alleles
A	0.2, 1.0, 1.8	6/28	15/28	7/28	
B	1.6, 2.0, 2.7, 3.0, 3.9, *N*	17/28	7/28	3/28	1/28
C	4.4, 4.9, 5.4	10/28	11/28	7/28	
D	6.2, 6.4, 6.5, 6.6, *N*	4/28	12/28	11/28	1/28

in Table II, which summarizes allelic variability for Esterase loci *A*, *B*, *C* and *D* in the parents of CCII.

Three alleles of the esterase *A* locus were represented in the 28 parents. Six of the 28 parents were monomorphic or fixed for one or the other of these three alleles. However, 15 of the parents were polymorphic with two alleles present and seven were polymorphic with three alleles present. At the *B* locus, six different alleles were represented in the parents. Seventeen of the 28 parents were monomorphic at this locus. The majority were fixed for the $B^{2.7}$ allele, which is expected because, as we saw earlier, this allele is very frequent on a worldwide basis. The 11 remaining parents were polymorphic for two, three, or four alleles at the *B* locus. The majority of the parents were also polymorphic for the *C* and *D* loci, including several which were polymorphic for three or four alleles. Considering all four loci simultaneously, only two of the 28 parents were monomorphic at all four loci, whereas three were polymorphic at one locus, five were polymorphic at two loci, nine were polymorphic at three loci and nine were polymorphic at all four loci. When more loci were included, no parent was entirely monomorphic.

Results are similar with the parents of CCV, and with a large number of other entries in the world barley collection which have been assayed electrophoretically. It is therefore apparent the entries in the world collection of barley are extensively polymorphic. In view of the high degree of self fertilization, and the long history of propagation of these entries in very small populations, this result is not consistent with adaptive neutrality. Adaptively neutral alleles are expected to become fixed very rapidly in such small populations. The extensive polymorphism observed, however, is consistent with certain types of balancing selection to be considered later.

3.3. *Changes in allelic frequencies.* CCII and CCV, in contrast to their parents, have been carried in very large populations in each generation. In populations of many thousands of individuals per generation, almost no drift will occur and *neutral* alleles are expected to remain constant in frequency, aside from changes due to mutation and migration. It can be stated with considerable confidence that in CCII and CCV neither mutation nor migration are factors of any consequence. Mutation can be eliminated on two counts: first, the mutation rate study mentioned earlier shows that mutation rates at the loci studied are too low to have much effect on the short term dynamics of the population; second, even though these loci are known from the study of worldwide variability in barley to be capable of mutating to many allelic forms, no alleles not present originally were found in any generation. Migration can be eliminated on this same basis; no alleles not present originally were found and such alleles are unlikely to have escaped detection had migration from the outside occurred into either CCII or CCV. Hence, if the molecular variants in these two populations are adaptively neutral, no change in allelic frequencies is expected in very large populations such as CCII and CCV. The observation that allelic frequencies remain constant in these populations, which have been grown in an environment which has fluctuated over generations, would therefore provide evidence in support of the proposition that the molecular variants are neutral. Conversely, the observation that gene frequencies change would provide evidence that they are affected by selection (or some yet undiscovered evolutionary factor producing effects parallel to those of selection).

Table III illustrates in terms of the esterase C locus, the sort of result that is obtained when allelic frequencies are monitored over generations in CCV. The parents contributed three alleles at this locus in CCV and the frequencies of these alleles, as inferred from their frequencies in the 30 parents, are 14.4 per cent for allele $C^{4.4}$, 24.7 per cent for allele $C^{4.9}$ and 60.9 per cent for allele $C^{5.4}$. The number of individuals assayed electrophoretically was large (from about 1000 up to nearly 4400 individuals) in each of the three early generations (4, 5, 6), four intermediate (14, 15, 16, 17), and three late (24, 25, 26) generations that were monitored. Standard errors for allelic frequencies are small (<0.01) so that changes of 0.02 in allelic frequency, or smaller, are significant. In several cases,

TABLE III

RELATIVE ALLELIC FREQUENCIES AT THE ESTERASE C LOCUS IN BARLEY COMPOSITE CROSS V [2]

The allelic frequencies for the initial generation are inferred from those of the 30 parents. Standard errors are < 0.01.

Generation	Number assayed	Allele $C^{4.4}$	$C^{4.9}$	$C^{5.4}$
Initial	4569	.144	.247	.609
4	1234	.033	.265	.702
5	1486	.049	.301	.650
6	1006	.074	.287	.639
14	1651	.034	.203	.763
15	2843	.082	.281	.637
16	2369	.050	.326	.624
17	2461	.077	.307	.616
24	4397	.102	.316	.582
25	3967	.211	.308	.481
26	3083	.279	.254	.467

for example, alleles $C^{4.4}$ and $C^{5.4}$ in transition from generation 24 to 25, changes in allelic frequency >0.10, that is, more than ten standard deviations, or larger occurred. Such changes are consistent with selection operating differentially in the drastically different environmental conditions to which this population was exposed in certain successive years. In a population of this size, they are not consistent with steps in a random walk by neutral alleles.

Table IV gives data for the same locus in CCII. Again significant changes in

TABLE IV

RELATIVE ALLELIC FREQUENCIES AT THE ESTERASE C LOCUS IN BARLEY COMPOSITE CROSS II [1]

The allelic frequencies for the initial generation are inferred from those of the 28 parents. Standard errors are < 0.01.

Generation	Number assayed	Allele $C^{4.4}$	$C^{4.9}$	$C^{5.4}$
Initial	3248	.112	.428	.460
7	1046	.193	.205	.602
8	1140	.190	.202	.608
9	948	.218	.165	.617
17	2398	.307	.275	.418
18	2094	.299	.282	.419
19	1903	.302	.249	.449
39	3472	.144	.010	.846
40	3075	.137	.015	.849
41	2868	.119	.018	.863

allelic frequencies occurred in certain single generation transitions, for example, for allele $C^{4.9}$ in transition from generation 8 to 9, and from generation 18 to 19. In addition, it is clear that longer term changes have also occurred. In the eight generation interval from the earlier generations (7, 8, 9) to the intermediate generations (17, 18, 19) alleles $C^{4.4}$ and $C^{4.9}$ both increased in frequency by about 0.10 at the expense of a loss in frequency of about 0.20 for allele $C^{5.4}$. In the generation interval between the intermediate generations (17, 18, 19) and the late generations (39, 40, 41) the trend reversed and allele $C^{5.4}$ gained 0.40 in frequency at the expense of allele $C^{4.4}$ and particularly at the expense of allele $C^{4.9}$, which was reduced to very low frequency in this population. Selection has not followed the same course for the C locus in the two populations. This is not surprising considering the different years in which the populations were grown and the different genetic backgrounds of the two populations.

Another feature of genetic change in Composite Crosses II and V is shown in Tables V and VI, which give the observed percentage of heterozygotes for three

TABLE V

Percentage of Heterozygotes for Esterase Loci A, B, and C in Barley Composite Cross II [1]

Generation	Locus A	B	C
7	5.36	1.05	6.31
8	3.67	0.78	2.28
9	3.37	1.26	2.64
17	2.21	0.33	2.71
18	1.39	0.77	2.01
19	0.96	0.21	1.95
39	1.44	0.69	1.10
40	0.98	0.56	0.94
41	0.49	0.14	0.83

representative loci in some early, intermediate and late generations. It can be deduced from the genotypes of the parents, and the sequence in which they were hybridized, that the F_1 generations of both CCII and V were highly heterozygous. In populations in which the mating system features more than 99 per cent of self fertilization, it is expected that about half of the initial heterozygosity will be lost per generation until an equilibrium level is approached in the fifth or sixth generation. The results given in Tables V and VI show that heterozygosity had been reduced to low levels in the earliest generations of CCII (generation 7) and CCV (generation 4) available for study and that no further consistent change occurred in the later generations. Thus, the pattern of change followed expected patterns, at least in a general way, in both populations.

The question which must now be asked is whether the observed changes in

TABLE VI

PERCENTAGE OF HETEROZYGOTES FOR ESTERASE LOCI *A*, *B*, AND *C* IN VARIOUS GENERATIONS OF BARLEY COMPOSITE CROSS V [2]

	Locus		
Generation	*A*	*B*	*C*
4	7.12	3.83	9.24
5	4.31	1.34	5.72
6	3.29	0.00	1.89
14	2.44	1.87	7.20
15	2.78	4.37	2.33
16	1.48	1.14	3.17
17	0.65	0.08	0.61
24	2.50	2.10	7.62
25	1.61	1.06	1.48
26	1.30	0.64	1.56

genotypic frequencies fit expectations for selectively neutral alleles. One way this question can be answered is to compute theoretical inbreeding coefficients F, which assume that the relationship between gene and genotypic frequencies is solely a function of mating system. These theoretical F values can then be compared with fixation indices $\hat{F}$, which are computed from observed genotypic frequencies, and hence measure the actual relationship between gene and genotypic frequencies, that is, the inbreeding actually realized. Computation of theoretical inbreeding coefficients requires precise estimates of the proportion of selfing *versus* outcrossing. Estimates were made by assaying electrophoretically about 18,000 progeny of plants taken from ten generations of CCV and 13,000 progeny of plants from CCII. These estimates are homogeneous over loci, generations, and years within the two populations. They also agree closely with earlier estimates of outcrossing made using morphological polymorphisms, and with general experience with barley. Hence, it seems safe to use the mean observed outcrossing rate of 0.78 and 0.57 per cent for CCII and CCV, respectively, to calculate theoretical inbreeding coefficients for the two populations.

Values of the theoretical inbreeding coefficient for CCV are given in Table VII. In populations such as CCII and CCV, which were synthesized by random crossing among diverse parents, F is expected to be zero in the original generation. In subsequent generations, with more than 99 per cent self fertilization, F is expected to follow approximately the series 0, ½, ¾, ⅞ $\cdots$ until in generation 6 or 7 it is expected to approach its equilibrium value of 0.989 (see footnote, Table VII). Table VII also gives a representative sample of fixation indices computed from the observed gene and genotypic frequencies. In virtually all cases the theoretical inbreeding coefficient is larger than the fixation index, which shows that there are consistent excesses of heterozygotes over levels expected on the basis of mating system alone. Thus, this result also does not

TABLE VII

THEORETICAL INBREEDING COEFFICIENTS F AND OBSERVED FIXATION INDICES $\hat{F}$ FOR REPRESENTATIVE GENOTYPES AND GENERATIONS IN BARLEY COMPOSITE CROSS V [2]

$$F^n = \frac{s}{1+t}[1 - (\tfrac{1}{2}s)^n],\ \hat{F} = 1 - \frac{H_{ij}}{2p_ip_j}$$

Generation	Theoretical F	Observed fixation indices				
		$A^{0.2}A^{1.8}$	$A^{1.0}A^{1.8}$	$B^{1.6}B^{2.7}$	$C^{4.4}C^{5.4}$	$C^{4.9}C^{5.4}$
4	.928	.830	.887	.779	.670	.795
5	.959	.890	.923	.830	.885	.876
14	.989	.971	.951	.903	.752	.817
15	.989	.947	.947	.978	.983	.948
25	.989	.965	.971	.970	.983	.975
26	.989	.955	.980	.983	.986	.966

conform to the hypothesis of neutral alleles. Again the simplest explanation for the excess of heterozygotes appears to be some sort of balancing selection which leads to a net advantage of heterozygotes in reproduction.

3.4. *Two locus interactions.* Before discussing expectations and results for pairs of loci considered simultaneously, it is necessary to define the two locus gametic and zygotic arrays in a population. For two loci with two alleles each (A, a and B, b) let allelic frequencies be p_1, q_1, and p_2, q_2, respectively, and let the frequencies of the four gametic types AB, Ab, aB, and ab be g_1, g_2, g_3, and g_4. There are ten possible genotypes with frequencies (f_i) as follows:

$$(1)\qquad \begin{array}{cccc} & AA & Aa & aa \\ BB & f_1 & f_4 & f_8 \\ Bb & f_2 & \begin{array}{c} f_5(AB/ab) \\ f_6(Ab/aB) \end{array} & f_9 \\ bb & f_3 & f_7 & f_{10}. \end{array}$$

Linkage equilibrium is defined as the condition in which the equilibrium frequencies of the gametic ditypes correspond to the products of the appropriate gene frequencies, that is,

$$(2)\qquad \hat{g}_1 = \hat{p}_1\hat{p}_2,\qquad \hat{g}_2 = \hat{p}_1\hat{q}_2,\qquad \hat{g}_3 = \hat{q}_1\hat{p}_2,\qquad \hat{g}_4 = \hat{q}_1\hat{q}_2.$$

In linkage equilibrium situations, gametic and zygotic frequencies thus correspond to the products of single locus gene frequencies. For nonequilibrium situations gametic frequencies are given by:

$$(3)\quad \hat{g}_1 = \hat{p}_1\hat{p}_2 + D,\qquad \hat{g}_2 = \hat{p}_1\hat{q}_2 - D,\qquad \hat{g}_3 = \hat{q}_1\hat{p}_2 - D,\qquad \hat{g}_4 = \hat{q}_1\hat{q}_2 + D,$$

where $D = \hat{g}_1\hat{g}_4 - \hat{g}_2\hat{g}_3$. Values of D range from -0.25 (all Ab and aB) to $+0.25$ (all AB and ab). When $D \neq 0$, gametic and zygotic frequencies do not correspond to the products of single locus gene frequencies.

The conditions necessary for the development and maintenance of linkage disequilibrium are the simultaneous existence of epistatic selection and certain combinations of tight linkage and/or inbreeding [5], [10]. Since selection is a requirement for linkage disequilibrium, pairs of adaptively neutral alleles originally in linkage equilibrium will not develop linkage disequilibrium ($D \neq 0$), or if $D \neq 0$ for some reason (for example, sampling effects due to the limited number of parents as in CCII and V), $D \rightarrow 0$ at a rate depending on the crossover value between the loci and/or the degree of inbreeding. This implies that, for neutral alleles, two locus zygotic frequencies should be predictable from the products of single locus frequencies, and that D should go to zero.

In illustrating the observed results in Composite Crosses II and V, it is convenient to start with the observed zygotic arrays for the esterase B and C loci in various generations of CCV, (Table VIII). There are three alleles at locus

TABLE VIII

Deviations from Products of One Locus Numbers for Esterase Loci B and C in Generations 6, 17, and 26 of Barley Composite Cross V [14]

	Locus B	Locus C $C^{4.4}C^{4.4}$	$C^{4.9}C^{4.9}$	$C^{5.4}C^{5.4}$
Generation 6	$B^{1.6}B^{1.6}$	+17	−5	−11
$N = 1006$	$B^{2.7}B^{2.7}$	−28	+1	+27
$\chi^2 = 124.9$	$B^{3.9}B^{3.9}$	+10	+4	−15
Generation 17	$B^{1.6}B^{1.6}$	+65	−37	−28
$N = 2461$	$B^{2.7}B^{2.7}$	−130	+45	+84
$\chi^2 = 1236.5$	$B^{3.9}B^{3.9}$	+65	−8	−56
Generation 26	$B^{1.6}B^{1.6}$	+387	−133	−247
$N = 3083$	$B^{2.7}B^{2.7}$	−492	+145	+354
$\chi^2 = 2187.8$	$B^{3.9}B^{3.9}$	+107	−9	−100

B and hence six possible genotypes at this locus. The same is the case for locus C. Considering both loci simultaneously, there are 36 possible genotypes, among which only the nine homozygous combinations are shown to keep the table within acceptable size. The values in this table are deviations of two locus numbers from numbers predicted from single locus frequencies. In generation 6 there is an indication that certain combinations of alleles at the two loci interact favorably with each other (for example, $B^{1.6}$ and $C^{4.4}$) in their homozygous combinations and that others interact unfavorably (for example, $B^{2.7}$ and $C^{4.4}$). By generation 17 the deviations from marginal frequencies have become very large and by generation 26 they have become larger still. This point is brought out by χ^2 values which show rapid increase over generations. Note that each of the three alleles at locus B interacts favorably in at least one of its homozygous combinations with each of the three alleles at locus C, and that each interacts unfavorably with at least one C locus allele.

Similar epistatic interactions, both favorable and unfavorable, also occur for the heterozygous combinations of alleles at the two loci, which are not shown here. This sort of epistatic interaction on the fitness scale leads to a balancing type of selection which can be shown to promote the development and maintenance of very stable polymorphisms at both loci. These data show in terms of the zygotic array, how the alleles at the two loci have gone from random associations in early generations, as expected in populations synthesized by random crossing between diverse parents, to very specific associations in later generations.

TABLE IX

LINKAGE DISEQUILIBRIUM (GAMETIC UNBALANCE) VALUES FOR THE TIGHTLY LINKED ESTERASE LOCI A, B, AND C IN BARLEY COMPOSITE CROSS V [14]

$D = g_1 \cdot g_4 - g_2 \cdot g_3$, where g_1, g_2, g_3 and g_4 are the frequencies of the gametic ditypes AB, Ab, aB, and ab, respectively.

Generation	Pairs of loci		
	A–B	A–C	B–C
5	.02	.02	.02
6	.02	.01	.03
16	.02	.05	.03
17	.02	.05	.03
25	.09	.08	.11
26	.10	.10	.11

Table IX shows the same results in terms of the gametic array for loci BC, and also for loci AB and AC. Initially, CCV was in near linkage equilibrium ($D = 0$) for all three pairwise combinations, but D gradually increased until it had become very large by generations 25 and 26. That such striking interaction systems have built up so rapidly implies that selection must have been of great intensity.

If the A, B, and C loci themselves are subject to selection, and if various alleles at these loci interact with one another in specific ways, it might be expected that the same interaction systems should develop whenever the same alleles occur together in the same population. All three alleles occur in both CCV and CCII and hence it is interesting to compare the zygotic arrays of these two populations. Deviations from expectations are very similar, in the two populations, the greatest difference being in the weaker interactions between allele $B^{3.9}$ and the C locus alleles in CCII (Tables VIII and X). The similarity of the two zygotic arrays is brought out most clearly by comparison of generation 17 in CCV and generation 18 in CCII. Note that correspondence between the populations in these comparable generations is identical in direction, and similar in magnitude, in each combination. For the zygotic array in either CCII or CCV to have progressed from linkage equilibrium to linkage disequilibrium as a result of a random walk of neutral alleles seems unlikely. For both popula-

TABLE X

DEVIATIONS FROM PRODUCTS OF ONE LOCUS NUMBERS FOR ESTERASE LOCI B AND C IN GENERATIONS 8, 18, AND 40 OF BARLEY COMPOSITE CROSS II [1]

	Locus B	Locus C $C^{4.4}C^{4.4}$	$C^{4.9}C^{4.9}$	$C^{5.4}C^{5.4}$
Generation 8	$B^{1.6}B^{1.6}$	+55	−13	−41
N = 1140	$B^{2.7}B^{2.7}$	−116	+25	+90
χ^2 = 413.0	$B^{3.9}B^{3.9}$	+8	0	−9
Generation 18	$B^{1.6}B^{1.6}$	+96	−37	−56
N = 2095	$B^{2.7}B^{2.7}$	−170	+66	+107
χ^2 = 452.2	$B^{3.9}B^{3.9}$	+19	−5	−14
Generation 40	$B^{1.6}B^{1.6}$	+153	−2	−149
N = 3075	$B^{2.7}B^{2.7}$	−214	+3	+219
χ^2 = 1268.6	$B^{3.9}B^{3.9}$	0	0	0

TABLE XI

DEVIATIONS FROM PRODUCTS OF ONE LOCUS NUMBERS FOR THE ESTERASE LOCI B AND D (UNLINKED) IN GENERATIONS 6, 17, AND 26 OF BARLEY COMPOSITE CROSS V [14]

	Locus B	Locus D $D^{6.4}D^{6.4}$	$D^{6.5}D^{6.5}$	$D^{6.6}D^{6.6}$	$D^{N}D^{N}$
Generation 6	$B^{1.6}B^{1.6}$	−5	−1	+6	+1
N = 1006	$B^{2.7}B^{2.7}$	+4	+2	−5	−1
χ^2 = 5.1 (NS)	$B^{3.9}B^{3.9}$	+1	−1	−1	−1
Generation 17	$B^{1.6}B^{1.6}$	+19	+4	−21	−2
N = 2461	$B^{2.7}B^{2.7}$	−46	−6	+17	+19
χ^2 = 49.5	$B^{3.9}B^{3.9}$	+8	+2	+4	−17
Generation 26	$B^{1.6}B^{1.6}$	+139	−27	−77	−29
N = 3083	$B^{2.7}B^{2.7}$	−166	+29	+62	+42
χ^2 = 240.8	$B^{3.9}B^{3.9}$	−2	−3	+10	−14

tions to go from random association to the same state of organization without the guiding force of selection seems even more unlikely.

Table XI gives, in terms of esterase loci B and D in CCV, an example of the build up of a specific interaction system between two unlinked loci. In generation 6 departures from expectations based on marginal frequencies are small and not significant, that is, the alleles at the two loci occur together at random. By generation 17 there is indication that certain alleles interact favorably in their homozygous combinations with each other (for example, $B^{1.6}$ and $D^{6.4}$) and others interact unfavorably (especially $B^{2.7}$ and $D^{6.4}$). Results in generation 26 confirm the reality of the earlier trends. Table XII gives D values for these pairwise comparisons of unlinked loci. Again the build up of D shows the change from random association to an organized state. Table XIII compares the zygotic

TABLE XII

LINKAGE DISEQUILIBRIUM (GAMETIC UNBALANCE) VALUES FOR ESTERASE LOCUS D WITH ESTERASE LOCI A, B, AND C IN BARLEY COMPOSITE CROSS V [14]

$D = g_1 \cdot g_4 - g_2 \cdot g_3$, where g_1, g_2, g_3 and g_4 are the frequencies of the gametic ditypes AB, Ab, aB, and ab, respectively.

Generation	Pairs of loci		
	A–D	B–D	C–D
5	.00	.00	.02
6	.01	.00	.01
16	.01	.02	.01
17	.01	.01	.02
25	.02	.06	.04
26	.02	.05	.03

TABLE XIII

DEVIATIONS FROM ONE LOCUS NUMBERS IN COMPARABLE GENERATIONS OF BARLEY COMPOSITE CROSSES II AND V [1]

	B Locus	Locus D			
		$D^{6.4}D^{6.4}$	$D^{6.5}D^{6.5}$	$D^{6.6}D^{6.6}$	$D^N D^N$
Composite Cross II	$B^{1.6}B^{1.6}$	+41	−3	−9	−30
Generation 18	$B^{2.7}B^{2.7}$	−61	−5	+14	+50
	$B^{3.9}B^{3.9}$	−3	+7	−1	−2
Composite Cross V	$B^{1.6}B^{1.6}$	+19	+4	−21	−2
Generation 17	$B^{2.7}B^{2.7}$	−46	−6	+17	+19
	$B^{3.9}B^{3.9}$	+8	+2	+4	−17

arrays for the same loci and alleles in comparable generations of CCII and CCV. Again the departures from random associations are the same in direction and magnitude in the two populations.

On the basis of results such as these we conclude that genes do not exist in populations in random backgrounds. On the contrary, the normal situation is probably existence in correlated blocks within chromosomes (as with loci A, B and C) and also between unlinked loci (as with loci AD, BD, and CD). Further, natural selection operates not only on single loci but also on the correlated state. Apparently the gametic and zygotic arrays, and evolutionary changes in these arrays, cannot be described adequately in terms of gene frequencies at single loci. Descriptions of the multilocus gametic and zygotic arrays apparently have to be in terms of larger units, such as linkage blocks, whole chromosomes, or even the entire population genotype, if they are to be consonant with the observations. In other words, before we can allow conclusions about rates of evolution based on single locus substitutions to rule out selection, we have to know much more than we do at present about interactions between loci at the level of the fitness scale.

4. Geographical variation in *Avena barbata*

Let us now turn to some observations on natural populations that are relevant to the random theory. Table XIV gives gene and genotypic frequencies for a

TABLE XIV

GENE AND GENOTYPIC FREQUENCIES IN A HILLSIDE POPULATION OF *Avena barbata* (SITE CSA) [4]

Locus	Genotype	Location 1	2	3	4	5
E_4	11	.129	.740	.734	.631	1.000
	12	.113	.016	.109	.062	.000
	22	.758	.194	.156	.308	.000
	q_2	.814	.202	.211	.338	.000
E_9	11	.564	.145	.156	.354	.000
	12	.097	.032	.078	.062	.000
	22	.339	.823	.766	.585	1.000
	q_2	.387	.839	.805	.615	1.000
APX_5	11	.190	.806	.859	.600	1.000
	12	.127	.032	.094	.062	.000
	22	.682	.161	.047	.338	.000
	q_2	.746	.177	.094	.369	.000

sample of three typical loci in a population of the Slender Wild Oat, *Avena barbata,* which occupies a site (CSA) about 200 feet wide and 400 feet long, extending up a hillside in the Coast Range near Calistoga, California. This site is mesic at the bottom of the hillside and it becomes progressively more arid up the hillside. Location 4 within the site departs from this ecological gradient in that it represents a flat area of deeper soil which is ecologically more like locations 1 and 2 at the bottom of the hillside than arid location 5 at the extreme top of the hillside.

Considering allelic frequencies at the esterase 4 and APX_5 loci, it can be seen that the faster migrating allele (allele 2) is in high frequency in the mesic location at the bottom of the hillside. Progressing up the hillside, the frequency of these alleles falls off in locations 2 and 3, increases again in location 4, and falls off to zero in location 5. The pattern for the esterase 9 locus differs only in that the slower migrating allele is in low frequency in the mesic locations and the faster migrating allele (allele 2) is in high frequency in the arid locations, becoming fixed in location 5. This progressive change in allelic frequencies, and in polymorphism, for these and for other loci, on this hillside in fact reflects the geographical variation which occurs throughout California [3].

In the mixed but generally mesic habitats of the Coast Range, from about Monterey northward, most populations of *A. barbata* are polymorphic with allelic frequencies falling generally within the range of those of locations 1 through 4 of site CSA. These populations are also polymorphic for many morphological characters and the extent of polymorphism for molecular and morphological traits is highly correlated [3], [12]. In the arid habitats east of the crest of the Coast Range, and in the foothills of the Sierra Nevada Mountains south to San Diego, all populations that have been analyzed are fixed for the genotype which is found in arid location 5 of the CSA site. The observation that only a single genotype occurs in the numerous isolated populations found over this very large geographical area is particularly difficult to explain by the random theory because the random theory predicts heterozygosity within populations, and it also predicts that different alleles will be present at different frequencies in different locations. In other words, it predicts the opposite of the observations.

There are also two aspects of the variability in the Coast Range that are difficult to explain by the random drift of neutral alleles. First, there is the observation that sharp geographical divergence, correlated with the details of the habitat, are maintained over very short distances (such as within site CSA), even though some pollen and some seed migration occurs between such locations. Even a little migration would homogenize allelic frequencies among populations if the molecular variants were neutral. Second, is the observation that the level of heterozygosity at all loci examined in all populations is higher than can be explained on the basis of mating system alone. An example is given in Table XV

TABLE XV

OUTCROSSING VALUES t, INBREEDING COEFFICIENTS F AND FIXATION INDICES $\hat{F}$ WITHIN SITE CSA [4]

Item	Location 1	2	3	4
Outcrossing Rate t	.027	.003	.013	.023
Inbreeding Coefficient F	.947	.994	.974	.955
	$\hat{F}$	$\hat{F}$	$\hat{F}$	$\hat{F}$
E_4	.697	.950	.673	.861
E_9	.796	.882	.752	.869
APX_5	.665	.890	.448	.867

which shows theoretical inbreeding coefficients and fixation indices for the four polymorphic locations within the CSA site. The fixation indices are substantially lower than theoretical inbreeding coefficients in all four locations. Table XVI expresses these excesses in terms of the selection that is necessary to maintain

TABLE XVI

SELECTIVE VALUES OF HOMOZYGOTES (ALLELE $ii = x$, ALLELE $jj = y$) RELATIVE TO HETEROZYGOTES TAKEN AS UNITY [4]

Locus	Location							
	CSA-1		CSA-2		CSA-3		CSA-4	
	x	y	x	y	x	y	x	y
E^4	.469	.541	.557	.549	.516	.459	.643	.632
E^9	.599	.587	.498	.519	.489	.533	.644	.652
APX_5	.495	.545	.521	.502	.503	.346	.649	.640

them. It can be seen that homozygotes have only about half the reproductive capacity of heterozygotes. The next step in this study will be to examine large enough samples within single populations to determine whether this excess of heterozygotes is due to epistatic interactions between alleles at different loci, similar to the situation in the experimental populations of barley.

5. Summary and conclusions

This discussion of enzyme variants in plant populations can be summed up in five main points.

First, there is extensive allelic variability within entries of the world collection of barley maintained by the U.S. Department of Agriculture. This variability appears to be much in excess of amounts that can be explained in such small populations on the basis of the drift of neutral alleles. However, it is consistent with certain types of rather strong balancing selection.

Second, the changes in allelic frequencies which occur from generation to generation in Composite Crosses II and V are much too large to be explained by genetic drift. The data also show that these changes cannot be due to mutation or migration, but that they are consistent with selection operating in different ways in the different environmental conditions of different years, or groups of years.

Third, comparisons of theoretical inbreeding coefficients and fixation indices show an excess of heterozygotes in Composite Crosses II and V, and in natural populations of *A. barbata*. This result is inconsistent with neutral alleles, but it is consistent with balancing selection.

Fourth, in early generations of Composite Crosses II and V, frequencies of two locus genotypes are generally in agreement with predictions based on single locus frequencies, that is, combinations of alleles at different loci are at random. However, within a few generations nonrandom associations develop which can be identified with favorable and unfavorable epistatic interactions between specific alleles at different loci. The same associations develop in the two populations. This result is consistent with very strong selection. It is not consistent with random walks of neutral alleles.

Fifth, the existence of a single genotype over a large part of the range of *A. barbata* in California is a most difficult observation to explain by the random theory, which predicts that different alleles should drift to fixation in different places.

We have studied all alleles which produce bands at migrationally different distances in our materials. None of the alleles appear to be neutral or physiologically irrelevant. Instead, all the alleles in our sample appear to be adaptive, but their effect on fitness seems to be expressed through apparently complex interactions with alleles at other loci. Until such interactions are better understood, we conclude that calculations based on single locus substitutions should be regarded cautiously, as should generalizations concerning evolutionary rates made from such calculations.

REFERENCES

[1] R. W. ALLARD, A. L. KAHLER, and M. T. CLEGG, "The dynamics of multilocus esterase polymorphisms in an experimental plant population," in preparation.

[2] R. W. ALLARD, A. L. KAHLER, and B. S. WEIR, "The effect of selection on esterase allozymes in a barley population," *Genetics*, in press.

[3] M. T. CLEGG, "Patterns of genetic differentiation in natural populations of wild oats," Ph.D. dissertation, University of California, Davis.

[4] J. L. HAMRICK and R. W. ALLARD, "Microgeographical variation in allozyme frequencies in *Avena barbata*," in preparation.

[5] S. K. JAIN and R. W. ALLARD, "The effects of linkage, epistasis and inbreeding on population changes under selection," *Genetics*, Vol. 53 (1966), pp. 633–659.

[6] A. L. KAHLER and R. W. ALLARD, "Genetics of isozyme variants in barley. I. Esterases," *Crop Science*, Vol. 10 (1970), pp. 444–448.

[7] ———, "Worldwide patterns of differentiation of esterase allozymes in barley (*Hordeum vulgare* and *H. spontaneum*)," in preparation.

[8] M. KIMURA, "Evolutionary rate at the molecular level," *Nature*, Vol. 217 (1968), pp. 624–626.

[9] J. L. KING and T. H. JUKES, "Non-Darwinian evolution," *Science*, Vol. 164 (1969), pp. 788–798.

[10] R. C. LEWONTIN and K. KOJIMA, "The evolutionary dynamics of complex polymorphisms," *Evolution*, Vol. 14 (1960), pp. 458–472.

[11] D. R. MARSHALL and R. W. ALLARD, "The genetics of electrophoretic variants in Avena," *J. Heredity*, Vol. 60 (1969), pp. 17–19.

[12] ———, "Isozyme polymorphisms in natural populations of *Avena fatua* and *A. barbata*," *Heredity*, Vol. 25 (1970), pp. 373–382.

[13] C. A. SUNESON, "An evolutionary plant breeding method," *Agronomy J.*, Vol. 48 (1956), pp. 188–190.

[14] B. S. WEIR, R. W. ALLARD, and A. L. KAHLER, "Analysis of complex allozyme polymorphisms in a barley population," *Genetics*, in press.

VARIATION IN FITNESS AND MOLECULAR EVOLUTION

W. F. BODMER
UNIVERSITY OF OXFORD
and
L. L. CAVALLI-SFORZA
STANFORD UNIVERSITY

1. Introduction

Molecular studies, especially of proteins and nucleic acids have added important new insights into evolutionary processes by providing new ways of investigating and measuring evolutionary rates over long periods of time. In particular, the estimation of the mean time necessary for an amino acid substitution (Zuckerkandl and Pauling [30]) has rightly generated much interest and has given considerable stimulus to further investigation into the mechanisms of evolution.

There seems, at the present time, to be substantial disagreement as to the meaning of the quantities observed and their interpretation in evolutionary terms (see, for example, Kimura and Ohta [16] who give citations to the relevant literature). Specifically, the analysis of data on molecular evolution has led to a revival of the old controversy concerning the relative roles in evolution of random genetic drift and selection.

In this paper, we shall extend some considerations that were made in a book that appeared recently. We shall also review some experiments on computer simulation of molecular evolution that were done some two years ago, and also review the molecular evidence from a variety of sources and organisms concerning the roles of random genetic drift and selection in evolution. The model of molecular evolution which we have used for computer simulation was designed to evaluate mean evolutionary time, both for neutral mutations and also for mutations which have an effect on fitness. It also provides an estimate of the extent of polymorphism for a given locus at any given time.

2. The computer model

Since the number of possible changes in a protein molecule is very large, we have used, as have others, a model in which every allele of a gene that can be

These experiments were supported in part by grant number GM10452-09 of the NIH and grant number GB 7785 of the NSF and AT(04-3)326PA33 of the USAEC.

produced by mutation is a new one, so that in practice there is an infinite number of alleles. This is very close to what is observed in molecular evolution, since with a protein of 100 amino acids and the possibility of twenty amino acids at each site, there are 20^{100} possible types, plus all other changes which do not involve a simple amino acid substitution. Many of these will, of course, be nonviable, but the number which are viable may still be very large.

A haploid population is used for our model, as is usually the case for genetic drift theories. The extension to diploids is easy as long as fitness is considered to be additive with respect to genotypes. The population is kept at a constant size N and mutation is allowed to occur with a constant rate μ per generation. Every new mutant is different. When fitness is allowed to vary, the mutant will have a fitness which may be different from that of the allele in which the mutant arises. The fitness of the mutant is assigned according to a chosen distribution of fitness values.

In our experiments, the fitness distribution was taken to be normal with arbitrary standard deviation σ_w and with a mean equal to that of the allele in which the mutation took place, plus a constant quantity Δw, which is zero if the average fitness of mutants is equal to that of the parental type. Checks were imposed to avoid negative fitness values. In such a system, one can, therefore, produce advantageous deleterious, neutral, or quasi neutral mutations in the desired proportions. All individuals present in the population were allowed to reproduce according to a Poisson distribution. The next generation was thus formed by giving to each type represented in the former generation an expectation of progeny equal to the number of individuals of that type times its fitness, and letting a Poisson variate represent its number of progeny. When the expectation computed in this way was above 20, then the computation of the number of descendants was simplified by replacing the Poisson distribution by a normal distribution having mean and variance equal to the expected number of progeny of that type. Under these conditions, the total number of individuals in the next generation also varies approximately according to a Poisson distribution with expectation N. In order to keep N constant, the realized population size was adjusted to its constant value by adding or eliminating individuals of the various types at random, that is, taking into account only the proportions of the various types. The constancy of N is a requirement which nearly always creates difficulties when setting up mathematical models. The program was adapted for the exact treatment (with a multinomial distribution) by Harry Guess and found to give undistinguishable results from those obtained with the above procedure. For N not very small the multinomial simulation requires more computer time than the Poisson approximation. In fact, it makes the computer time proportional to N (times the number of generations) while with the Poisson approximation the computer time is proportional to the number of alleles present, which is a function of the product $N\mu$ (times the number of generations).

We are grateful to Harry Guess who pointed out an error in the computer program used in the simulation.

3. Some results of the model

Table I shows an example of a simulation with $N = 30{,}000$ and $\mu = 10^{-5}$. The fitness w of the original type at the beginning of the experiment was 1. The variation in fitness had a standard deviation σ_w of 0.01 and the average decrease in viability of new mutants Δw was $= 0.01$. Newly produced mutants were thus mostly deleterious, having on average a fitness which was one standard deviation below the fitness of the type in which they were produced. But because of the normal distribution of fitness values, about 15 per cent of new mutants had fitnesses which were higher than that of the parental type. The table shows the composition of the population at various times. Each mutant is identified by its fitness as well as its birth date, which is the generation in which it arose. Each mutant is also associated with a count of the *number of mutational transitions* which it has undergone since the beginning of the experiment. Thus if, for example, a new mutation arises in an allele produced by a mutation from the allele which was present in all individuals in the original population, this has undergone two mutational transitions, and so on. This quantity, the number of

TABLE I

AN EXPERIMENT OF EVOLUTION BY COMPUTER SIMULATION

$N = 20{,}000$, $\mu = 10^{-5}$

Each column refers to one of the mutant alleles present in the population at the time given. There are as many columns as alleles.

Mutant born at generation 1,320 was fixed by generation 3,750.

Generation 1,000							
No. individuals	19,833	5	162				
Fitness	1	0.9962	0.9987				
Birth date	1	610	809				
No. mutational transitions	0	1	1				
Generation 2,000							
No. individuals	16,794	3,206					
Fitness	1	1.0036					
Birth date	1	1,320					
No. mutational transitions	0	1					
Generation 10,000							
No. individuals	10,521	8,151	1,024	274	23	5	2
Fitness	1.0244	1.0221	1.0176	1.0147	1.0147	1.0140	1.0412
Birth date	8,859	7,887	9,783	9,965	9,965	9,982	9,997
No. mutational transitions	4	3	4	4	5	4	5

mutational transitions, had to be introduced in order to deal with the statistics of evolutionary rates. The original purpose of the simulations was to estimate the time taken to fix new mutations. It soon became evident, however, that unless the mutation rate was much lower than the reciprocal of the population size, no mutant, or at least very few mutants, ever really became fixed.

The general consequences of this model, which seem quite close to reality, were rather that there are usually several alleles present in a population which may have undergone different numbers of mutational transitions from the original allele, which was assumed to have a frequency of 1 at time zero. The mean number of mutational transitions for the alleles present in a population can be calculated at each time point. The time taken for this mean number to increase by 1 is the reciprocal of the rate of gene substitution. The mean evolutionary time estimated from amino acid substitutions should correspond to this number. In fact, the number of amino acid differences between two proteins is, assuming an almost infinite number of alleles, proportional over a wide range to the number of mutational transitions. The proportionality constant is somewhat less than one because of reverse mutation (a rare event), the complications arising from the degeneracy of the genetic code, and other sources.

Part of the experiment shown in Table I is illustrated in Figure 1. Here the mutation rate is less than $1/N$ and the *effective* mutation rate, that is, the rate of production of mutants that have a fitness above neutrality, is very low, being about 1/30 of $1/N$. In this example, a few mutants do get fixed. Two were actually fixed during the first 10,000 generations (see Figure 1). A third mutant was not fixed because at the time its frequency was approaching 100 per cent, it was supplanted by a new mutant with a higher fitness that had meanwhile developed from it.

That very few mutants ever get fixed, is more clearly illustrated in Figure 2, which gives an experiment with the same values of N and μ as before, but with only neutral mutations. Because there are no fitness differences, there are considerable short term fluctuations in the frequencies. Only one of the many mutants indicated in the figure became fixed. The frequency of this particular mutant, which underwent two mutational transitions, is also indicated in its descent phase to emphasize its long persistence in the population. It should also be noticed that around generation 7,500, for instance, mutants that differ by more than one mutational substitution may be present with appreciable frequencies at the same time, in one population.

This suggests that the variance of the number of mutational transitions undergone by mutants present in a given population at the given time may be an indication of the evolutionary forces at work. Our simulation experiments are still inconclusive on this point, but it may be worth remembering that Prager and Wilson [23] reported the coexistence in a population of two alleles differing by at least six mutational transitions.

Table II gives data from another experiment in which the variation of fitness was so small that most mutations can be thought of as almost neutral ("quasi

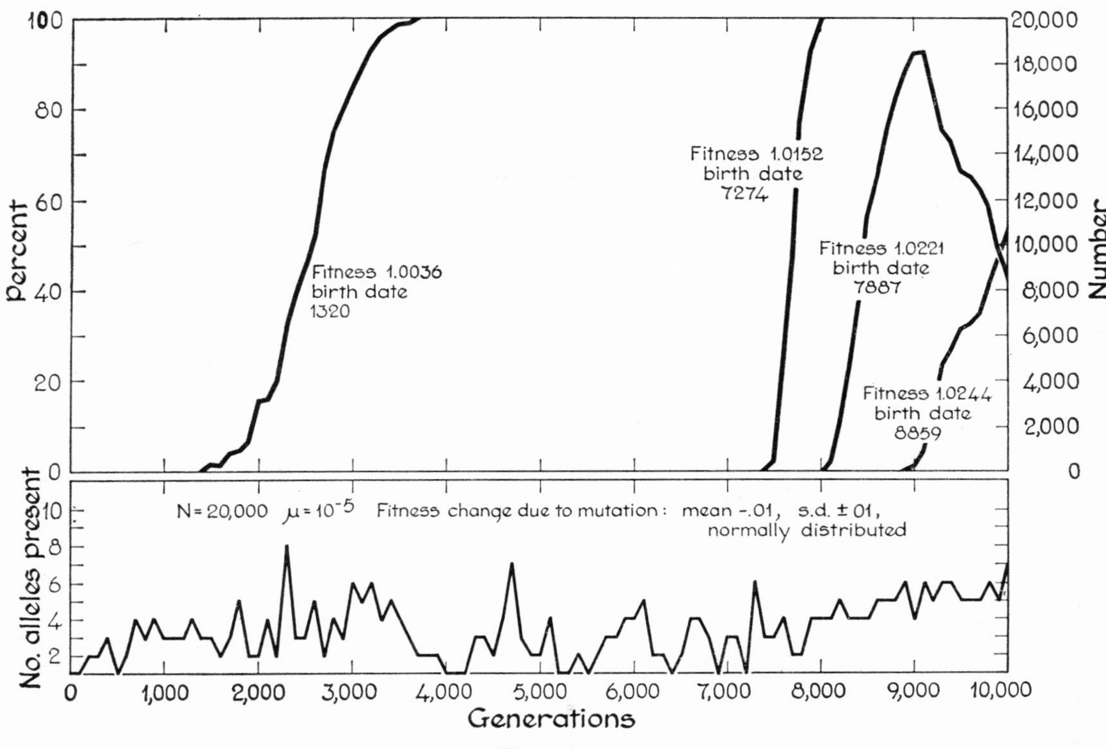

FIGURE 1

The experiment of Table I plotted.

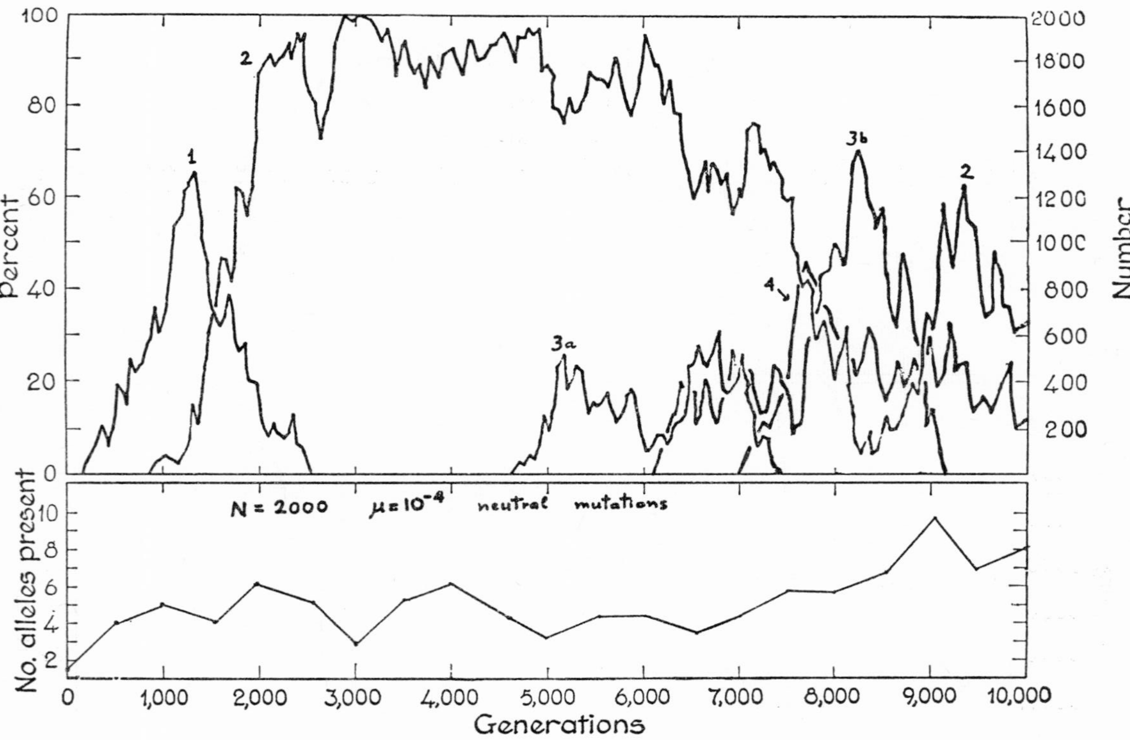

FIGURE 2

An experiment with neutral mutations.

The numbers 1, 2, 3a, 3b, and 4 indicate mutations, sampled because they reached substantial frequencies whose histories are traced throughout all generations. One of them (2) was fixed for a short time and was the only one to reach fixation.

TABLE II

COMPUTER SIMULATION OF EVOLUTION WITH QUASI NEUTRAL MUTATIONS

$N = 2{,}000$, $\mu = 10^{-4}$

Each column refers to one of the mutant alleles present in the population at the time given. There are as many columns as alleles.

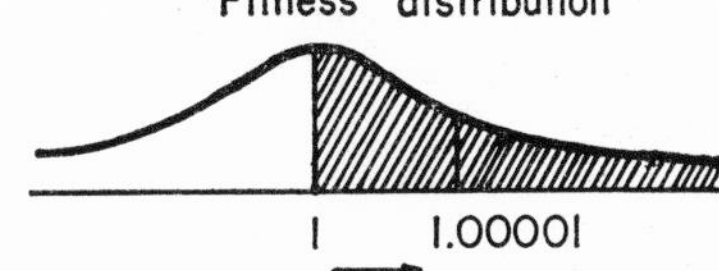

Generation 10,000				
No. individuals	1,959	38	3	
Fitness (minus 1, % s.d.)	0.0	2.9	1.14	
Birth date	5,617	9,694	9,949	
No. mutational transitions	1	2	2	
Generation 20,000				
No. individuals	1,108	724	134	4
Fitness	0.0	0.67	1.72	1.14
Birth date	5,617	18,360	19,764	19,990
No. mutational transitions	1	2	2	3
Generation 30,000				
No. individuals	1,908	72	16	4
Fitness	2.19	2.0	3.24	2.38
Birth date	29,866	27,844	29,991	29,998
No. mutational transitions	5	6	6	6
Generation 40,000				
No. individuals	1,676	241	42	41
Fitness	2.19	2.67	1.43	2.48
Birth date	36,469	38,609	39,021	38,249
No. mutational transitions	8	9	9	9

neutral" following Kimura's definition). In such experiments, the mean evolutionary time is close to that expected for neutral mutations, but a small increase in fitness is observed and "positive" mutants are eventually preferred; thus, even if the advantages are very small, they cannot be neglected. From observations obtained from a number of similar experiments, it appears that the mean fitness increases by an amount that tends to be smaller than that expected, the smaller the variation in fitness is with respect to $1/N$. In other words, an increase in the relative importance of drift decreases the expectation of the rate of increase in fitness. One might thus visualize a possible generalization of Fisher's fundamental theorem of natural selection which included terms that represent a reduction in the expected rate of increase of fitness due to drift.

Some data on the mean observed number of substitutions and other quantities of interest obtained in various experiments are given in Table III. The mean substitution time was computed by dividing the number of generations the experiment was run by the mean observed number of mutational transitions (NMT). The first 1,000 generations were not included to avoid possible effects of initial conditions. All populations are started at time zero with only one type. Standard errors of NMT and other quantities are computed on the basis of the

TABLE III

RESULTS OF SOME COMPUTER EXPERIMENTS SIMULATING MOLECULAR EVOLUTION
The number of mutational transitions (NMT) is given per 1,000 generations, and its expectation for neutral changes ($\sigma_w = 0$) is $1{,}000\mu$. The mean substitution time is 1,000/NMT.

Population size N (haploid)	Mutation rate μ	Variation of fitness	NMT ($\times$ 1,000 gen.) obs.	NMT exp.	Mean substitution time (generations) obs.	Mean substitution time exp.	Mean F obs.	Mean F exp.	Average no. of alleles
100	0.01	$\sigma_w = 0$ (neutral)	10.01 ± 1.04	10	99.9	100	.368	.333	6.7
100	0.003	$\sigma_w = 0$	$3.36 \pm .29$	3	297.6	333	.656	.769	3.2
100	0.001	$\sigma_w = 0$	$0.85 \pm .16$	1	1,176.5	1,000	.879	.833	1.55
500	0.01	$\sigma_w = 0$	$11.44 \pm .86$	10	87.4	100	.102	.091	33.67
100	0.01	$\sigma_w = 0.05$	13.38 ± 1.77		74.7		.389		7.1
100	0.01	$\sigma_w = 0.02$	14.10 ± 1.29		70.9		.366		7.4

variation of estimates of NMT obtained every 1,000 generations (from 9 to 22 such observations for each mean). In general, the number of mutational transitions is found to be equal to expectation; that is, equal to $1/\mu$ and independent of N for neutral mutations (Kimura [14], Cavalli-Sforza and Bodmer [6]). It is higher when selection is involved ($\sigma_w > 0$, the last two lines of Table III) even though in the experiments presented in Table III ($\Delta w = 0$) half of all the mutations have fitness lower than the parental type and are constantly discarded.

The mean F value ($\sum p_i^2$, where p_i is the frequency of each existing mutant) corresponds well to its expectation $1/(1 + 2N\mu)$ (see Kimura and Crow, [15]), where we have $2N\mu$ instead of $4N\mu$, the population being haploid. It was observed, however, that F values have an extremely high variance. This corresponds to expectation according to theoretical work (unpublished) by Ewens. Also the average number of alleles observed is given in Table III.

4. Form of the fitness distribution

Two examples of approximate distributions illustrating the variation in fitness of new alleles, assumed in our computer model are shown in Figure 3. In both cases the majority of mutations are deleterious. Such mutations practically never get fixed unless the population is extremely small, and so can safely be neglected. Thus, the mutation rate that must be considered is that to advantageous and neutral mutations. The latter are shown in the figure as corresponding to the approximate range $1 \pm 1/2N$. Our experiments confirm the prediction by Kimura that, when the variation in fitness is of this order of magnitude, the mean number of transitions is practically the same as that observed with strictly neutral mutations. In the upper distribution the fraction of advantageous mutations which cannot be considered neutral is relatively large, while in the lower distribution it is small. The lower distribution, therefore, corresponds more

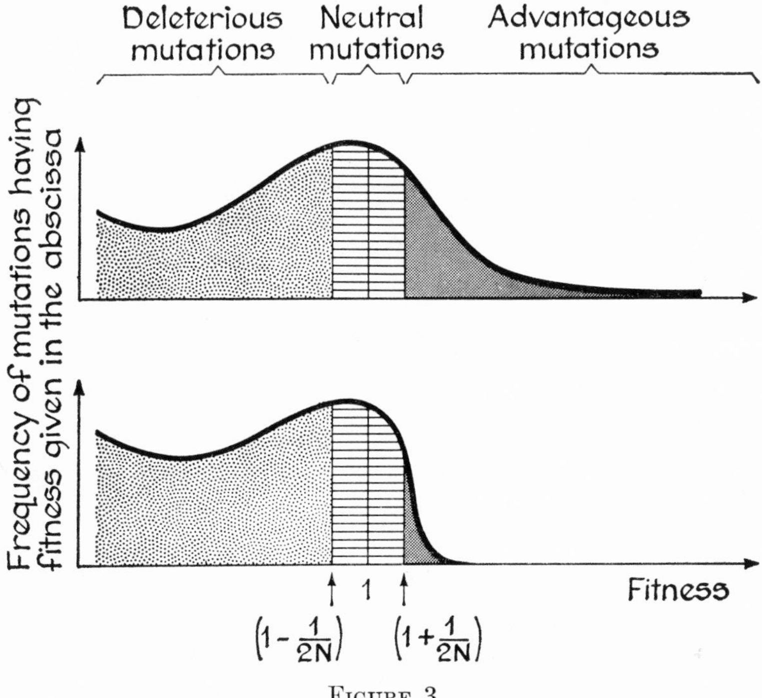

FIGURE 3

Approximate distributions illustrating the variation in fitness of new alleles.

closely to the model suggested by Kimura for molecular evolution, in which most mutations are neutral or almost so.

The picture suggested in Figure 3 is of course an over simplification which can at best be valid for haploids. In diploid organisms, the situation is further complicated by the fact that we must consider the fitnesses both of the homozygote and of the heterozygote. Figure 4 shows a suggested distribution of fitness values in mutant homozygotes and heterozygotes, taking the fitness of the normal homozygote as equal to 1. It is perhaps reasonable to assume that most mutations will be distributed around the line indicating additive fitnesses, that is, the situation in which the homozygote has a fitness $1 + 2s$ when the heterozygote has fitness $1 + s$. The distribution indicated in the figure has, for illustrative reasons, a variance which is much larger than appropriate for the actual distribution. There is actually little, if any, data for individual mutations that can be used to give this distribution.

Perhaps the best indication from published data is given by observations of Dobzhansky, Holz, and Spassky (see Hadorn [11] and Figure 5). These data, however, refer only to homozygotes and not to heterozygotes and so give only one marginal distribution that would be obtained from the surface given in

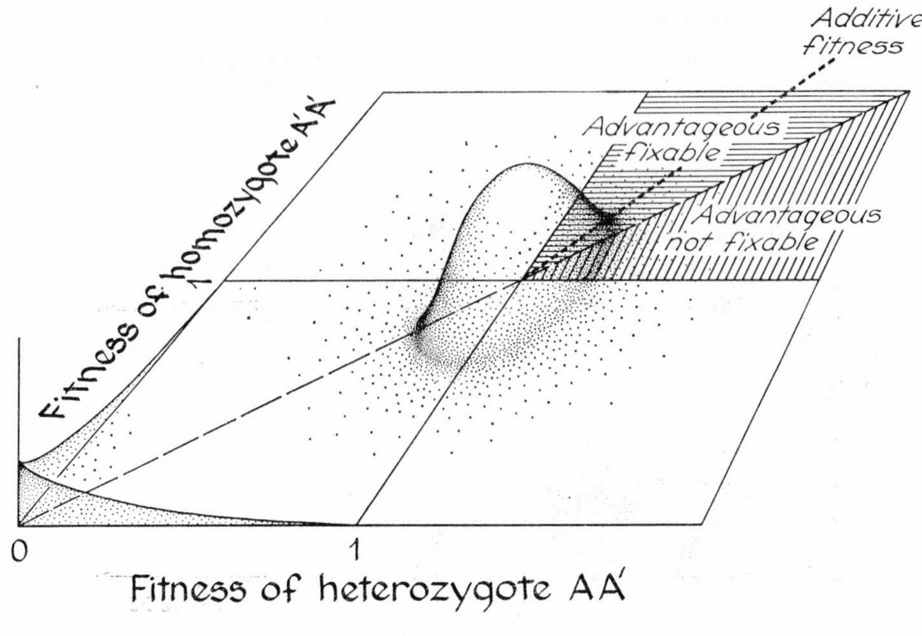

FIGURE 4

Distribution of fitness values in mutant homozygotes and heterozygotes, taking the fitness of the normal homozygote as equal to one.

Figure 4. The viabilities computed by these workers were for entire chromosomes. Therefore, they refer to the sum of a large, unknown number of different mutations located on these chromosomes. The standard deviation for fitness in the part of the distribution which peaks around normal fitness is approximately 0.05. This may correspond to the sum of hundreds, possibly thousands or more, of different mutations that were heterozygous in the population that was analyzed. It may be, therefore, that the average fitness of each of the individual mutations is exceedingly small so that a large fraction of them lie within the range $\pm 1/2N$ of quasi neutral mutations. These are not, however, *new* mutations, but a sample of mutations that has already been tested by natural selection, because they have been found in wild populations. A distribution which may be closer to that appropriate for new mutations was given by Käfer (see [11]) who studied X-ray induced mutations. The fraction of deleterious mutations is then increased, but the general shape of the distribution remains the same as that shown in Figure 5. This is, perhaps, surprising because in the irradiation experiment only a relatively small number of mutations should be induced on each chromosome. This type of observation is, however, subject to a large experimental error which may

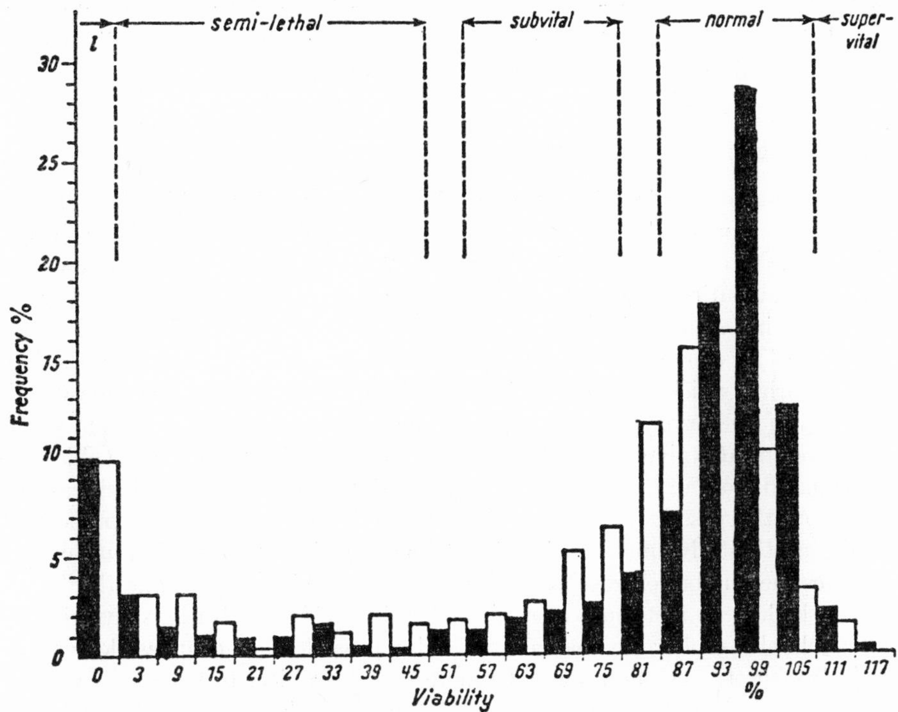

FIGURE 5

Distribution (possible) for individual mutations (from [11], page 118). Viability spectra for factors from wild populations of *Drosophila pseudoobscura*. Black = distribution of relative viabilities of homozygotes for 326 second chromosomes, white = the same for 352 fourth chromosomes, l = lethal. (Recalculated and illustrated after data by Dobzhansky, Holz, and Spassky, 1942.)

obscure the actual variation in fitness values. Small fitness differences are extremely difficult to measure, especially in higher organisms, and it is very difficult to measure satisfactorily fitness differences that are less than 0.01 (see below). If many mutations have fitness differences less than 0.01, the problem of estimating the distribution of fitnesses associated with new mutations, especially the part that matters for the present discussion, may be exceedingly difficult.

Even if it were possible to obtain actual data giving the distribution surface illustrated in Figure 4, it would still have to be remembered that this surface would refer to a specific environment. The variety of environments with which an organism might be confronted would complicate the interpretation of such surfaces still further. Most organisms of course live in a great variety of environments that are heterogeneous in time as well as space, even perhaps over quite small distances. Fitnesses estimated in natural populations, however, for example in man, do generally represent average values that may be valid over a wide range of different environments.

5. The fitting of theory to observation

There are three major observable evolutionary quantities which have to be explained by our models. The first, which we have already discussed extensively, is the mean rate of gene substitution or the mean time taken for the average number of evolutionary transitions to increase by one. This is estimated from data on amino acid substitution. The second is the observed degree of polymorphism. This can be expressed in a variety of ways such as the overall fraction of time during which a gene is polymorphic, or $1 - F$, where F is the overall frequency of homozygotes for the gene in question, or also the mean number of alleles present at a given time. The mean number of alleles and the F value can be estimated from data on electrophoretic variation for enzymes which can be stained or otherwise identified on gels following electrophoresis. This procedure permits us to study unselected loci but has the disadvantage that it underestimates the number of existing alleles by a factor which may be one third and possibly higher. In fact, only one third of amino acid substitutions give rise to observable electrophoretic changes. It is also possible that changes detectable by electrophoretic techniques may be more usually subject to selective pressure than mutational changes which do not determine a charge difference and are, therefore, usually not detectable by electrophoresis. The third observable is the variation between different populations in different environments in the level of polymorphism for a given locus. This is usually expressed as the variance of the gene frequencies from the various populations. For existing theories to be applicable to the data, effective migration rates between the populations must be neglibible, or at least their intensity should be known.

Our computer model is based on four main parameters: N the population size, μ the mean mutation rate per locus, Δw the mean difference in fitness between a new mutant and its immediate ancestor, and σ_w the variance of the distribution (assumed normal) of the fitnesses of new mutants. The problem, in principle, is the estimation of these four parameters, if possible, from data on the three major observable evolutionary quantities. The issue, for example, that has been raised by Kimura [14], by King and Jukes [18], and by others is whether the observables are compatible with values of Δw and σ_w inside the range $\pm 1/2N$. Since the population size and mutation rate can in principle be estimated using quite different sorts of information from that we are considering, there should be adequate scope for estimating Δw and σ_w and even for testing the goodness of fit of the model using the third degree of freedom in the observables.

There are, however, at least two major complicating factors in this apparently simple approach. The first is that there is no universal agreement on what are the appropriate values for N and especially for μ. The second, and perhaps more important, is that a single normal distribution with parameters Δw and σ_w is not enough to describe adequately the distribution of fitness values for new mutants. Apart from anything else, as already pointed out, this model can only apply to diploid organisms on the assumption of additive fitness values.

The important features of the distributions illustrated by Figures 3 and 4 are the proportions of deleterious, neutral, heterotic, and fixable alleles. The heterotic and fixable parts of the distribution can be further subdivided according to whether they apply to all environments or only to some environments. This distinction is especially important in the consideration of observed variations in the level of polymorphism, when different populations are compared (our third observable above). If we characterize the distribution of fitness values of new mutants by these six subdivisions (equivalent to considering six different mutation rates according to the fitnesses of the newly derived genotypes), we have, with N, seven rather than only four parameters for our theoretical model. We may not, however, even with six parameters, have adequately catered for variations in the environment changing the shape of the fitness distribution. Even accepting independent estimates of N and μ (the overall average mutation rate), we are now no longer in a position to be able to estimate from observed data, all the parameters of the model, let alone test the goodness of fit. The best that can now be done is to see whether the observed data rule out any significant regions of the parameter space defined by the values of N, μ, and the describers of the fitness distribution. A schematic summary of the effects of increases in the seven parameters defined above on the three major observable evolutionary quantities is shown in Table IV.

We shall now review briefly published data on three major observable evolutionary quantities starting with variations in the level of polymorphism between different populations. Apart from man, the best studied mammal is the mouse. A paper by Petras, Reimer, Biddle, Martin, and Linton [22] has shown that relatively unrelated populations of *Mus musculus* can show quite similar distributions of polymorphisms. This is more in agreement with selectively balanced polymorphism than with neutrality of the mutants present in a population. Little is, however, known about migration in the mouse so that populations that seem to be widely isolated geographically may in fact be more interconnected by migration than one might expect *a priori*. If this were true, the similarity of polymorphism found at a great distance might also be compatible with the theory of neutral mutation. The authors of this study also mention the possibility that the observed similarity of polymorphisms in widely separated geographical isolates may represent transient polymorphism due to selection following the introduction of new pesticides.

Prakash, Lewontin, and Hubby [24] have found even more extensive similarities in the polymorphism exhibited by many loci in *Drosophila pseudobscura* from quite different geographical origins. Here, again, population sizes, mutation rates, and migration rates are generally not well known, though the similarity in the distribution of polymorphisms encountered in widely separated localities is

TABLE IV

EFFECTS OF INCREASES IN SEVEN PARAMETERS ON THREE OBSERVABLE EVOLUTIONARY QUANTITIES

See text for further explanation.
Parentheses indicate effects are limited to some environments.

Parameters (which increase)	Observable evolutionary quantity: Mean evolutionary time	Average level of polymorphism	Variation in level of polymorphism
N	increase (only in presence of selection)	increase	no effect
Mutation rate to deleterious alleles	no effect	no effect	no effect
Mutation rate to neutral alleles	decrease	increase	no effect
Mutation rate to heterotic alleles:			
In some environments	(some contribution)	(increase)	increase
In all environments	small contribution	increase	decrease
Mutation rate to fixable alleles:			
In some environments	(decrease)	(increase)	increase
In all environments	decrease	increase	no effect

certainly surprising. It would be difficult not to conclude with the authors that the simplest explanation is that polymorphisms showing such a remarkable similarity in the frequency of the various genes in different populations represent the consequence of balancing selection. The identification of an allele purely on the basis of electrophoretic mobility is not, however, generally sufficient, and identity should be shown by further molecular analysis. A number of hemoglobins previously believed to be identical on the basis of identical electrophoretic mobility were later shown to be different alleles when fingerprinting and sequencing were carried out. It should also be emphasized that it may be very hard to distinguish the direct selective effects of an identifiably polymorphic locus from those of other so far unidentified but closely linked loci. Weak selective interaction between closely linked genes may make an important contribution to the overall maintenance of polymorphism (see, for example, Bodmer and Parsons [4], Bodmer and Felsenstein [3], and Franklin and Lewontin [10]). Even in the absence of selection, close linkage to a selectively maintained polymorphic locus can also in finite populations contribute to the overall level of polymorphism (see, for example, Sved [26], [28]). The results presented by Ayala at this conference extend considerably the range of the original observations by Prakash, Lewontin, and Hubby [24], but do not alter the conclusions above.

In man, the average frequency of polymorphisms is similar to that so far observed in other species. Population sizes and migration rates are, on the whole, more easily ascertained in man than in other species. This makes it possible to compare the observed level of geographic variation of polymorphisms with that expected on the basis of relevant demographic quantities. The migration matrix method (Bodmer and Cavalli-Sforza, [2]) has been used in various studies of rural populations from various parts of the world (partly unpublished, see [6]). This approach allows one to compare observed with expected variation in gene frequencies for given migration rates and population sizes. In all these cases the observed variation, computed as an f value (variance of gene frequencies divided by $\bar{p}$ $(1 - \bar{p})$, where $\bar{p}$ is the mean gene frequency) is in "semiquantitative" agreement with that expected under the balance of drift and migration and in the absence of selection.

These results thus suggest selection played a minor role in generating the observed variation between populations. In each case, however, only variation at a microgeographic scale was measured. The studies were also based on areas selected to have low population numbers or lower migration and thus relativey stronger drift effects so that they cannot be considered to represent the species as a whole. When variation is analyzed at a wider geographic level—for example, by comparing broad ethnic groups, then the effect of selection becomes apparent. The criterion used is a simple one. If drift alone were responsible for the observed variation, then every locus should show the same amount of variation in gene frequency between populations. Thus, we know that for genes that are polymorphic, or more precisely, that are not maintained by the balance of mutation and selection under drift alone, f should be the same for all genes, being a function only of N and of migration rates. The observed f values in interracial comparisons vary greatly from gene to gene (over a range of at least 10 fold, see Cavalli-Sforza [5]). This clearly suggests that selection is operating at this level of comparison. Selection may be disruptive for genes having relatively high values of f, in which case the genes are responding differently to selection in different environments. Selection may, on the other hand, be balancing for those genes giving low f values. In this case, similar balancing selection in different environments is presumably reducing the level of variation in comparison with that expected from drift alone. Unfortunately, however, the analysis of interracial variation cannot yet be carried to the level of comparing observed with expected f value, as in the case of the analysis of microgeographic variation. This is because we know too little about the demographic conditions that prevailed during the formation of races and this information is needed to compute the expected values of f.

On the whole, these analyses of the variation in polymorphic gene frequencies between different populations in mouse, *Drosophila*, and man, do suggest the existence of detectable differences due to selection.

Let us now consider the data derived from amino acid sequences on the rate of gene substitution which lead to a comparison of the observed and expected rate of evolution under different assumptions. We want values of N and μ, the latter possibly subdivided according to the selection effects of the mutational change. In general, the relevant value of N depends on the population which is sampled. For molecular evolution this is the whole species and so at the upper limit of possible values of N. In all the cases of variation discussed so far, such as interracial comparisons, or the analysis of variation at a microgeographic level, the values of N involved were smaller as implied by the populations being sampled. We will limit our discussion to man as this is the species for which this quantity can be estimated most satisfactorily.

We should, of course, not consider the present world population as the basis for evaluating N for man. Very large increases in population size have occurred just during the last 10,000 or so years, that is, since the domestication of plants and animals has augmented the carrying capacity of the land for man. Most of our evolution, however, took place before this, while man was still a hunter and gatherer. The relevant estimates of population size which have been suggested, for example, 125,000 by Deevey [9], seem far too low. Today, there are still people who live with a hunting and gathering economy, such as, for example, the African Pygmies. These alone number over 100,000 and occupy a very small portion of the African continent, at a density of about 0.2 per Km^2 (see [6]). On this basis, a minimum estimate of the total human population size throughout the Paleolithic must be of the order of 10^6 to 10^7. Reduction of N to N_e the effective population size, involves two factors: (1) overlapping generations, which reduces N by a factor of about one third [6] and (2) isolation. With respect to the latter, a theorem by Moran [21] states that, if a population of N individuals is separated into k groups amongst which exchange of individuals takes place, and each group receives from the other groups k individuals per generation, then the effect of the subdivision on the drift experienced by the population as a whole is practically negligible. That is, the effective size of the whole group is still close to N. It would seem that the effective size of the human population as a whole should therefore, not be taken as less than 10^5 and is probably nearer to 10^6.

Mutation rates have been estimated in man using pedigree data and mutation-selection balance theory, but an important source of bias in these estimates has apparently so far been overlooked. Average published mutation rates are generally about 3×10^{-5} per gene per generation. These estimates, however, generally ignore the fact that mutations at the particular locus for which they were derived were known to occur before they were studied. This implies that the particular loci studied must have been selected at least to some extent on the basis of their mutation frequency. A simple statistical computation shows that this can lead to a considerable bias in the estimated mutation rate. If one assumes, as a first approximation, that the probability of a mutation being included in a survey is proportional to its mutation rate, it can be shown that the unselected average mutation rate is equal to the harmonic mean of the observed selected mutation rates [6]. The results of the calculations show that the average mutation rate, because of the extreme variation in mutation rates between different loci, is 3×10^{-7}, two orders of magnitude lower than the values given before.

These mutation rate estimates in man refer only to deleterious alleles. The proportion of all mutations that are deleterious in man is not known though attempts have been made to estimate it in other organisms. It at least seems unlikely that the order of magnitude of the mutation rate to neutral and to advantageous alleles is higher than that to deleterious alleles.

Data on amino acid differences between proteins of different species suggest a median rate of evolution corresponding to 10^{-9} amino acid substitutions per year per amino acid position [30], [18]. In other words, the average time between amino acid substitutions at a given position in a protein is 10^9 years. When multiplied by three, to allow for the fact that three nucleotide pairs are needed to code for one amino acid, this gives 3×10^9 years as the mean time taken for the number of mutational transitions, as given by our computer model, to increase by one. As already discussed, the mean expected rate of gene substitution per generation, assuming only neutral mutations, is the mutation rate μ. Since the amino acid substitution data comes mainly from mammals, the relevant generation time should be an average for mammals, which can reasonably be taken to be four years. The molecular data thus suggests a mutation rate of $4/3 \times 10^9$ or 1.3×10^{-9} per nucleotide pair per generation, on the assumption that all or most mutations are neutral. If we assume that the mutation rate to neutral alleles is equal to that to deleterious alleles, and that there are on average about 1,000 nucleotide pairs per gene, then using the mutation rate estimate to deleterious alleles of 3×10^{-7} per gene, we obtain a neutral mutation rate per nucleotide pair of $3 \times 10^{-7}/1{,}000 = 3 \times 10^{-10}$. This is three times less than that suggested by observations on amino acid substitutions assuming neutrality of all mutations. At face value, this would argue against the idea suggested by Kimura, King and Jukes, and others, that most observed amino acid substitutions are due to neutral or quasi neutral mutations. However, the fact that Kimura can come to an opposite conclusion, using similar arguments and published data should stand as a warning against taking these numerical data too seriously as evidence either for or against neutrality. The figures involved are known with insufficient accuracy to make precise statements.

Consider now the situation when there can be both neutral and advantageous mutations. Assume that a proportion p_n of all mutations are effectively neutral (that is, lead to fitness differences in the range $\pm 1/2N$) and a proportion p_a are advantageous, that is, lead to fitness differences greater than $1/2N$. Since there will also be a fraction of mutants that are deleterious,

$$p_n + p_a < 1. \tag{1}$$

For the neutral mutants, the rate of gene substitution is simply obtained from the mutation rate to neutral changes, μp_n. For the advantageous mutants, the rate will be $k\mu p_a$, where k, a factor greater than one, represents the average effects of selection on the rate of gene substitution. The overall rate of substitution, taking into account both neutral and advantageous mutants, is therefore given by

$$M = \mu(p_n + kp_a) \quad (2)$$

(see [6]). Clearly, M can be much greater than μ (even by a factor of ten or more) depending on the magnitudes of k and p_a, that is, depending on the distribution of fitness values among mutants. Thus, even reducing the number of variables from seven in Table IV to a minimum of three, as we have now done, the expected mean evolutionary times, based on population genetic models, are compatible with practically any reasonable observed rate of evolution.

The order of magnitude of N in man determines the order of magnitude of a selection differential that can be considered neutral, namely, $< 10^{-5}$ or even $< 10^{-6}$. The estimation of selection coefficients is in practice, however, very difficult. Selective differentials for advantageous mutations have only been estimated in a few cases mainly limited to malarial environments, such as for sickle cell anaemia heterozygotes, and for the G6PD gene. These two are both of the order 0.1 and even selection coefficients of this order of magnitude already require for their estimation the detailed examination of a considerable number of individuals. In most experimental situations, it is difficult or impossible to estimate selective coefficients smaller than 0.01. Only in very special situations has it proved possible to estimate small selection coefficients. Thus, the relative advantage of *ABO* alleles that protect against duodenal ulcer is of the order of 10^{-4} in males and 10^{-5} in females [6]. These estimates, however, depend on the assumption that differential mortality from ulcer is the sole cause of selection. Many such small selective differences could exist, usually unmeasurable, that could account for an observed rate of gene substitution which is higher than that expected for only neutral mutations.

Kimura ([14] and later) has suggested, following Haldane's earlier work on the cost of natural selection [12], that most mutations that eventually become substituted in a population must be neutral, because the genetic load implied by substitution at the rate indicated by observations on amino acid differences between species would be excessive. His computations are based, however, on the somewhat arbitrary assumption of independent action of different loci at the level of fitness. If a threshold model for selection is assumed, as has been suggested by Sved, Reed, and Bodmer [29], King [17], and Milkman [20] for heterotic polymorphisms, then the apparently excessive substitutional load disappears. It has actually been shown by Sved [27] that, assuming a threshold model, the observed rates of gene substitution can be readily accommodated with relatively minimal selective loads.

A number of other arguments, not based on the theoretical considerations we have discussed so far, have been put forward by King and Jukes [18] and Kimura ([14] and other papers) in favor of neutrality of most new mutations or "non-Darwinian evolution," as it has been called by King and Jukes. These arguments concern, for example, the distribution of the number of amino acid substitutions per amino acid position in a protein, the question of "synonymous" substitutions, the apparent equivalence, according to some protein chemists, of different amino acid substitutions at many positions in many proteins and the apparent uniformity of the rate of evolution of some proteins over a wide evolutionary time span. Though we do not propose to elaborate further on these questions in this paper, we do not find any of these arguments particularly convincing as has been discussing by Richmond [25] and Clarke [7], [8].

The work of Lewontin and Hubby [19] in *Drosophila*, and Harris [13] in man has indicated average heterozygosity level per locus of 10 to 30 per cent corresponding to F values of from 0.7 to 0.9. If we take the minimal suggested value of N, namely 10^5, and use a minimal value of $\mu = 10^{-7}$ for the mutation rate to neutral alleles, then the formula used by Kimura and Crow [15] to evaluate F on the assumption of only neutral mutations, namely,

$$F = \frac{1}{1 + 4N\mu} \tag{3}$$

gives $F = 0.96$ which is almost certainly too high. If, on the other hand, we take $N = 10^6$ and $\mu = 10^{-6}$, this gives $F = 0.2$ which is clearly too low. Moreover, the high variance of F which was already mentioned makes the test insensitive. As already mentioned, it has been shown by Ewens (unpublished) that F is a poor statistic. Thus, observed levels of polymorphism could, in principle, be accounted for by neutral mutations, but the test is a weak one. This of course says nothing about the extent to which selection for fixable alleles is actually involved in maintaining observed levels of polymorphism. In Table III, we can notice that the introduction of selection does not alter the mean F values where F and μ are the same.

It seems worth recalling that in microorganisms, situations are available in which the rate of formation of advantageous mutants can be measured with some precision as illustrated by early work by Atwood, Ryan, and Schneider [1]. These authors noticed that asexual bacterial populations in which the equilibrium between a specific mutant and the rest of the population due to mutation selection balance was being investigated, occasionally underwent significant shifts in the relative frequency of the mutant in the population. These shifts could be interpreted on the hypothesis that new mutations with an increased fitness had occurred somewhere in the bacterial genome in one individual of the population, usually not of the original mutant type. These new advantageous mutations then wiped out the original mutant type whose equilibrium was being investigated. Once these new fitter types have replaced the old types, the specific mutant being investigated can reappear among the fitter types and return slowly to its former equilibrium. The estimate of the rate of mutation to such advantageous types under these conditions, was extremely low, namely, of the order of 10^{-12}, leaving plenty of scope for neutral or quasi neutral mutations. This system, however, only uncovers mutations with an increase in fitness that is above a certain threshold and this may account for their extremely low rate of appearance. Though it is, of course, clear that results with bacteria and other microorganisms cannot readily be extrapolated to higher organisms, it does seem likely that further studies of the kinetics of such selective processes in microorganisms, both at a theoretical and at an experimental level, might well be rewarding.

6. Conclusions

The main point of constructing and describing our model has been to try and clarify the issues involved in matching population genetic theory to observed data on evolutionary rates and polymorphism. The results, perhaps unfortunately, are so far inconclusive, though we hope that further elaboration of the models and data will lead to a clearer understanding of the problems and, in particular, of the relative importance of neutral *versus* advantageous gene substitution. Although it appears that there is no major discrepancy between theory and data, the data do not yet clearly indicate what should be the prevailing values of N, μ, and the fitness differences to account for the observed properties of evolving populations. The major question of the extent to which new mutants are or are not associated with selective differences is, apparently, no nearer resolution today than it was well before the recent revival of discussion about "non-Darwinian" evolution.

REFERENCES

[1] K. C. Atwood, L. K. Schneider, and F. J. Ryan, "Selective mechanisms in bacteria," *Cold Spring Harbor Symp. Quant. Biol.*, Vol. 16 (1951), pp. 345–355.

[2] W. F. Bodmer and L. L. Cavalli-Sforza, "A migration matrix model for the study of random genetic drift," *Genetics*, Vol. 59 (1968), pp. 565–592.

[3] W. F. Bodmer and J. Felsenstein, "Linkage and selection: theoretical analysis of the deterministic two locus random mating model," *Genetics*, Vol. 57 (1967), pp. 237–265.

[4] W. F. Bodmer and P. A. Parsons, "Linkage and recombination in evolution," *Advan. Genet.*, Vol. 11 (1962), pp. 1–100.

[5] L. L. Cavalli-Sforza, "Population structure and human evolution," *Proc. Roy. Soc. London B Biol. Sci.*, Vol. 164 (1966), pp. 362–379.

[6] L. L. Cavalli-Sforza and W. F. Bodmer, *The Genetics of Human Populations*, San Francisco, Freeman, 1971.

[7] B. Clarke, "Darwinian evolution of proteins," *Science*, Vol. 168 (1970), p. 1009.

[8] ———, "Selective constraints on amino-acid substitutions during the evolution of proteins," *Nature*, Vol. 228 (1970), pp. 159–160.

[9] E. S. Deevey, "The human population," *Sci. Amer.*, Vol. 203 (1960), pp. 194–204.

[10] I. Franklin and R. C. Lewontin, "Is the gene the unit of selection?" *Genetics*, Vol. 65 (1970), pp. 707–734.

[11] E. Hadorn, *Letalfaktoren*, Stuttgart, Georg Thieme Verlag, 1955.

[12] J. B. S. Haldane, "The cost of natural selection," *J. Genet.*, Vol. 55 (1957), pp. 511–524.

[13] H. Harris, "Enzyme and protein polymorphism in human populations," *Brit. Med. Bull.*, Vol. 25 (1969), pp. 5–13.

[14] M. Kimura, "Evolutionary rate at the molecular level," *Nature*, Vol. 217 (1968), pp. 624–626.
[15] M. Kimura and J. F. Crow, "The number of alleles that can be maintained in a finite population," *Genetics*, Vol. 49 (1964), pp. 725–738.
[16] M. Kimura and T. Ohta, "Protein polymorphism as a phase of molecular evolution," *Nature*, Vol. 229 (1971), pp. 467–489.
[17] J. L. King, "Continuously distributed factors affecting fitness," *Genetics*, Vol. 55 (1967), pp. 483-492.
[18] J. L. King and T. H. Jukes, "Non-Darwinian evolution," *Science*, Vol. 164 (1969), pp. 788–798.
[19] R. C. Lewontin and J. L. Hubby, "A molecular approach to the study of genetic heterozygosity in natural populations. II. Amount of variation and degree of heterozygosity in natural populations of *Drosophila pseudoobscura*," *Genetics*, Vol. 54 (1966), pp. 595–609.
[20] R. D. Milkman, "Heterosis as a major cause of heterogeneity in nature," *Genetics*, Vol. 55 (1967), pp. 493–495.
[21] P. A. P. Moran, *The Statistical Processes of Evolutionary Theory*, Oxford, The Clarendon Press, 1962.
[22] M. L. Petras, J. D. Reimer, F. G. Biddle, J. E. Martin, and R. S. Linton, "Studies of natural populations of *Mus*. V. A survey of nine loci for polymorphisms," *Can. J. Genet. Cytol.*, Vol. 11 (1969), pp. 497–513.
[23] E. M. Prager and A. C. Wilson, "Multiple lysozymes of duck egg white," *J. Biol. Chem.*, Vol. 246 (1971), pp. 523–530.
[24] S. Prakash, R. C. Lewontin, and J. L. Hubby, "A molecular approach to the study of genic variation in central, marginal and isolated populations of *Drosophila pseudoobscura*," *Genetics*, Vol. 61 (1969), pp. 841–848.
[25] R. C. Richmond, "Non-Darwinian evolution: A critique," *Nature*, Vol. 225 (1970), pp. 1025–1028.
[26] J. A. Sved, "Linkage disequilibrium and homozygosity of chromosome segments in finite populations," *Theor. Pop. Biol.*, Vol. 3 (1971), pp. 125–141.
[27] ———, "Possible rates of gene substitution in evolution," *American Naturalist*, Vol. 102 (1968) pp. 283–293.
[28] ———, "The stability of linked systems of loci with a small population size," *Genetics*, Vol. 59 (1968), pp. 543–563.
[29] J. A. Sved, T. E. Reed, and W. F. Bodmer, "The number of balanced polymorphisms which can be maintained in a natural population," *Genetics*, Vol. 55 (1967), pp. 469–481.
[30] E. Zuckerkandl and L. Pauling, "Evolutionary disease, evolution and genic heterogeneity," *Horizons in Biochemistry* (edited by M. Kasha and B. Pullman), Chicago, Academic Press, 1965.

EVOLUTIONARY INDICES

LILA L. GATLIN
UNIVERSITY OF CALIFORNIA, BERKELEY

1. Introduction

As *Homo sapiens* we have always believed that we are higher organisms. After all, we are more complex, more differentiated, more highly ordered than lower organisms. As thermodynamicists we recognize these words and realize that the concept of entropy must somehow enter into the explanation of this.

We have always had the vague notion that as higher organisms have evolved, their entropy has in some way declined because of this higher degree of organization. For example, Schröedinger made his famous comment that the living organism "feeds on negative entropy." We reason that this decreasing entropy of evolving life, if it exists, does not in any way, violate the second law of thermodynamics which states that the entropy of an isolated system never decreases. The living system is not isolated and the reduction in entropy has been compensated for by a correspondingly greater increase in the entropy of the surroundings. It does not violate the letter of the second law, and yet something about it seems to make us uneasy. Why should the evolution of the living system constantly drive in the direction of increasing organization while all about us we observe the operation of the entropy maximum principle, which is a disorganizing principle? I know of no other system except the living system which does this.

First of all, can we establish that the entropy has, in fact, declined in higher organisms? No one has ever proved this quantitatively. In fact, one can argue that it is impossible to establish this thesis by classical means because of the uncertainty principle in its broadest sense. In particular, if we were to make the precise and extensive measurements necessary to determine accurately the entropy difference between a higher and a lower organism, these measurements would disturb the living systems so much that they would kill them. So, it is impossible by classical means to even establish this proposition in which almost all of us seem to believe.

When concepts break down like this they are of little use to us. I think that our classical notions of entropy are totally inadequate in dealing with the living system. This does not mean that there is anything mysterious, supernatural, or vitalistic about the living system. It simply means that our classical notions of entropy are inadequate, just as the laws of Newtonian mechanics were inadequate in dealing with the interior of the atom.

I shall extend the entropy concept primarily through the apparatus of information theory, but I shall extend this also. Shannon [10] gave the most general definition of entropy to date and I shall extend the concept of Shannon. Specifically, I shall show that the entropy function which Shannon called the redundancy is composed of two parts which I call D_1 and D_2. We must characterize the redundancy of a sequence of symbols by two independent numbers, one describing the amount and the other the kind of redundancy of the sequence. I can state this in terms of entropy. I shall show that phrases like, increasing entropy or decreasing entropy, are not completely definitive. We must ask, in what way the entropy has increased or decreased or what kind of entropy is it? We do not encounter such questions in either classical thermodynamics or information theory. I shall develop a theory which can answer these questions.

In classical thermodynamics we dealt with the ordering of three dimensional aggregates of matter, but in information theory we begin to grapple with the concept of ordering of one dimensional sequences of symbols. This is very significant because we now know that the DNA molecule is a linear sequence of symbols which stores the primary hereditary information from which the entire living organism is derived just as a set of axioms and postulates stores the primary information from which a mathematical system is deduced. Therefore, if we wish to investigate the organization of living systems, we must investigate the ordering of the sequences of symbols which specify them.

DNA stores the hereditary information in a sequence of symbols from an alphabet of four letters, the four DNA bases, A, T, C and G. DNA stores its information in the particular sequential arrangement of these four letters just as any language. Therefore we are dealing with language in general although we will apply it to DNA in particular.

2. Theory

We must first define the alphabet. We let

$$S_1 = \{X_i: i = 1, a\}, \tag{1}$$

where the X_i are the letters of the alphabet and a is the number of letters. For DNA, $a = 4$.

With each X_i there is associated a probability $0 \leqq P_i \leqq 1$;

$$\sum_i P_i = 1. \tag{2}$$

Thus, S_1 is a finite probability space. The entropy of S_1 according to Shannon [10] is

$$H_1 = -K \sum_i P_i \log P_i. \tag{3}$$

When $K = 1$ and the logarithm base is 2, the units of H_1 are bits. It can be shown under very reasonable postulates (Khinchin [6]) that (3) is unique and takes on its maximum value, $\log a$, if and only if all the P_i are equal. Thus, $H_1^{\max} = \log a$.

The maximum entropy state for a sequence of symbols is characterized by equiprobable, independent single letter elementary events. This statement is not difficult to justify. Almost any game situation illustrates that the most "random" state is characterized by equiprobable, independent events.

We all know that in any language the single letter frequencies diverge from equiprobability. For example, in the English language the letter e occurs more frequently than any of the others. The divergence from the maximum entropy state due only to this divergence from equiprobability is given by

$$D_1 \equiv \log a - H_1, \tag{4}$$

where the P_i in H_1 are the experimentally observed values for a given language. Biologists call the distribution of the P_i on S_1 the "base composition" of DNA.

We are interested in the sequential arrangement of the letters in a sequence. Therefore, we define a space of n tuples:

$$S_n = \{X_i X_j \cdots X_n; i, j \cdots n = 1, a\}. \tag{5}$$

There are a^n n tuples in S_n.

If the letters in the sequence are independent of each other,

$$H_n^{\text{Ind}} = -\sum_i \sum_j \cdots \sum_n P_i P_j \cdots P_n \log P_i P_j \cdots P_n \tag{6}$$

or

$$H_n^{\text{Ind}} = nH_1. \tag{7}$$

Let m be the memory of a Markov source. If $m = 1$, the probability of occurrence of a given letter depends only on the letter immediately preceding it in the sequence. Then the entropy of S_n is given by

$$H_n^{\text{Dep}} = -\sum_i \sum_j \cdots \sum_n P_i P_{ij} \cdots P_{(n-1)n} \log P_i P_{ij} \cdots P_{(n-1)n}, \tag{8}$$

where P_{ij} is the one step Markov transition probability from letter i to letter j. Utilizing the summations

$$\sum_j P_{ij} = 1 \tag{9}$$

and

$$\sum_i P_i P_{ij} = P_j, \tag{10}$$

equation (8) reduces to

$$H_n^{\text{Dep}} = H_1 + (n-1)H_M^1, \tag{11}$$

where

$$H_M^1 = -\sum_i \sum_j P_i P_{ij} \log P_{ij}. \tag{12}$$

This is just the well-known form for the entropy of a first order Markov source. If $m = 2$,

$$H_n^{\text{Dep}} = -\sum_i \sum_j \cdots \sum_n P_i P_{ij} P_{ijk} \cdots P_{(n-2)(n-1)n} \log P_i P_{ij} P_{ijk} \cdots P_{(n-2)(n-1)n}, \tag{13}$$

$$H_n^{\text{Dep}} = H_1 + H_M^1 + (n-2)H_M^2, \tag{14}$$

where

$$H_M^2 = -\sum_i \sum_j \sum_k P_i P_{ij} P_{ijk} \log P_{ijk}. \tag{15}$$

Following this same pattern, we generalize for an mth order Markov source:

$$H_n^{\text{Dep}} = H_1 + H_M^1 + H_M^2 + \cdots H_M^{(m-1)} + (n - m)H_M^m. \tag{16}$$

If the sequence of symbols diverges from the maximum entropy state due only to a divergence from independence of the symbols, this divergence must be a function of the difference between H_n^{Ind} and H_n^{Dep}.

Since $H_M^{m+1} \leqq H_M^m$ (this is a generalized form of Shannon's fundamental inequality; its proof is in Khinchin [6]),

$$H_n^{\text{Ind}} \geqq H_n^{\text{Dep}}, \tag{17}$$

and since both H_n^{Ind} and H_n^{Dep} are monotonically increasing functions of n, I define the divergence from independence

$$D_2 \equiv \lim_{n\to\infty} \frac{1}{n} (H_n^{\text{Ind}} - H_n^{\text{Dep}}). \tag{18}$$

From (7),

$$\lim_{n\to\infty} \frac{1}{n} H_n^{\text{Ind}} = H_1, \tag{19}$$

and from (16) if we impose the condition that $m \ll n$,

$$\lim_{n\to\infty} \frac{1}{n} H_n^{\text{Dep}} = H_M^m. \tag{20}$$

Therefore,

$$D_2 = H_1 - H_M, \tag{21}$$

where the order of H_M is understood.

Our condition that $m \ll n$ holds for DNA. DNA molecules are very long. For human DNA, $n \cong 4 \times 10^9$ and for even a small bacteria such as *Eschericia coli,*

$$n \cong 4 \times 10^6. \tag{22}$$

At the present time there is no conclusive evidence that the m is any greater than one for DNA. There are good theoretical arguments for expecting future evidence that it is greater than this, but it is highly unlikely that m will be of any greater magnitude than a small integer. Therefore, $m \ll n$ will almost certainly hold for DNA. If one knows m for any given language, one can always impose the condition $m \ll n$ simply by considering sequences of sufficient length.

The total divergence from the maximum entropy state is $D_1 + D_2$, which we shall call the information density of the sequence I_d,

$$I_d = D_1 + D_2. \tag{23}$$

I previously called this quantity the "information content" of DNA (Gatlin [2]), but these are poorly chosen words for a number of reasons.

Now let us show the relationship between this quantity I_d and the redundancy of Shannon. According to Shannon's [10] definition, the redundancy of a sequence of symbols is given by

$$R \equiv 1 - \frac{H_M}{\log a}. \tag{24}$$

From (21) and (4),

$$R = \frac{D_1 + D_2}{\log a} \tag{25}$$

or

$$R = \frac{I_d}{\log a}. \tag{26}$$

The redundancy of Shannon is just the information density expressed as a fraction of its maximum value, log a. This entire picture is illustrated in Figure 1 which is an entropy scale.

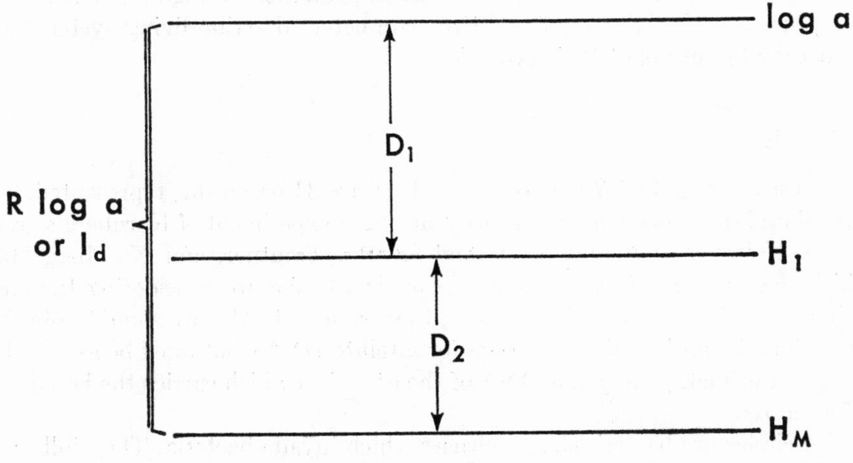

FIGURE 1

The entropy scale.

It is clear from the entropy scale that a given value of I_d may be achieved with different relative contributions from D_1 and D_2. Hence, I define the divergence indices or simply the D indices to characterize this relative contribution:

$$RD1 \equiv \frac{D_1}{D_1 + D_2} = \frac{D_1}{I_d} = \frac{D_1}{R \log a} \tag{27}$$

or

$$RD2 \equiv \frac{D_2}{D_1 + D_2} = \frac{D_2}{I_d} = \frac{D_2}{R \log a}. \tag{28}$$

There is, of course, only one independent index being defined and

$$RD1 + RD2 = 1. \tag{29}$$

Thus, we have in R and $RD1$ or $RD2$ two independent parameters, both dimensionless fractions with a range of zero to one. The R tells us how much the system has diverged from the maximum entropy state and the D index tells us in what way this divergence has taken place, that is, primarily through D_1 or D_2.

Let us now observe how these two parameters describe living systems and what they tell us about their evolution.

3. Results

Figure 2 is a plot of R versus $RD2$. There are 34 organisms represented. The basic data is obtained from the nearest neighbor experiment of Kornberg's group (Josse, Kaiser, and Kornberg [4] and Swartz, Trautner, and Kornberg [13]) which measures the basic P_{ij} for a given DNA. We are considering therefore only a first order Markov dependence, that is, $m = 1$. Also we should note that these data do not include any values for satellite DNA's but must be assumed to represent primarily the main DNA of the organism which carries the hereditary information.

The circles are bacteriophage, viruses which invade bacteria. They follow an empirical functional dependence. The squares are bacteria and follow a similar empirical curve. The vertebrates, however, do not exhibit a similar functional behavior between R and $RD2$ but fall into a rather restricted domain. R lies between about 0.02 to 0.04 and $RD2$ lies between about 0.6 to 0.8. There are some lower organisms with R values as high as or even higher than vertebrates, but whenever this occurs the $RD2$ value invariably drops quite low. This means that whenever lower organisms achieve R values in the vertebrate range they do so primarily by increasing D_1, the divergence from equiprobability of the DNA symbols (or bases). This confirms a well established experimental fact that the base composition of lower organisms, particularly bacteria, has a wide variational range from almost 20 to 80 per cent cytosine plus guanine while the base composition of vertebrates lies within the restricted range of about 40 ± 4 per cent $(C + G)$ (Arrighi, Mandel, Bergendahl, and Hsu [1]). Therefore, vertebrates have achieved their higher R values by holding D_1 relatively constant and in-

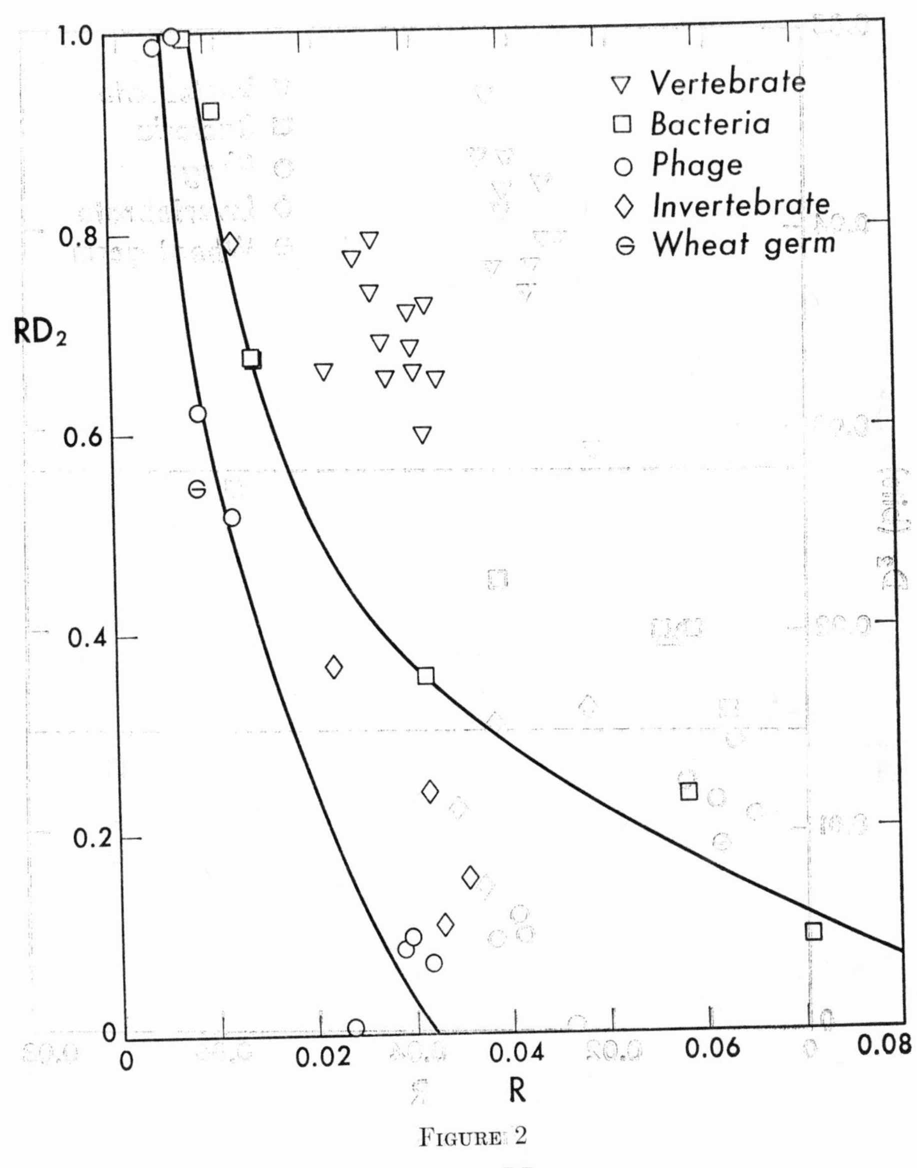

FIGURE 2

R versus RD_2

creasing D_2 whereas lower organisms use D_1 as the primary variable. The mechanism is fundamentally different.

If this is the case, we should expect to find the vertebrate D_2 values higher in general than for lower organisms. This is what we observe. Figure 3 is a plot of R versus D_2.

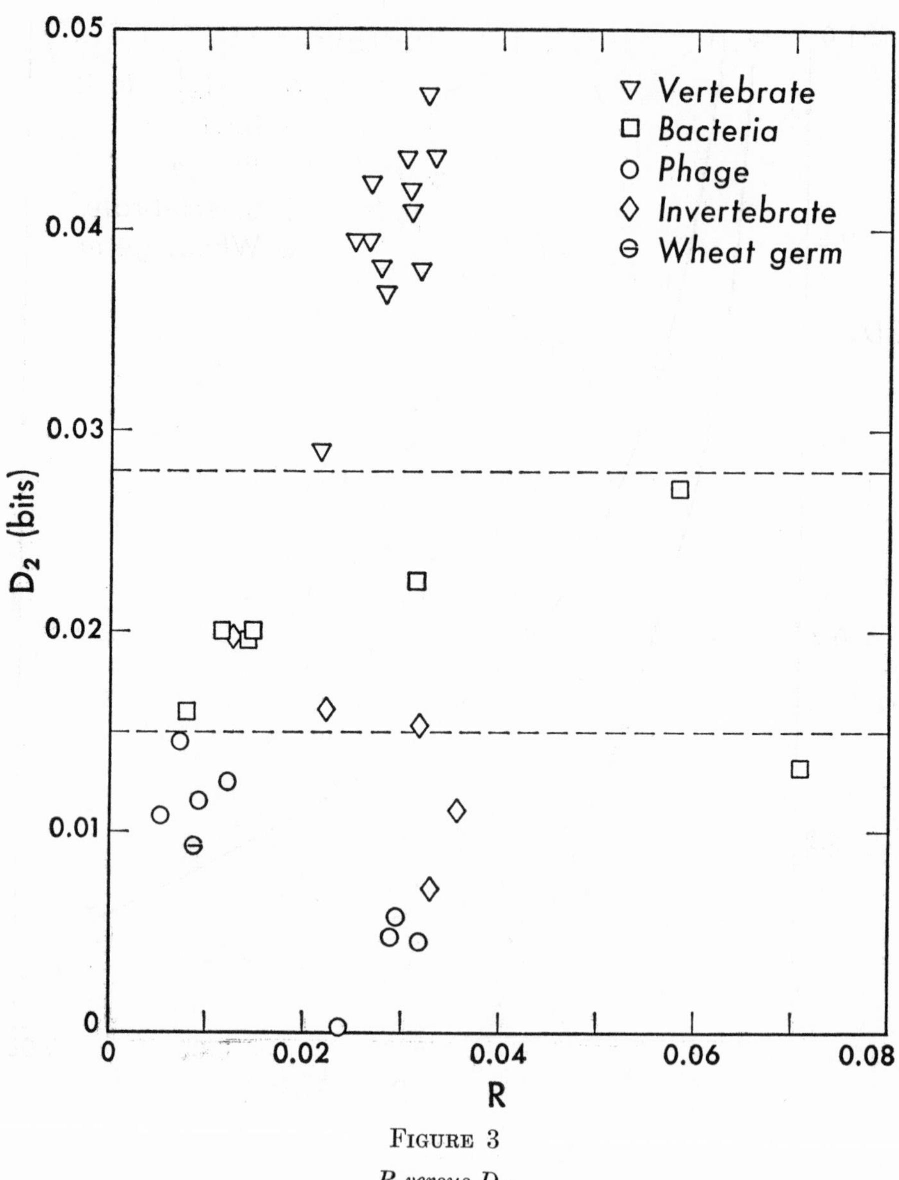

FIGURE 3

R versus D_2.

The vertebrates are characterized by the highest absolute magnitudes of D_2, the divergence from independence of the bases. We might say that D_2 is an evolutionary index which separates the vertebrates from all other "lower" organisms. In terms of entropy, vertebrate DNA does not necessarily have the lowest values of H_M, but they have the lowest values of H_M relative to H_1, that is, they have the highest values of D_2 which is a more important measure of the ordering of a sequence of symbols than any *single* entropy value.

This situation can be described in terms of game theoretic limits. Table I lists

TABLE I

GAME THEORETIC LIMITS OF D_1 AND D_2

The † indicates min-max, the asterisk indicates max-min.

		D_1	D_2
Phage	max	.059	.015
	min	.000	.000
Bacteria	max	.129	.027
	min	.000	.013
Vertebrates	max	.026†	.047
	min	.011	.029*
Invertebrates	max	.211	.020
	min	.005	.007

the maximum and minimum values of D_1 and D_2 for the groups of bacteriophage, bacteria, invertebrates and vertebrates. The vertebrates display a max-min of D_2 and a min-max of D_1. From the relations we have derived and from an inspection of the entropy scale this can be stated in terms of entropy. The vertebrates display a max-min of H_1 and a min-max of H_M. These game theoretic limits are indicative of an optimization between the opposing elements of variety versus reliability which must occur in any sophisticated language (Gatlin [3]).

4. Other evolutionary indices

A statistician might be interested in whether or not the result that vertebrates have the highest values of D_2 could be duplicated by classical statistical procedures. After all, D_2 is a measure of the "deviation from random" of the base sequence in DNA. Usually when we speak of a "random" sequence, we mean one where there has been no divergence from independence of the symbols as separate and distinct from the divergence from equiprobability.

Figure 4 is a plot of R versus σ, where

$$\sigma = \{\sum_i \sum_j \tfrac{1}{16} (P_iP_j - P_iP_{ij})^2\}^{1/2}. \tag{30}$$

Therefore, σ is the standard deviation from the random of the base sequence in DNA taking into consideration only a first order Markov dependence. This should be a classical counterpart of our D_2 measure. However, σ cannot begin to duplicate the results of the D_2 index. There is significant overlap of the bacterial and vertebrate domains.

It is possible to define arbitrarily other classical evolutionary indices using the standard root mean square form with a slightly different base. T. F. Smith [11]

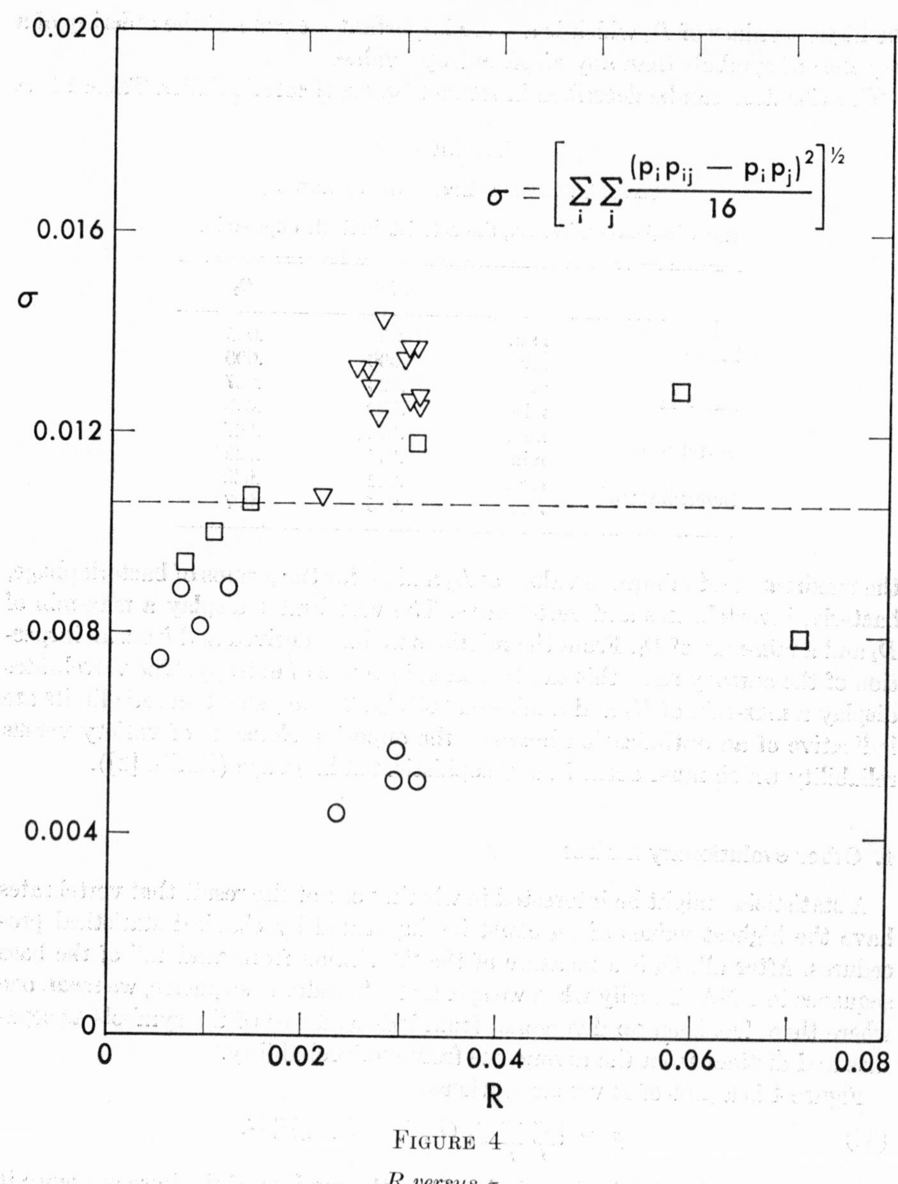

FIGURE 4

R versus σ.

has defined the following evolutionary index. Figure 5 is a plot of R versus e, where

$$e = \{\sum_i \sum_j (P_{ij} - P_j)^2\}^{1/2}. \tag{31}$$

Here the base of the index is the transition matrix element P_{ij} minus the base composition value P_j. The results are much better. There is separation of the

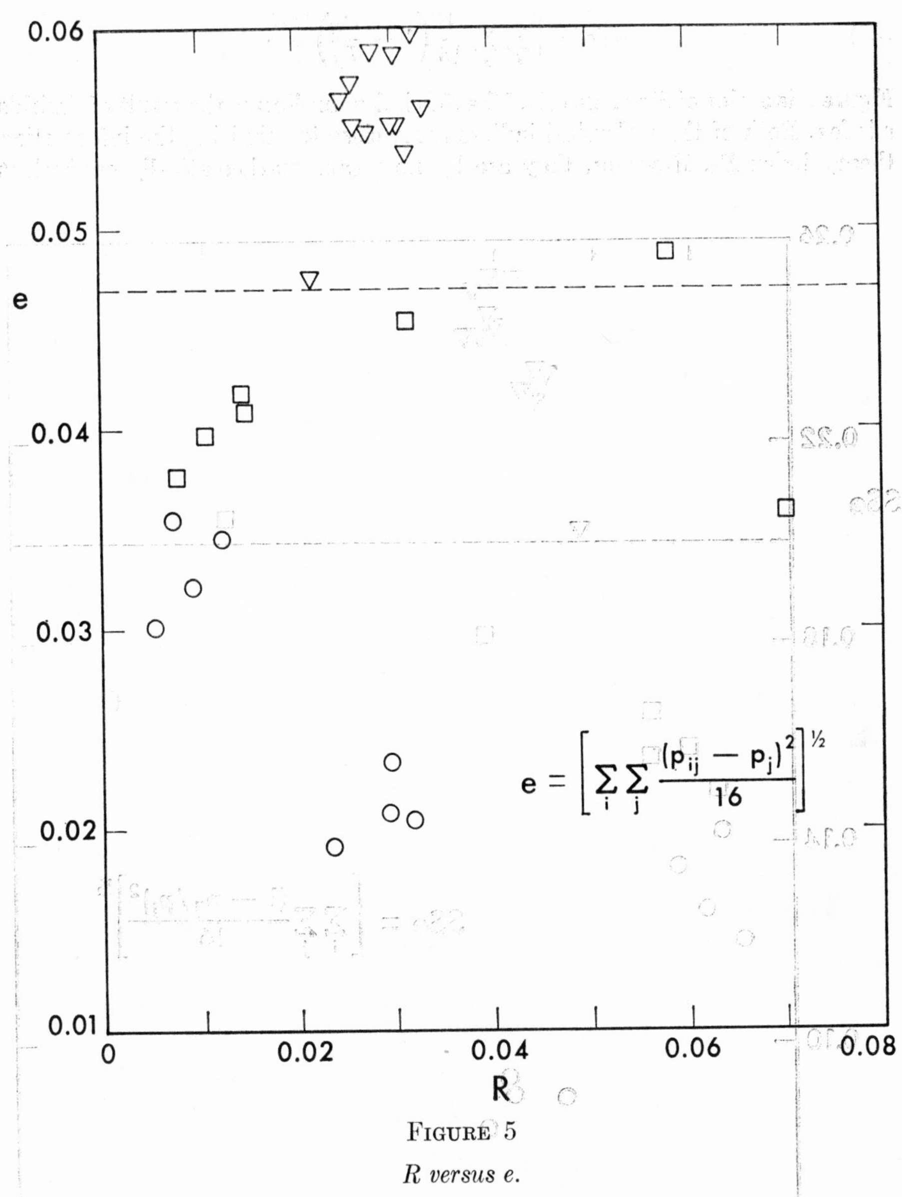

FIGURE 5

R versus e.

vertebrate and lower organism domains with only a very slight overlap at the boundary.

The experimentalist, Subak-Sharpe [12], who has worked extensively with the nearest neighbor data, has an intuitive, algorithmic procedure by which he analyzes the data. I have summarized his algorithm and defined the following evolutionary index:

$$SSe = \left\{\sum_i \sum_j \frac{1}{16}\left(1 - \frac{P_{ij}}{P_j}\right)^2\right\}^{1/2}. \tag{32}$$

Figure 6 is a plot of R versus SSe. The SSe index duplicates the result of Smith's e index. Both of these classical indices come close to mimicing the information theory index D_2. However, they are by no means mathematically equivalent

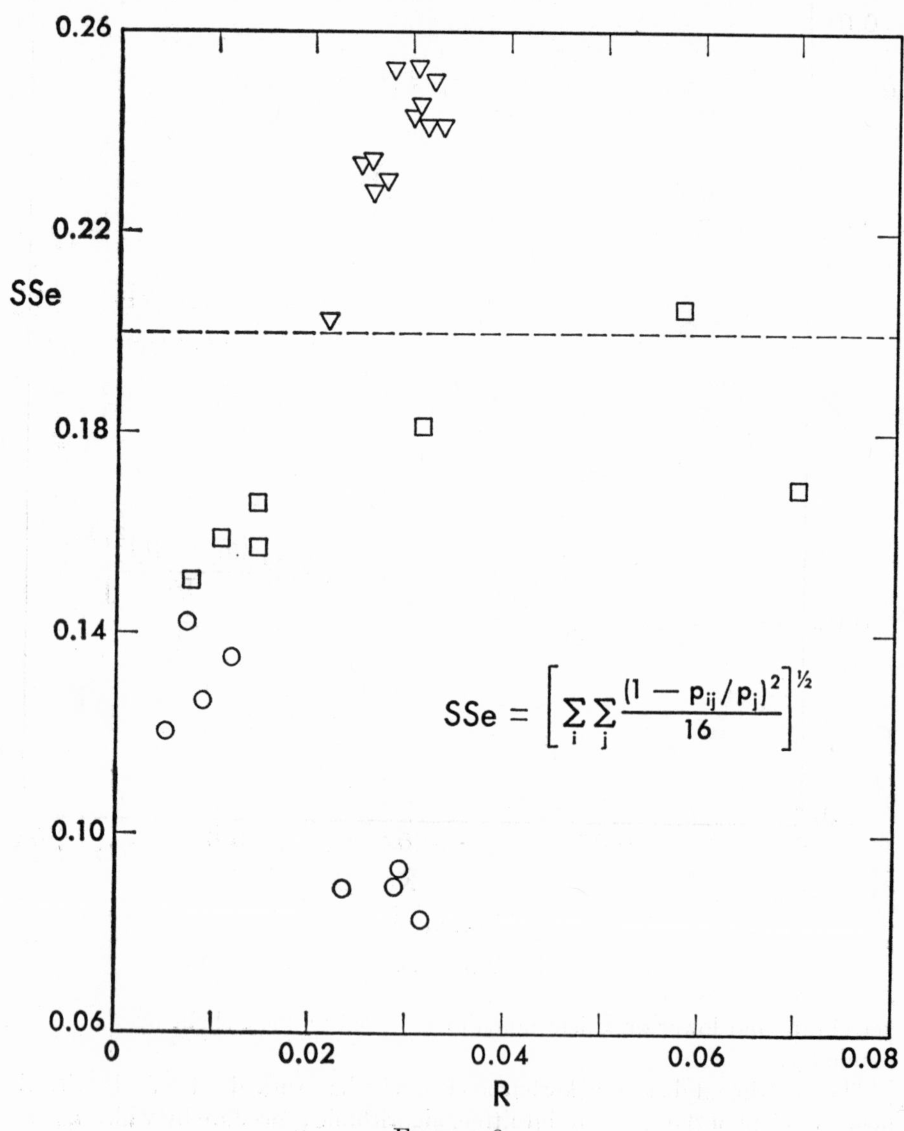

FIGURE 6

R versus SSe.

because nowhere either in the definition of e or SSe does the concept of entropy enter in, along with its inevitable logarithmic functional form.

The index D_2 is slightly better quantitatively than either of the classical indices but this is not the primary reason why D_2 is vastly superior as an evolutionary index. It is an entropy function and because of the structure with which it endows the entropy concept we are left not with just an isolated arbitrary result as we would have been with the classically defined indices, but with an explanation of our result and a workable theory which allows us to explore a vast and new conceptual area.

5. Shannon's second theorem

Let us speculate on the evolutionary implications of our observations. Let us assume that the first DNA molecules assembled in the primordial soup were random sequences, that is, D_2 was zero, and possibly also D_1. One of the primary requisites of a living system is that it reproduce itself accurately. If this reproduction is highly inaccurate, the system has not survived. Therefore, any device for increasing the fidelity of information processing would be extremely valuable in the emergence of living forms, particularly higher forms.

Redundancy, or information density, is a measure of all the constraints placed upon a sequence which make possible error detection and correction. Therefore, redundancy is in this sense a measure of the fidelity of a message. Lower organisms first attempted to increase the fidelity of the genetic message by increasing R primarily by increasing D_1, the divergence from equiprobability of the symbols. This is a very unsuccessful and naive technique because as D_1 increases, the potential message variety, the number of different words that can be formed per unit message length, declines. This is not difficult to show (Gatlin [3]) and in the limit at the maximum divergence from equiprobability, we would have the distribution where one of the p_i is one and all the rest are zero. This is a monotone, a sequence of only one letter which has no message variety at all. Hence, the lower organisms which have achieved R values in the vertebrate range or above have purchased them at the expense of a reduction in potential message variety. This is why they have remained "lower" organisms.

A much more sophisticated technique for increasing the accuracy of the genetic message without paying such a high price for it was first achieved by vertebrates. First they fixed D_1. This is a fundamental prerequisite to the formulation of any language, particularly more complex languages. We observe it in human languages. The particular distribution of the single letter frequencies in human language is so stable and characteristic of a given language that this is a fundamental tool used by cryptographers in decoding messages. When a cryptographer is faced with an unknown message, he first begins to count the single letter frequencies. If the message is in English, the letter *e* will always be the most frequently occurring providing the text is of sufficient length. The distribution of the P_i on S_1 is stable and characteristic of a given language. The vertebrates were the first living organisms to achieve the stabilization of D_1, thus laying the foundation for the formulation of a genetic language. Then they increased D_2 at relatively constant D_1. Hence, they increased the reliability of the genetic message without loss of potential message variety. They achieved a reduction in error probability without paying too great a price for it, and an information theorist would recognize this as the utilization of Shannon's second theorem, the coding theorem of information theory.

This kind of sophisticated reduction in the error of a message was first set forth in the second theorem of Shannon [10] which states that under certain conditions it is possible to reduce the error of a message to an arbitrarily small value even in a noisy channel and without reduction in transmission rate provided that the message has been properly encoded at the source. This statement still reflects the jargon of the communications engineer, but the second theorem principle is a broad, fundamental principle which can be stated in many ways. Let us state it in the language of the biologist.

It is possible within limits to increase the fidelity of the genetic message without loss of potential message variety provided that the entropy variables change in just the right way, namely, by increasing D_2 at relatively constant D_1. This is what the vertebrates have done. They have utilized Shannon's second theorem. This is why we are "higher" organisms.

6. Language in general

In review, the theory upon which the definition of D_1 and D_2 is based is perfectly general and could be applied to language in general. Then we observed in the genetic language the increase of D_2 at constant D_1 as a fundamental mechanism for increasing the fidelity of the genetic message. Now I ask the question: Is this mechanism a general mechanism for increasing the fidelity of any message? Is it used anywhere in human language? It is.

The human mind is an information processing channel, the most complex in the universe, and like any channel possesses a certain capacity, an upper limit to the rate at which it can receive and process information. If information is transmitted at a rate which overloads this capacity, the result is *not* that an amount of information up to the channel capacity is received and processed and the rest "spills over." The result of overloading the channel is utter confusion and chaos. Any good teacher knows this, and very carefully and with deliberation lays a firm foundation of fundamentals before increasing the rate of transmission of information to the student. It is extremely important in the initial stages of this process that error is held to an absolute minimum. Therefore, any device for increasing the fidelity of a message is extremely useful.

One of the most important learning processes which the human mind undergoes is when a little child learns to read the written language. He has spoken it for several years before he learns to read it and this is a major advancement. It is obvious that any safeguards against error in the early critical stages of this learning process would be invaluable.

I shall now show that the writers of children's textbooks intuitively utilize the basic device of increasing D_2 at constant D_1 to increase the fidelity of the message. I selected a series of well-known children's readers beginning with the primer and continuing through the sixth grade. The series selected is the Ginn Basic Reader series (Ginn and Company, New York). I calculated the redundancy of each book by taking texts of increasing length until the R value stabilized taking into consideration only the first order Markov effect. Figure 7 is a plot of R versus the grade of the reader. The R value is quite high in the primer and follows a very smoothly declining curve as the grade of the reader increases. In Figure 8, I have taken the R value apart into D_1 and D_2. It is very apparent that the high R value in the early readers has been achieved by increasing D_2 at constant D_1, just like the vertebrates. Therefore, this appears to be a fundamental mechanism, a general mechanism, for increasing the fidelity of a message.

7. Second theorem selection

We must now inquire into the detailed evolutionary mechanisms whereby the vertebrates have achieved a DNA message with higher R of the high $RD2$ type. The fundamental underlying mechanism is natural selection; but it is a different type of selection than we have considered previously. To define this type of selection we must be more explicit about the jargon of the communications engineer. This is diagrammed in Figure 9.

Here we have an information processing channel. The source or transmitter is just any mechanism for generating a sequence of symbols. The encoding of a message in a particular language occurs at the source. The channel is simply any medium over which the message is transmitted and finally received at the output of the channel. Conceptually, it is just anything one regards as intermediate between the transmitter and receiver, and hence may be sometimes somewhat a matter of definition. I define the base sequence of DNA as the encoded message at the source of the living channel and the amino acid sequence of proteins as the message which is finally received at the output. This is, of course, in a different language. The channel consists of the entire mechanics of protein synthesis which we know a great deal about today due to the massive experimental efforts expended in this area.

All evolutionary thought to date has focused its attention primarily upon the output of this channel, the protein. Natural selection acts because of the sequence of amino acids in proteins. Even the so-called "non-Darwinian" theories of evolution which have arisen recently still focus their attention on the output of the channel and it is here that they search for the reason why a mutation is selectively neutral, the ultimate reason being that the amino acid in the protein is not critical to the function of the protein.

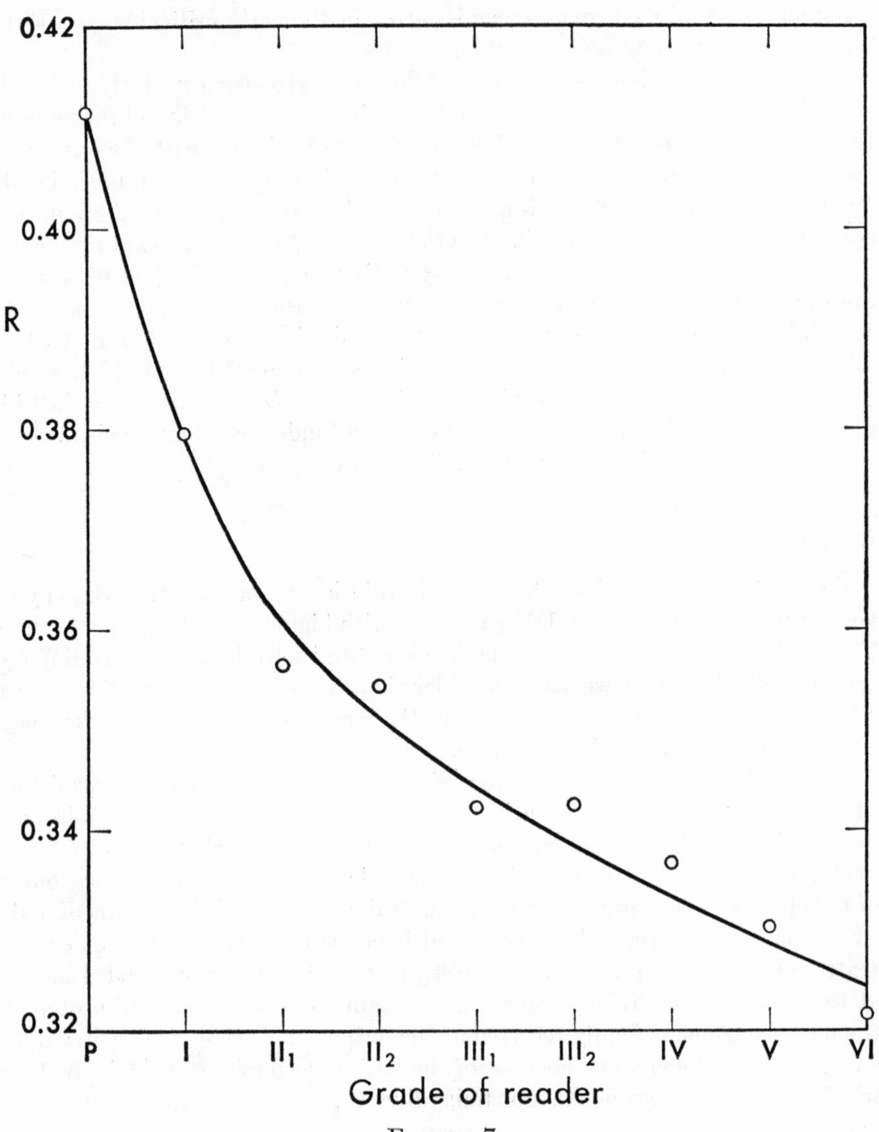

FIGURE 7

R versus grade of reader.

One can pick up any paper in the evolutionary literature, particularly the more recent ones, and confirm this preoccupation with the output of the channel. For example, I quote from Ohta and Kimura [9]: "From the point of view of survival probability, the amino acid substitution between a particular pair has a certain average probability of being accepted by natural selection." Even survival

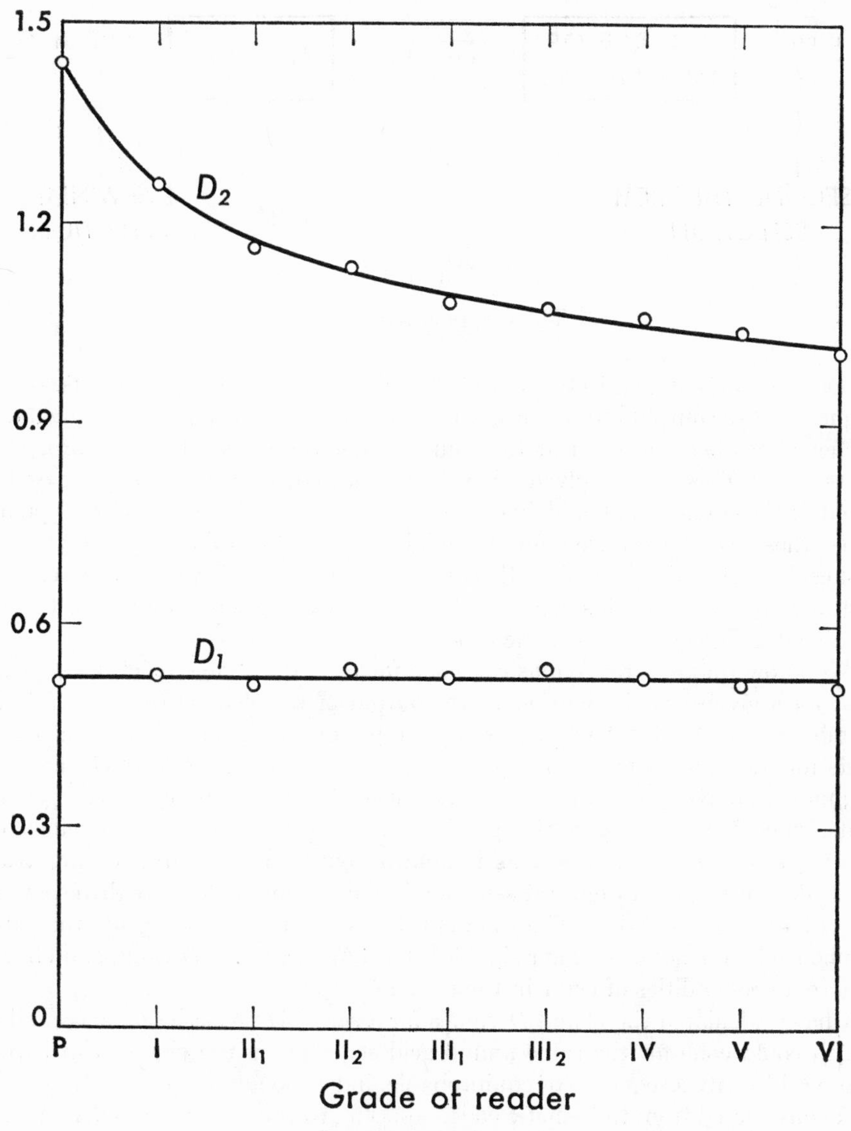

FIGURE 8

D_1 and D_2 versus grade of reader.

probabilities are conceived of in terms of amino acid substitutions in the protein at the output of the channel.

I wish to consider a new type of selection which I shall call second theorem selection because this is the basic principle under which it acts. Second theorem selection directs our attention for the first time to the input of the channel. I

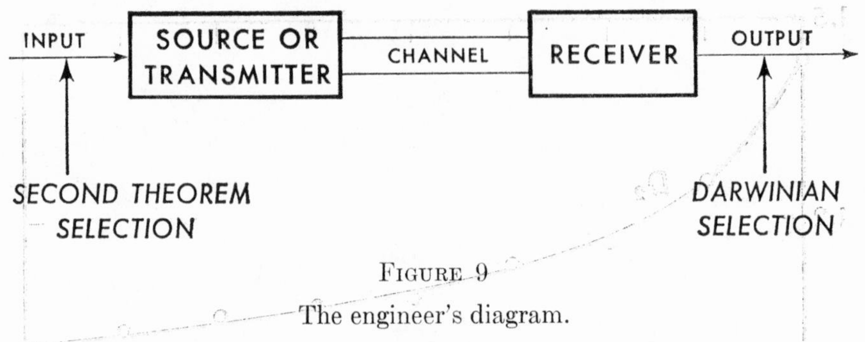

FIGURE 9

The engineer's diagram.

define second theorem selection as natural selection which acts not because of the sequence at the output but because of the informational efficiency with which this sequence has been encoded at the source under the second theorem principle. As vertebrates have evolved, they have selected for DNA sequences at the input of the channel with a higher information density of the high $RD2$ type because these sequences have a lower probability of error in the information processing channel, and they achieve this higher measure of fidelity without paying an excessive price for it. This type of selection is made possible because of the extensive degeneracy of the genetic code.

We know that several codons can code for the same amino acid. This means that for a given protein message at the output of the channel there are a large number of possible DNA sequences at the input of the channel all of which could code for it. Under current concepts these sequences are all selectively neutral. I quote from King and Jukes [8]: "Because of the degeneracy of the genetic code, some DNA base-pair changes in structural genes are without effect on protein structure. . . . As far as is known, synonymous mutations are truly neutral with respect to natural selection." This is not the case with respect to second theorem selection. The different DNA sequences coding for the same amino acid sequence could have significantly different R and D values, and hence different probabilities of error in the channel.

I have calculated the R and D values for a set of DNA base sequences, all of which could code for the same amino acid sequence in protein (Gatlin [2]). I chose arbitrarily a sequence of equiprobable, independent amino acids in protein and constructed from the genetic code, a dozen arbitrary types of DNA base sequences, all of which could code for this same amino acid sequence. For this small sample of DNA sequences the R value ranged from 0.021 to 0.224, a variation of 20.3 per cent of the entire theoretical range of R. This is a very significant variation. The $RD2$ values ranged from 0.58 to 0.97 which is very close to the vertebrate range of $RD2$ values. Thus, there is adequate variation for second theorem selection to act upon. Therefore, second theorem selection can distinguish between different DNA base sequences all of which give rise to the same amino acid sequence in protein. This is a new concept in evolutionary thought.

Let us go on a step further. It is possible that second theorem selection can distinguish between different DNA sequences which code for slightly different amino acid sequences which are selectively neutral in the Darwinian sense.

We now believe that selectively neutral mutations are fixed by random drift (Kimura [7]). However, there are certain discrepancies between this random model and the experimental data (Jukes [5]). If we impose the concept of second theorem selection as a constraint upon this random model, perhaps this will improve the agreement. This possibility is totally unexplored.

In conclusion, we see that D_2 is not *just* another evolutionary index which can distinguish between vertebrates and lower organisms. It is an entropy function. It extends the entropy concept and endows it with structure. It defines a fundamental mechanism for increasing the fidelity of a message which we observed in the genetic language and in human language. We are led into the consideration of a new evolutionary principle which is the confluence of Darwin's principle of natural selection and Shannon's second theorem. This is an organizing principle in contrast to the disorganizing principle of thermodynamics. And finally, we are left with the rather satisfying explanation that it is the second theorem of information theory rather than the second law of thermodynamics which has given the evolution of life its unique direction.

REFERENCES

[1] F. E. Arrighi, M. Mandel, J. Bergendahl, and T. C. Hsu, "Buoyant densities of DNA of mammals," *Biochem. Genet.*, Vol. 4 (1970), pp. 367–376.

[2] L. L. Gatlin, "The information content of DNA. II," *J. Theor. Biol.*, Vol. 18 (1968), pp. 181–194.

[3] ———, *Information Theory and the Living System*, New York, Columbia University Press, 1972, in press.

[4] J. Josse, A. D. Kaiser, and A. Kornberg, "Enzymatic synthesis of deoxyribonucleic acid. VIII. Frequencies of nearest neighbor base sequences in deoxyribonucleic acid," *J. Biol. Chem.*, Vol. 236 (1961), pp. 864–875.

[5] T. H. Jukes, "Comparison of polypeptide sequences," *Proceedings of the Sixth Berkeley Symposium on Mathematical Statistics and Probability*, Berkeley and Los Angeles, University of California Press, 1972, Vol. 5, pp. 101–127.

[6] A. I. Khinchin, *Mathematical Foundations of Information Theory*, New York, Dover, 1967.

[7] M. Kimura, "Evolutionary rate at the molecular level," *Nature*, Vol. 217 (1968), pp. 624–626.

[8] J. L. King and T. H. Jukes, "Non-Darwinian evolution," *Science*, Vol. 164 (1969), pp. 788–798.

[9] T. Ohta and M. Kimura, *Nature*, 1971, in press.

[10] C. Shannon and W. Weaver, *The Mathematical Theory of Communication*, Urbana, University of Illinois Press, 1949.

[11] T. F. Smith, "The genetic code: information density and evolution," *Math. Biosci.*, Vol. 4 (1969), p. 179.

[12] H. Subak-Sharpe, R. R. Burk, L. V. Crawford, J. M. Morrison, J. Hay, and H. M. Keir, "An approach to evolutionary relationships of mammalian DNA viruses through

analysis of the pattern of nearest neighbor base sequences," *Symp. Quant. Biol.*, Vol. 31 (1966), p. 737.

[13] M. N. Swartz, T. A. Trautner, and A. Kornberg, "Enzymatic synthesis of deoxyribonucleic acid. XI. Further studies on nearest neighbor base sequences in deoxyribonucleic acids," *J. Biol. Chem.*, Vol. 237 (1962), pp. 1961–1967.

THE AMOUNT OF INFORMATION STORED IN PROTEINS AND OTHER SHORT BIOLOGICAL CODE SEQUENCES

THOMAS A. REICHERT
CARNEGIE-MELLON UNIVERSITY

1. Introduction

These remarks were made to the conference assembly in a context not unlike that of a surprise witness for the defense. This work was so hot off the press that there had been no time to communicate it before the conference itself. I am grateful to Dr. Lila Gatlin for the opportunity to make this presentation. All of the work to be discussed here has been done in collaboration with A. K. C. Wong, also of the Biotechnology Program at Carnegie-Mellon University.

In the last year, we have developed a measure of the amount of information required to perform genetic mutations, together with an algorithm utilizing these measures, for aligning amino acid and RNA code sequences [9], [11]. In the process of this development, we attempted to calculate the amount of information which was stored in a protein's amino acid sequence. We had, at the time, only the tools of the conventional communications form of information theory. Thus, we attempted the calculation using the two expressions:

$$H = -\sum_{i=1}^{a} p(i) \log p(i) \tag{1}$$

or

$$I_{\text{self}} = -\sum_{i=1}^{a} n_i \log p(i). \tag{2}$$

Equation (1) is the expression for the entropy of a discrete information source operating with an alphabet of a letters. This quantity is also interchangeably called the information content of such a source. Since information and entropy are more nearly opposites than synonyms, this equivalence has always been confusing. Indeed, the values of H obtained for a set of cytochrome c sequences, by allowing the amino acid frequencies in each sequence to determine the alphabet character probabilities used in equation (1), displayed the then embarrassing trend of higher information content with lower organism complexity. Since the method used to estimate these probabilities is not generally known, let me describe it here.

It was Laplace, I believe, who first noted that the frequency limit estimate of the probability of an event's occurrence was applicable only in the limit of infinite frequency. In the anecdote reported by many biographers concerning the probability of the morrow's sunrise, he showed the importance of the real difference from this limit. Laplace's formula for the probability of the sun's not rising was $p = 1 - (x + 1)/(n + 2)$, where x is the number of times the sun has previously risen and n is the number of times a new day has presented itself (x apparently $= n$).

R. A. Christensen [2] has formally generalized this estimate. Probabilities so constructed are called relatively unbiased probabilities and are defined by the expression

$$p(i) = \frac{x_i + t}{n + t + f}, \tag{3}$$

where x_i is the number of occurrences of the event i (successes), n is the total number of occurrences of any event (trials), t is the number of possible different realizations of the event i, and f is the number of events in the event space which are not realizations of the event i.

In the absence of any experimental data, this expresssion reduces to the condition of maximum ignorance $p(i) = t/(t + f)$, and assumes the frequency limit form as the data acquisition proceeds to infinity.

This formula has proved especially useful in our case because occasionally one or more amino acids would be completely absent from a particular sequence, and the resulting zero frequency would otherwise have been difficult to handle in a fashion not arbitrary.

Equation (2) is the sum of the self information associated with each character in a sequence. We reasoned that since H is simply the average value of the self information of the particular alphabet in use, $I = -\log p(i)$; then, if we were to use the entire ensemble of homologous sequences to determine the character probabilities, the total self information for each sequence would give us the elements of the distribution having H as its first moment. The set of ensemble based probabilities characterize what we call the "super source" for the particular protein. The values of I_{self} so obtained were, if anything, more blatant in the inverse correlation of information content with intuitive notions of complexity. Faced with this apparently incomprehensible result, we took the only course open to conscientious scientists, and placed the results in an appendix to an overlong paper where it was referred to only obliquely.

2. Formulae

The matter hung thus in limbo until we discovered the work of L. L. Gatlin [6], [7], [8]. She has explained how, in her formulation, the information storage ability of an information source is measured by the deviations of its entropy from the theoretical maximum. For a source whose next emission depends only on the last character emitted, a first order Markov source, the information density I_d is given by $I_d = H_{\max} - H_{\text{Markov}}$. This difference she has further decomposed into two components

$$D_1 = H_{\max} - H_1, \quad (4)$$
$$D_2 = H_1 - H_{\text{Markov}},$$

where D_1 is the deviation of the source entropy from equiprobability and D_2 is its deviation from independence.

DNA sequences are of enormous length, effectively infinite, so that the source information density and the average information per sequence character are essentially identical quantities. The short sequences of proteins are, however, another matter. You will remember that H_{Markov} is given by

$$\sum_i p(i) \sum_j p(j/i) \log p(j/i), \quad (5)$$

where $p(j/i)$ is the probability of occurrence of the $(i - j)$th pair of characters.

The probabilities, both marginal and conditional, are determined by the entire ensemble of homologous sequences so that H_1, H_{Markov}, and thereby I_d would, utilizing the two expressions given above, have the same value for every such sequence. The key to this dilemma lies in recalling that H_1 and H_{Markov} are both averages of the form $\langle f(i)\rangle = \sum_i p(i)f(i)$. Replacing this formulation by that for obtaining a simple mean of the correspondent self information measures, we obtain, for equations (1) and (5),

$$\bar{I}_{\text{self}_1} = -\sum_i \frac{n_i}{\sum_i n_i} \log p(i) = -\frac{1}{N} \sum_i n_i \log p(i),$$

$$\bar{I}_{\text{self}_M} = -{\sum_i}' \frac{n_i}{\sum_i' n_i} \sum_j \frac{n_{ij}}{\sum_j n_{ij}} \log p(j/i) \quad (6)$$

$$= -\frac{1}{N} {\sum_i}' \frac{n_i}{\sum_j n_{ij}} \sum_j n_{ij} \log p(j/i),$$

where n_{ij} is the number of $i \to j$ directed pairs in the sequence. The $\sum'$ indicates that only those residues which can form pairs are to be included in the average. The last residue in the sequence has no following residue. Thus, $\sum_i' n_i = \sum_i n_i - 1 = N'$, and

$$\bar{I}_{\text{self}_M} = -\frac{1}{N} {\sum_i}' \sum_j n_{ij} \log p(j/i), \quad (7)$$

where $\bar{I}_{\text{self}_1}$ and $\bar{I}_{\text{self}_M}$ are the short sequence analogs of Gatlin's H_1 and $H_{M=\text{Markov}}$. Thus, we may define the analogous deviations

$$D_1 \triangleq \log a - \bar{I}_{\text{self}_1}, \quad (8)$$
$$D_2 \triangleq \bar{I}_{\text{self}_1} - \bar{I}_{\text{self}_M}.$$

3. Applications

3.1. *Cytochrome c.* When these measures are assembled for the set of available cytochrome c sequences (the ensemble presently contains 33 sequences), the values in Table I are obtained. Figure 1 is a plot of D_2, the deviation of the amino

TABLE I

INFORMATION MEASURES FOR CYTOCHROME *c* SEQUENCES

D_2 and R are calculated only for the 27 sequence ensemble.

Cytochrome *c*	Length	$\bar{I}_{\text{self}_1}$	$\bar{I}_{\text{self}_M}$	D_2	R	TD_2	TI
1. Human	104	3.998	3.017	1.008	0.3104	105.1	138.8
2. *Rhesus* monkey	104	3.993	2.985	1.042	0.3195	107.9	142.1
3. Pig and bovine	104	3.958	2.902	1.105	0.3431	112.7	150.6
4. Horse	104	3.956	2.994	1.025	0.3253	103.1	141.1
5. Donkey	104	3.969	2.94	1.093	0.3377	110.	146.7
6. Dog	104	3.956	2.861	1.138	0.3508	116.7	154.8
7. Rabbit	104	3.987	2.899	1.138	0.3433	116.	150.9
8. California gray whale	104	3.964	2.88	1.131	0.3472	115.6	152.9
9. Kangaroo	104	3.935	2.99	0.9671	0.3166	101.2	141.5
10. King penguin	104	4.007	2.962	1.116	0.3329	111.7	144.4
11. Chicken and turkey	104	4.016	3.001	1.077	0.3216	107.6	139.4
12. Pekin duck	104	4.015	2.948	1.137	0.3354	113.9	145.8
13. Pigeon	104	3.999	2.997	1.059	0.3217	107.1	140.7
14. Snapping turtle	104	3.955	3.003	0.983	0.3144	102.	140.1
15. Dogfish	104	4.05	3.14	0.9303	0.2773	97.8	126.1
16. Pacific lamprey	104	4.044	3.147	0.9318	0.2799	96.3	125.3
17. Silkworm moth	107	4.049	3.216	0.8433	0.256	92.4	121.6
18. Tobacco horn worm moth	107	4.071	3.2	0.8938	0.2626	96.3	123.2
19. Fruit fly	107	4.018	3.122	0.9254	0.2839	99.	131.5
20. Screw-worm fly	107	4.024	3.063	0.9926	0.2981	105.9	137.8
21. Wheat	112	4.110	3.44	0.4475	0.1454	78.5	102.2
22. *Neurospora crassa*	107	4.024	3.725	0.2615	0.1294	35.7	67.6
23. Baker's yeast	108	4.077	3.695	0.3914	0.1465	45.	71.4
24. *Candida krusei*	109	4.146	3.862	0.2952	0.1047	34.8	54.
25. Rattlesnake	103	4.071	3.219	0.8905	0.2631	91.	116.8
26. Tuna	104	3.996	3.153	0.8555	0.2746	90.8	124.7
27. Bullfrog	104	4.013	3.056	0.9951	0.3029	102.6	134.7
28. Mung bean	111	4.114	3.533			68.	91.1
29. Castor bean	111	4.087	3.499			68.8	94.9
30. Sesame	111	4.112	3.483			73.3	96.6
31. Sunflower	111	4.099	3.407			80.2	104.9
32. *Physarum*	108	4.132	3.531			68.4	88.9
33. *Euglena*	105	4.127	3.997			17.7	38.2

Hemoglobin α	Length	TD_2	TI
Human	141	123.3	185.5
Gorilla	141	117.6	180.9
Rhesus monkey	141	121.	184.3
Mouse C57 strain	141	107.7	168.1
Mouse NB strain	141	105.7	165.5
Rabbit	141	109.5	152.4
Horse (slow sequence)	141	118.2	183.6
Donkey	141	110.2	175.2
Sheep α strain	141	107.1	172.3
Sheep δ strain	141	105.8	170.9
Carp	142	48.3	77.1
Horse (fast sequence)	141	118.5	180.8
Bovine	141	106.1	176.6
Chicken	140	87.	125.7
Kangaroo	129	78.8	125.3

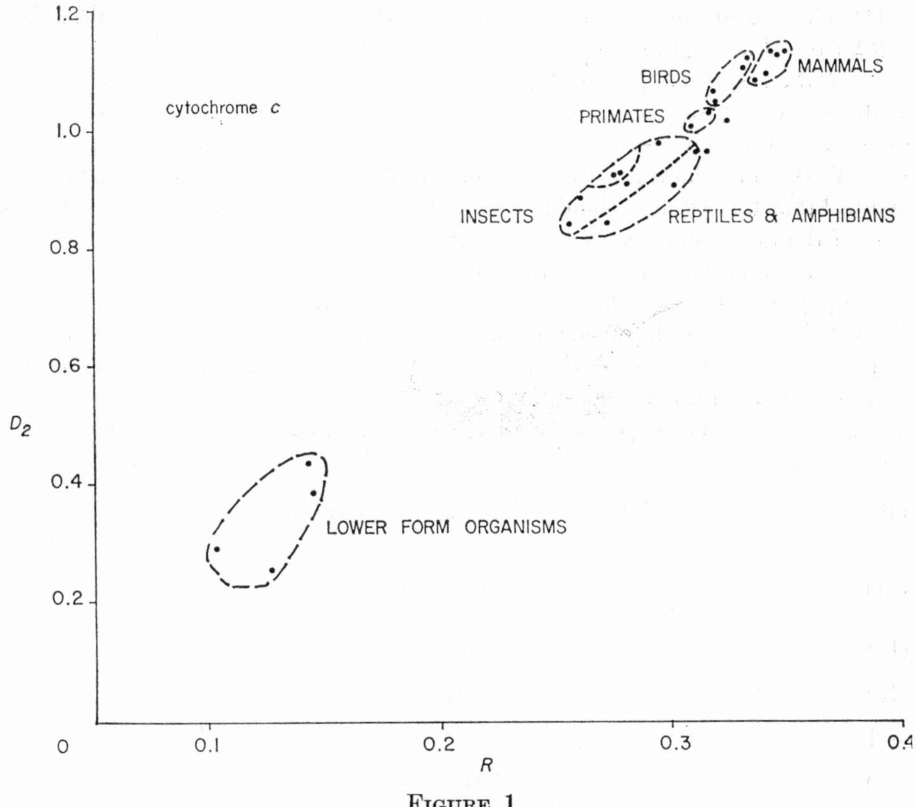

FIGURE 1

The average deviation of the amino acid sequence average self information from independence, D_2, *versus* the redundancy of that sequence for 27 species relative to the ensemble super source based on the same set of sequences.

acid code of each sequence from independence, *versus* R, the Shannon redundancy of the code based on an ensemble comprised of the first 27 cytochrome c sequences enumerated in Table 1 of [3]. The Shannon redundancy

$$R = \frac{D_1 + D_2}{\log a} \tag{9}$$

is proportional to the short sequence average information density. This plot is the short sequence analog of the figure just presented by Lila Gatlin for DNA sequences [8]. In her figure, you will remember that the DNA for the vertebrate was characterized by higher values of D_2 than that of bacteria which was, in turn, higher in D_2 than the DNA for phages. The same correlation with complexity is apparent in Figure 1. What are commonly called higher organisms exhibit higher values of D_2. There are, however, additional features present in Figure 1 which are absent in the DNA plot.

(1) The relationship appears to be linear with an intercept of zero. Unlike the DNA case, then, higher organisms are higher both in D_2 *and* in I_d.

(2) Data from many more higher organisms are available for the protein analysis so that a finer screening is possible. We note that it is possible to enclose nearly all of the members of established taxonomic groupings with convex boundaries. We hesitate, however, to ascribe taxonomic significance to these boundaries and feel that these values should be looked on more as the cytochrome *c* coordinate of the taxonomic vector of these organisms.

Since we have acknowledged that the length of our sequences is not infinite, it seems proper to ask if, in view of the fact that the average information per character is higher in higher organisms, the total amount of information stored is also different. The length of all of these sequences is tabulated in Table I. Note that lower form sequences are longer than vertebrate sequences by a significant amount. The expressions for the total information stored in such a sequence are

$$TI_1 = -\sum_{i=1}^{a} n_i \log p(i) = N\bar{I}_{\text{self}_1}, \tag{10}$$

$$TI_M = -\sum_{i=1}^{a}\sum_{j=1}^{a} n_{ij} \log p(j/i) = N'\bar{I}_{\text{self}_M}, \tag{11}$$

$$TD1 = N \log a - T1_1, \tag{12}$$

$$TD2 = TI_1 - T1_M, \tag{13}$$

and

$$TI = TD1 + TD2. \tag{14}$$

A plot of $TD2$ *versus* TI for all 33 sequences is presented in Figure 2. Note that TR is not suitable since it is length independent as seen in the analog expression $TR = TI/N \log a$.

The clusterings here become a bit more tortuous to construct. Particularly troublesome are the set of insect sequences. On the other hand, the data is even more highly linear, indicating that our earlier caution against the use of the information stored in a single protein as a monophyletic basis for taxonomy is probably well advised. The addition of the *Euglena* sequence [4] not only provides the lowest point thus far, but also makes the straight line a poor extrapolation to low values. One obvious interpretation of Figure 2 is that evolution appears to be a process of the acquisition of information. The curve best fitting the data would indicate that the lowest life forms have virtually all of their information stored as $TD1$. Thus, life would appear to have arisen merely from a properly asymmetrical distribution of amino acids with little or no interresidue correlation. This correlation, however, apparently developed quite rapidly after function first evolved. At the level of the slime mold, it would appear that the nature of the process changed, perhaps to refinement rather than development. In this mode both $TD1$ and $TD2$ are augmented at the same rate.

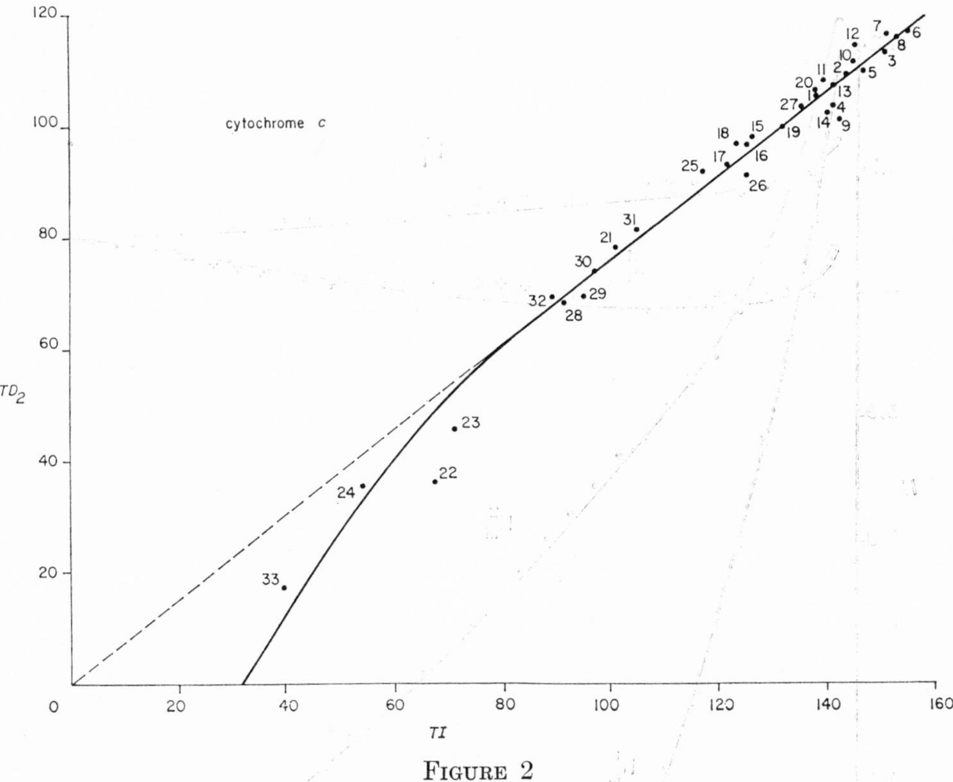

FIGURE 2

The total deviation from independence *versus* the total amount of information stored in the amino acid code of the 33 cytochrome *c* sequences of Table I. The numbers refer to the elements of the table. The slope of the limiting line is 0.75.

Figure 2 contains 33 sequence points based on the 33 sequence super source. Figure 1 contains 27 points based on the ensemble source of the first 27 sequences. This was intentional because it is necessary to examine the measures used for stability and bias.

Figure 3 is a plot of the ensemble entropies H_1 and H_M as a function of the number of sequences included in the ensemble. The sequences were added in two different orders. The approach to some stable value for H_M is clear only in the first case. From the figure, we note that H_I stabilizes much more quickly than H_M. A curious fact relative to these measures is that those elements (both single residues and amino acid pairs) which are *most* common to the ensemble contribute most to the stored information. Thus, those sequences with the highest TI would utilize the elements most common to the ensemble super sources most often. Since the source entropies are, as noted above, the first moments of the distributions characterizing the super source, these high TI sequences will contribute the most to the final frequency distribution. Sequences, then, should be added to the

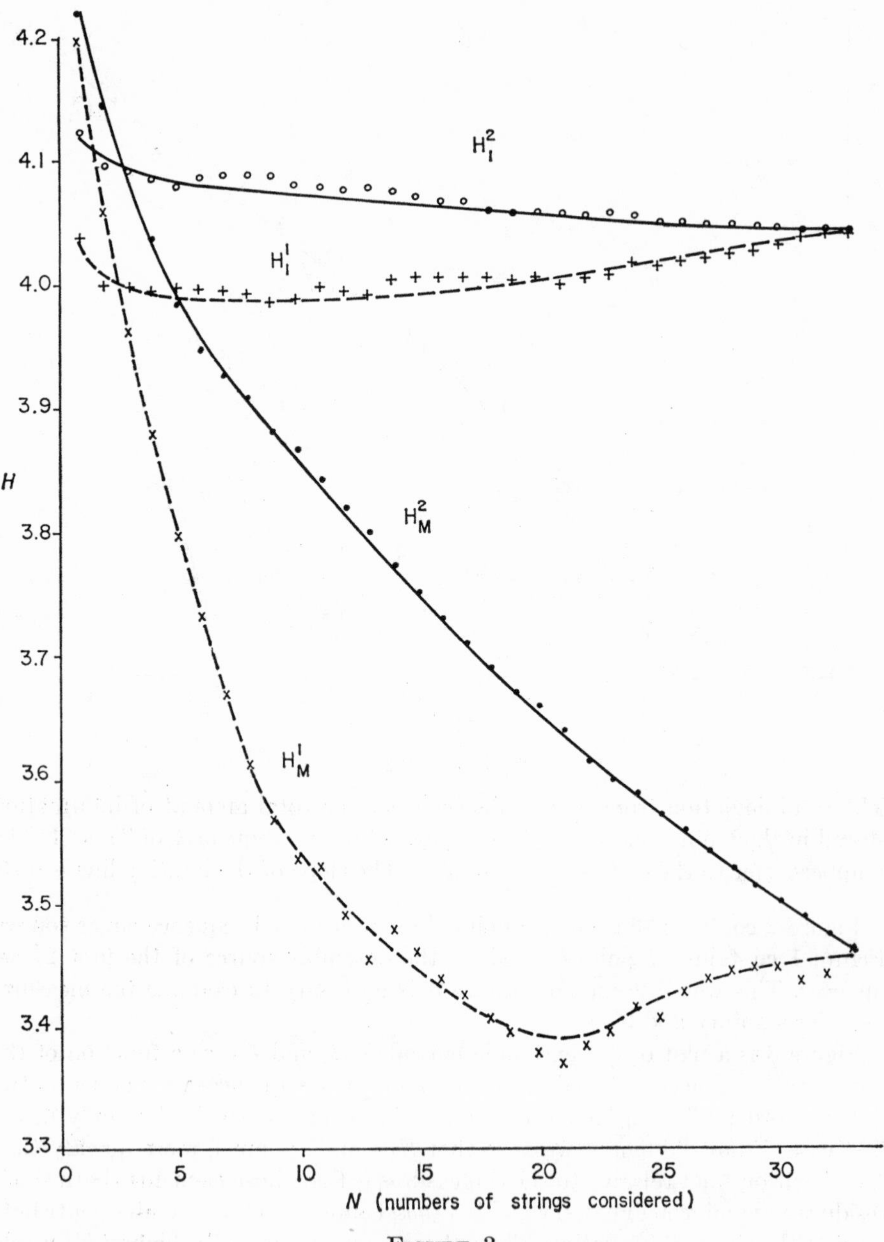

FIGURE 3

The development of the source entropies for cytochrome *c* with ensemble size showing the effect of altering the order of addition of sequences.

ensemble in order of decreasing TI to obtain the most rapid approach to the limiting values. This has been done for H_M^1 for the 33 sequence ensemble source appearing as the dashed line in Figure 3. The second ordering is the inverse of the first, that is, sequences were added in order of increasing TI. No limiting value is in evidence for this case; whereas, for the first ordering, the source entropy was within two to three per cent of its apparent limiting value after ten sequences. It would appear then, that high information sequences are the most important in establishing a protein super source.

The feature, more common-more information, is entirely in conflict with the established notion of the communications form of information theory in which information value and the amount of surprise the message element generates in the receiver are somehow correlated. Because of this feature, one might suspect that the high values accorded some of the taxonomic groupings might be due solely to their overrepresentation in the ensemble. Among the species represented by the first 27 sequences of Table I, there are nineteen vertebrates: 8 mammals, 1 marsupial, 4 birds, 6 reptiles, fish, or amphibians; and eight nonvertebrates: 4 insects, 1 higher plant, 3 lower organisms. Certainly the sample appears to be mammalian biased although the less well represented birds fare at least as well as the mammals. In the ensemble used in Figure 2, the added sequences are four higher plants [1], one slime mold and one protozoan [4]. The mammalian-vertebrate bias is mitigated and certainly the wheat value is substantially altered without changing materially the other orderings. (Typical changes are on the order of three per cent for $TD2$—the same order as the alteration in the source entropy H_M.)

3.2. *Hemoglobin.* The set of hemoglobin sequences which have been completely elucidated number 26 (15 α chains, 8 β chains, 2 δ chains and 1 γ chain). Only the entropies for the α chain source appear to be suitably stable, although it is possible to lump all of the hemoglobin sequences together in an all hemoglobin ensemble. We have already dealt with this possibility in another paper [10]. In Figure 4 we have presented the stability plot for the α chain source in the ordering of the entries in Table I. Figure 5 is a view of the same parameters presented earlier for cytochrome *c*. The limiting linearity is again in evidence. The slope of the line characterizing cytochrome *c* development in Figure 1 is nearly identical to that of a similar plot for hemoglobin. The limiting slopes of Figures 2 and 5 are, however, different. The values for β hemoglobins would fall at lower values along the same line in Figure 5.

4. Discussion

4.1. *Evolution.* The obvious interpretation of Figures 1, 2, and 5, in complete agreement with intuition, that evolution proceeds in the direction of increasing information, has a strong bearing on the construction of ancestor sequences. Fitch [5] has shown that a divergent evolution may be obtained using an algo-

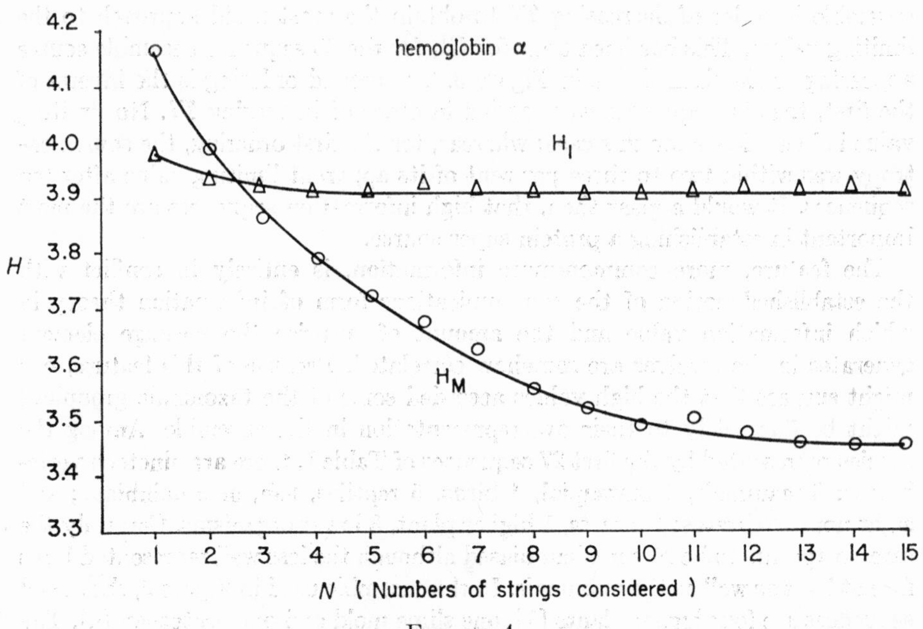

FIGURE 4

The development of the source entropies for hemoglobin α. The order of sequence addition is that of Table I.

rithm for constructing ancestor sequences which has as its basic element the following rule.

Given two homologous sequences which differ at a single site, the ancestral sequence will contain, at that site, that residue, of the two possible, which is most common in the remainder of the sequence.

From the discussion here, an ancestral sequence should possess less information than its progeny. Thus, the residue which would lower *TI* for the sequence should be incorporated in the ancestral sequence. This would be the *rarer* of the two residues and/or that which formed the rarer pairs—where the lowering of *TI* was the preeminent feature. Fitch's sequences are thus more nearly descendant than ancestral.

If *TI* for a cytochrome *c* sequence is calculated using the super source probabilities of another protein, very low, generally negative values of some of the parameters are obtained. The same is true of some myoglobin sequences relative to the hemoglobin α source. The effect is not unlike looking up a particular word, say an English word, in a dictionary for a language other than English. If the language is different enough from that of the word or set of words, no meaning will be assignable. It would seem, for example, that one cannot "say" cytochrome *c* in hemoglobin. This is, of course, very close to an operational definition of homology.

4.2. *The meaning of information.* That higher organisms store more information in homologous molecules, seems to be the message so far. Can we conclude that a plethora of this stuff is better? To begin to answer this question, we examined the set of variants of human hemoglobin α which differ from the normal α chain only by a single amino acid substitution. The informational parameters of this group relative to the ensemble of hemoglobin α sequences are presented in Figure 6. There are three subsets of variants of particular interest demonstrated in the figure.

The first contains the two variants which have TI greater than the normal sequence while $TD2$ is lower. Bearers of these deviant hemoglobins are apparently clinically normal (cn).

The second cluster is the group of variants with $TD2$ and TI closest to the normal sequence. Of these five are clinically normal variations and *one*, M Boston, produces severe complications.

The third cluster is the group of *four* variants which are the lowest in both $TD2$

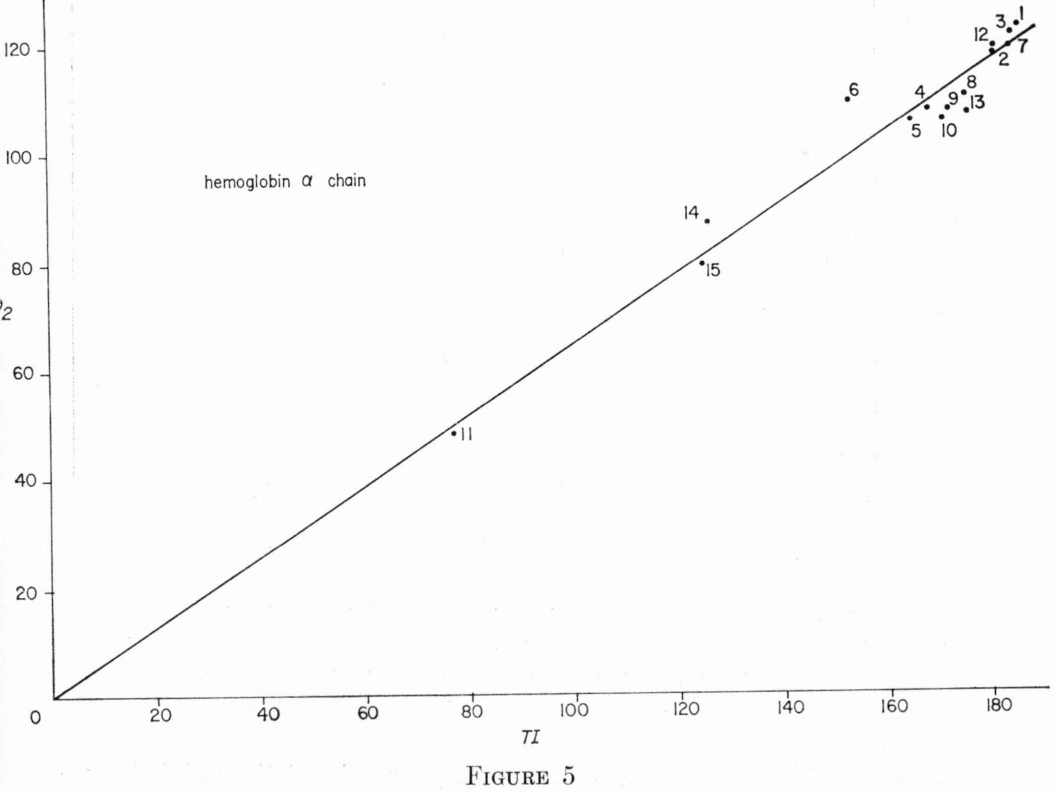

FIGURE 5

A plot of the informational parameters $TD2$ *versus* TI for the α chain of hemoglobin for an ensemble of 15 sequences. The slope of the limiting line is 0.65.

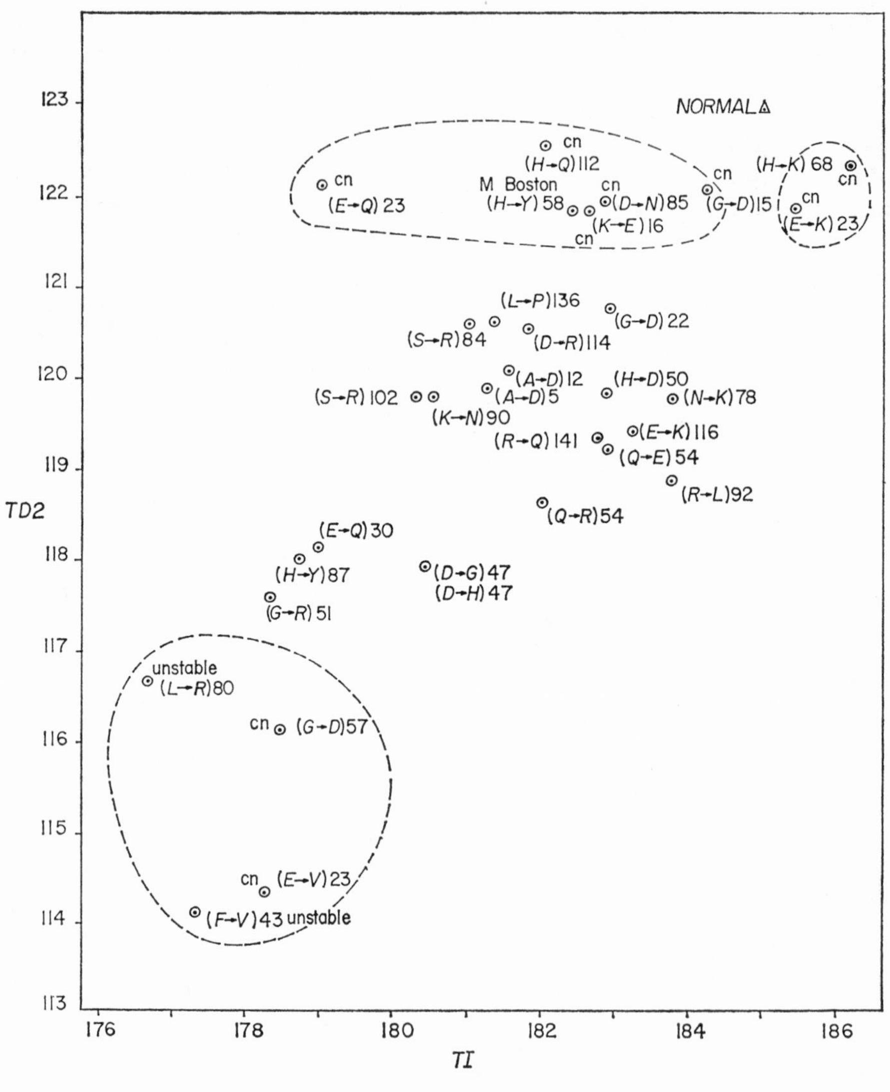

FIGURE 6

A plot of the information stored in the selection of amino acid pairs, $TD2$, *versus* the total amount of information stored in the sequence for all known single substitution variants in the α chain of human hemoglobin. The letters cn mean clinically normal in the heterozygous condition.

and TI. Of this group, *two* represent clinically normal variations while *two* are decidedly pathological. Thus, we have some evidence that indeed "more is better." A careful study of the function associated with each variant site should even further pigeonhole this "stuff," information.

We have recently completed an analysis [12] of the correlation of the informational parameters developed here with the known structural and functional significance of each amino acid in the normal hemoglobins (both α and β) of humans and horses. We find that the residues which are ascribed structural/functional significance are highly optimized, that is, any substitution at that site will produce a decrease in TI. The sites accorded no such importance, on the other hand, often have several possible information-improving substitutions.

By further refinements in these studies, we may hope to further localize this property heretofore spoken of only in its form as an average. The molecular biological format provides a testable basis for significance, and may even lead us, quite incidentally, to the meaning of meaning.

We have limited our discussion here to proteins. We have made some attempt to treat *t*-RNA premethylation sequences in a similar fashion. The results were not well defined, however, and it is clear that the modified bases must, in some way, be put into the formulation.

REFERENCES

[1] D. Boulter, E. W. Thompson, J. A. M. Ramshaw, and M. Richardson, "Higher plant cytochrome *c*," *Nature*, Vol. 228 (1970), pp. 552–554.

[2] R. A. Christensen, "A general approach to pattern discovery," University of California, Berkeley, Computer Center, Technical Report No. 20, 1967.

[3] M. O. Dayhoff, *Atlas of Protein Sequence and Structure*, Silver Spring, Md., National Biomedical Research Foundation, 1969.

[4] W. M. Fitch, Personal communication.

[5] ———, "Distinguishing homologous from analogous proteins," *Syst. Zool.*, Vol. 19 (1970), pp. 99–113.

[6] L. L. Gatlin, "The information content of DNA," *J. Theoret. Biol.*, Vol. 10 (1966), pp. 281–300.

[7] ———, "The information content of DNA II," *J. Theoret. Biol.*, Vol. 18 (1968), pp. 181–194.

[8] ———, "Evolutionary indices," *Proceedings of the Sixth Berkeley Symposium on Mathematical Statistics and Probability*, Berkeley and Los Angeles, University of California Press, 1972, Vol. 5, pp. 277–296.

[9] T. A. Reichert and A. K. C. Wong, "An application of information theory to genetic mutations and the matching of polypeptide sequences," submitted to *J. Theoret. Biol.*

[10] ———, "Toward a molecular taxonomy,'" *J. Molec. Evol.*, Vol. 1 (1971), pp. 99–111.

[11] A. K. C. Wong, T. A. Reichert, and B. Aygun, "A generalized method for aligning unambiguous code sequences," submitted to *J. Computers Biol. Med.*

[12] A. K. C. Wong and T. A. Reichert, "The structure and function of hemoglobin reflected in code sequence optimization," submitted to *J. Mol. Biol.*

HIERARCHIES OF CONTROL PROCESSES AND THE EVOLUTION OF CONSCIOUSNESS

RICHARD BELLMAN
UNIVERSITY OF SOUTHERN CALIFORNIA

1. Introduction

The purpose of this paper is to sketch an application of some ideas in the mathematical theory of control processes to biological phenomena such as instinct, learning, curiosity, adaptation, and, finally, consciousness. We shall employ the language and methodology of the theory of dynamic programming. Detailed accounts of the mathematical ideas will be found in [1], [2], [3]. For a quite different approach to consciousness, see [6].

2. Control processes

We begin with the idea of a process. Consider a system described by a point p in a space S. Let $T(p)$ be a transformation with the property that $p_1 = T(p)$ belongs to S whenever p is in S. We call the pair $[p, T(p)]$ a process. More precisely, this is a particular description of a process.

When the transformation is repeated, yielding a sequence of states, $p_1, p_2, \cdots$, where $p_1 = T(p)$, $p_2 = T(p_1)$, $\cdots$, we call it a multistage process. This is an abstract version of a dynamic process.

Assume next that $T(p)$ is replaced by a transformation of the form $T(p, q)$ having the property that for any p in S and any q in a decision space D, the point $p_1 = T(p, q)$ is in S. A choice of vectors (decisions) $q_1, q_2, \cdots$, then yields a sequence of states $p_1 = T(p, q_1)$, $p_2 = T(p_1, q_2)$, $\cdots$. We call this a multistage decision process. It is also an abstract version of a control process. From the mathematical point of view, control and decision processes are equivalent; see [2], [3].

A determination of the q_i may be effected by maximizing a criterion function which depends on the history of the process $K = K(p, p_1, \cdots ; q_1, q_2, \cdots)$. In many important cases this has a separable structure $K = k_1(p, q_1) + k_2(p_1, q_2) + \cdots$, in other words an accumulation of single stage effects. A criterion function is a measure of the effectiveness of a control process.

The maximizing q_i will be functions of the states $p_1, p_1, \cdots$. In the most important cases the q_i which maximizes depends only upon the present and past history of the process $q_i = q_i(p, p_1, \cdots, p_{i-1})$, and frequently only upon the current state, $q_i = q_i(p_{i-1})$.

Let us call any function of this type a policy, reserving the term optimal policy for a policy which maximizes K. The study of control and decision processes may then be considered to be the study and effective determination of optimal policies. Unfortunately, in many of the control processes of greatest significance either no criterion function exists or there are too many of them. (The criterion function is a measure of the sequence of decisions.) This makes the application of mathematical theories such as the calculus of variations and dynamic programming quite difficult; see, however, [5].

Nevertheless, the concept of policy remains meaningful. Furthermore, the powerful and flexible theory of simulation can be used to study the complex processes associated with animate and human systems; see [3].

3. Instinct

Let us now identify instinct as a policy, a policy which controls the behavior of an organism in a particular situation. We consider instinctive behavior a precise automatic response to a signal or stimulus acting on an organism. The evolutionary value of instinct is clear since in critical circumstances there may not be time for conscious behavior. The response must be preprogrammed to ensure survival.

The point we wish to emphasize, however, is that instinct is not solely a low level intellectual activity. What seems to be the case is that there are levels of instinctive behavior, intermingled with conscious behavior. We shall return to this point below.

4. Levels of control processes

When the original system does not cope in a satisfactory style, survival of the species depends upon the development over time of control mechanisms to improve the performance. At the lowest level these are feedback control devices. These control systems, however, themselves require supervision and modification. Gradually, then, we see the development of control systems for control systems, and so on. The operation of any complex system requires a hierarchy of control systems.

We can identify this hierarchy of control systems with levels of consciousness. Indeed, if we wish to discuss consciousness in a meaningful fashion, we must utilize the concept of levels of consciousness. Similarly, to study thinking, we must consider levels and kinds of thinking. A confusion of levels leads to paradoxes and other difficulties; see [4], [7].

5. Instructions

Let us pursue this subject of hierarchies in another direction. Suppose that an organism contains certain types of mechanisms for conveying instructions. Some of these instructions may well be instructions for issuing instructions, and so on, again a hierarchal concept.

An error in one kind of instruction is thus seen to be far more critical than an error in another type of instruction. An error in the instructions for issuing instructions will lead to an infinitely greater number of errors than an inaccurate instruction which leads to the development of one faulty organ, for example. We see then that numerical probabilities of mutation are not inherently meaningful. We must take the structure of the organism into account and examine the responsiveness of the structure to a change in one component. These are questions of mathematical stability and sensitivity.

6. Uncertainty

So far we have considered only deterministic processes, processes where cause and effect is assumed to hold. Let us now consider processes where uncertainty plays a major role.

One way to construct mathematical models of uncertainty is to introduce random variables. This leads to the concept of a stochastic transformation $T(p, q, r)$. The point $p_1 = T(p, q, r)$ belongs to S whenever p is in S, q is in D and r is a random variable in R; see [2].

A choice of a sequence $q_1, q_2, \cdots$, and a selection of random variables $r_1, r_2, \cdots$, leads as before to a sequence of states $p_1, p_2, \cdots$. An optimal policy may be determined in a process of this nature by maximizing the expected value of a criterion function.

This, however, takes care of only first level processes. There are far more complex levels of uncertainty. In the foregoing we assumed that the probability distributions for the random variables were known. Frequently, this is not the case. Instead, we may possess some initial clues as to the nature of the uncertainty and then proceed to discover more of the structure of the process using observation of the events that occur. We call this an adaptive control process, and equate this operation with learning. One instinct associated with learning is curiosity. There are, however, levels of learning. We learn, we learn how to learn, and so on; see [5].

7. Higher level instinct

A mathematical technique for studying adaptive control processes is the theory of dynamic programming. In animals, however, it seems that adaptive control involves higher level instincts. If so, this plausibly explains why it is so difficult to carry out mathematical studies of pattern recognition, language translation, and human communication. If these are instinctive, they are not based upon the mathematical principles of the last 5000 years, but instead upon techniques developed by evolutionary selection over hundreds of millions of years.

In other words, we possess in our brain very complex, genetically determined, internal mechanisms specifically designed for particular tasks. Existing mathematical methods and computers cannot yet compete with these consequences of selective breeding.

8. Theory of types

It is apparent that what has proceeded has been considerably influenced by the Russell theory of types and the Liouville-Ritt theory of elementary functions. See the expository discussion in [7], as well as [4] for an analysis of one type of humor in terms of paradoxes.

REFERENCES

[1] R. Bellman, *Dynamic Programming*, Princeton, Princeton University Press, 1957.

[2] ———, *Adaptive Control Process; A Guided Tour*, Princeton, Princeton University Press, 1961.

[3] ———, "Humor and paradox," *A Celebration of Laughter* (edited by W. Mendel), Los Angeles, Mara Books, 1970, pp. 35–45.

[4] ———, "Adaptive process and intelligent machines," *Proceedings of the Fifth Berkeley Symposium on Mathematical Statistics and Probability*, Berkeley and Los Angeles, University of California Press, 1966, Vol. 4, pp. 11–14.

[5] R. Bellman and P. Kell, *Simulation in Human Systems: Decision-Making Applied to Psychotherapy*, to appear.

[6] E. Neumann, *The Origins and History of Consciousness*, Princeton, Princeton University Press, 1954.

[7] A. Tarski, "Truth and proof," *Sci. Amer.*, Vol. 220 (1969), pp. 63–77.

THEORETICAL FOUNDATIONS OF PALEOGENETICS

RICHARD HOLMQUIST
Space Sciences Laboratory
University of California, Berkeley

1. Introduction

This paper presents mathematical reflections that were completed during the Spring of 1966 (Holmquist [9]). In light of increasing knowledge and interest in the evolutionary significance of the primary amino acid or nucleotide base sequences of homologous (see Section 3.2 for definitions) proteins or DNA's, both within and among various phylogenetic species, these calculations are given here in the hope that they may prove useful to a wider audience. The mathematics, though straightforward, is complex. Therefore, a conscientious effort has been made to relate the mathematical equations to concrete physical phenomena so that the paper may be more readable to mathematicians for whom the historical jargon of molecular biology may be unfamiliar as well as to the biologists, biochemists, and anthropologists who may want to use the mathematics as a tool for interpreting their experimental data.

The problems in molecular evolution that are soluble by a study of protein sequences and homologies may for convenience be divided into three classes: (a) the construction of phylogenetic trees; (b) the deduction of the primary amino acid sequences of the common ancestral proteins at the branch points of the phylogenetic tree; and (c) the assignment of a time scale to each leg of the phylogenetic tree. Historically, three concepts have been extremely useful in solving these problems: amino acid differences between homologous proteins (Zuckerkandl and Pauling [28]; Needleman and Wunsch [18]), the minimum mutation distance between homologous proteins (Jukes [10]; Fitch and Margoliash [4], [5]), and the path of least information or maximum entropy (Reichert and Wong [21]) between two homologous proteins.

It has seemed plausible to attempt to correlate the amino acid or base differences between two homologous proteins or nucleic acids with a time of origin, measured from the present, of a "common ancestor" protein or nucleic acid which is homologous to both. In favorable cases a possible primary sequence for the common ancestral molecule can be deduced. Proceeding in this way, one can build up a biochemical tree of life which can be compared to those

The preparation of this manuscript was supported by NASA Grant NgR 05-003-020, "The Chemistry of Living Systems."

evolutionary and phylogenetic relationships that are already known from classical biology. The name *paleogenetics* or *paleobiochemistry* has been suggested for that branch of science which concerns itself with molecular restoration studies of the above kind (Zuckerkandl and Pauling [28]; Pauling and Zuckerkandl [20]; Zuckerkandl [27]). Among the proteins for which such studies have been made, one may mention the hemoglobins of a great many species (see above references), the cytochromes *c* of various organisms (Margoliash [15]), the A and B fibrinopeptides (Doolittle and Blombaeck [2]), and the ferredoxins (Matsubara, Jukes, and Cantor [17]). In each case, the primary amino acid sequences of the proteins were used as base data. Wilson and Sarich [26] have used quantitative immunological techniques to study albumin and transferrin. Paleogenetic studies have been made on deoxyribonucleic acids by Martin and Hoyer [16] and by Kohne [14]. These investigators used hybridization, kinetics of renaturation, and thermal denaturation techniques to establish the degree of similarity of various mammalian DNA's and the evolutionary relationships between them. A more complete tabulation of primary sequence data may be found in Dayhoff's *Atlas of Protein Sequence and Structure* [1].

The success of the above methods is attested to by the general concordance of their results with the paleontological fossil record.

Infrequently, the evolutionary relationships revealed by studies of the above type differ radically (Doolittle and Blombaeck [2]; Wilson and Sarich [26]) from those that have been deduced from a great body of classical biological and paleontological evidence. The resolution of such differences remains a current problem.

Nevertheless, with the exception of the paper by Reichert and Wong, theoretical, quantitative, justification of the computational methods employed with each of these approaches has been lacking. It has been no more than fortuitous that for most of the proteins which have been studied to date, the mutation rate has been relatively low. The situation is particularly confusing when the mutation rate is moderate, for then the low rate approximations are being used at or above the limits of their validity; however, the protein sequences still haven't been completely randomized, so that evolutionary information is still present. The problem is "how much," and how to extract it, with confidence in the final result.

In order to interpret the experimental data, which consists of (in decreasing order of information content) the comparison of the primary nucleotide sequences, the primary amino acid sequences, the number and kind of amino acids, the amino acid compositions, the number and kind of nucleotide bases, and the nucleotide base compositions, a more sound theoretical foundation is needed. Two contributions to this theory may be found in the papers by Jerzy Neyman [19] and by Reichert and Wong [21]. Neyman particularly clearly defines, and emphasizes the complexity of, the statistical, topological, and temporal aspects of paleogenetics; and he makes a mathematically rigorous quantitative beginning towards solving them. Reichert and Wong approach these problems from the viewpoint of set theoretic, informational, and thermodynamic principles. Fitch and Margoliash [4] have utilized the concept of "minimum mutation distance" to construct phylogenetic trees. Although this parameter is not mathematically the "best," and gives incorrect results when the mutation rate is high (it ignores, for example, multiple hits at the same nucleotide site and back mutations, and requires not generally correct *a priori* assumptions about the time sequence of mutational events), it has been extremely useful in making sense out of the mass of data now available. Gatlin [6] has suggested that living systems "may utilize the principle of Shannon's Second Theorem (Shannon and Weaver [23]) which states that it is possible to reduce transmission error without undue sacrifice of message variety or rate by properly encoding the message. These coding devices may form a quantitative basis for evolution and differentiation." King and Jukes [13] have pointed out the fact that evolution is fundamentally non-Darwinian in character: "natural selection is the editor, rather than the composer, of the genetic message."

There exist two theoretical approaches for relating the observed number of nucleotide base or amino acid differences between two homologous nucleic acids or proteins: minimizing the energy required to effect these changes and maximizing the entropy change. The concept of "minimum mutation distance," mentioned above, belongs to the first class. However, the energies required to interconvert one base to another are all of the same general order of magnitude, and it is therefore entropic factors that are usually the determinants of these differences. Also, the "minimum mutation distance" is sometimes a gross underestimate of the true number of primary mutagenic events that have occurred, is related in no simple way to these events, and has no firm theoretical foundation, except that it does state a minimum below which the number of primary mutagenic events cannot go. Although it may be useful in establishing the approximate topology of a phylogenetic tree, it does not suffice, particularly when the mutation rate is relatively high, to accurately establish the length of the legs of these trees: accurate values of these lengths are an absolute necessity if macromolecules are to be used as evolutionary clocks. In the latter respect, two summers ago I had the privilege of visiting the Olduvai gorge. Considering the landscape and the vastness of the African continent, one cannot help but admire the monumental and scientifically productive efforts of Dr. and Mrs. Leakey and their colleagues. As molecular evolutionists we do well to remember that the absolute time scales that we use come directly from the fossil record and not from macromolecules.

Whether the "ticks" of the evolutionary clock be electromagnetic, thermal, chemical, or other, in origin, the molecular quantity most closely and simply related to these "ticks" is the number of one step nucleotide base changes within DNA. At the observational level, these primary events are reduced by multiple hits at the same base site, back mutation, and the chance coincidence of having the same base at a given site in two homologous nucleic acids. At the protein level additional correction must be made for multiple hits within the same codon, codon degeneracy, and the possibility that two homologous sites will have the same amino acid there by chance. The purpose of this paper is to show how to make corrections for these phenomena accurately.

The methods presented here differ in three important aspects from earlier approaches. First, they are more general: the only starting assumption is that the mutagenic events occur randomly along the polynucleotide sequence which codes for a particular protein. The mutation rate may have any numerical value and any temporal dependence; the nucleic acid segment, or the protein for which it codes, may be of any length. If, in fact, the mutagenic events occur nonrandomly along the polynucleotide sequence, it is inherent in the method to detect such nonrandomness. Second, they are more exact: the phenomena of multiple hits at the same nucleotide site, back mutation, and accidental identity at the same site are quantitatively accounted for without approximation. The effect of amino acid codon degeneracy and of multiple nucleotide base changes at different base sites within the codon triplet are evaluated. The formulas developed herein thus include as special cases those methods based on a low mutation rate. And third, the probability of obtaining *exactly* a given value of a parameter of interest is calculated so that not only the average and most probable values of these parameters are known, but also their variances. This latter fact establishes quantitative objective criteria for the significance of any computed or observed values of those parameters.

Before we proceed to the mathematical theory and derivations, the results of these calculations and their significance for paleogenetics will be summarized and discussed in the next section so that the more important points do not become lost in the necessarily lengthy mathematical development. Following these results, the abstract formulas from which they were obtained will be derived; and finally, an Appendix illustrating the formulas by numerical example, based on actual experimental data, will conclude the paper, so that others may be able to do similar calculations themselves on experimental data of their own choosing.

2. Results

Let us examine the first row in Table I. The first column shows that we are considering, for illustration, a DNA segment of $L = 18$ base residues which codes for a hexapeptide (second column $T = 6$). This DNA segment evolves to two present day homologous DNA's. The number of one step base changes or hits (See Section 3.2 for definitions) which separates each of these contemporary DNA's from the ancestral DNA is 9 (third column), or in an alternative viewpoint, the number of one step base changes which separates the two contemporary DNA's from each other is 18 (fourth column). However, some of the base changes will occur at the same base site, so that the number of different base sites hit in each homologue is less than 9, namely, 7.24 (fifth column). Because a base site hit more than once may revert (back mutate) to the same

TABLE I

AMINO ACID DIFFERENCES

L = number of nucleotide base sites ($L = 3T$),
T = number of amino acid sites,
X = primary mutagenic events (hits),
$N(x)$ = average number of base sites hit at least once,
$N'(x)$ = average number of base sites different from corresponding sites in the ancestral DNA,
$N(D)$ = average number of base differences between two present day homologous DNA's,
$N(A)$ = average number of amino acid sites different from corresponding sites in ancestral fibrinopeptide fragment,
$N(d)$ = number of amino acid differences between two present day homologous fibrinopeptide fragments.

Numbers in parentheses are calculated for restricted mutation: purine to purine; pyrimidine to pyrimidine.

L	T	X	$2X$	$N(x)$	$N'(x)$	$N(D)$	$N(A)$	$N(d)$
18	6	9	18	7.24	6.74	10.12	3.96	5.14 ± 0.86
					(5.88)	(7.91)	(3.28)	(4.65 ± 1.02)
9	3	4.5	9	3.70	3.46	5.15	2.28	2.62 ± 1.51
					(3.04)	(4.03)	(2.12)	(2.40 ± 1.38)
18	6	2	4	1.94	1.94	3.61	1.33	2.35 ± 1.20
					(1.94)	(3.46)	(1.23)	(2.19 ± 1.18)
9	3	1	2	1	1	1.85	0.76	1.32 ± 0.86
					(1)	(1.78)	(0.65)	(1.16 ± 0.84)

base as in the ancestral DNA, the number of base sites in each homologue which differ from the homologous sites in the ancestral DNA is still less, or 6.74 (sixth column). The number of base differences between the two present day DNA's will be twice 6.74 less that number of homologous sites, which though differing from the ancestral site, are the same by chance coincidence. The net result is that the two contemporary DNA's will differ in 10.12 (seventh column) homologous sites, on the average. Now the number of amino acid differences, 3.96 (eighth column), between the ancestral hexapeptide coded for by the ancestral DNA, and either of the two present day homologous peptides will be less than 6.74, because some of the differing base sites may fall within the same codon triplet, and also because some amino acids are coded for by more than one triplet. The number of amino acid differences between the two contemporary homologous hexapeptide is twice 3.96 less that number due to chance identity, or a total of 5.14 differences (ninth column), on the average.

The other rows in Table I are interpreted similarly and are discussed more exhaustively in the analysis of the fibrinopeptide A sequences in Section 6.3.

The first row of Table I demonstrates that the expected number of amino acid differences between two homologues may be less than the number of primary mutagenic events by as much as a factor of 3.5. Table I also clearly brings out the fact that the number of amino acid differences is not only not proportional to the number of mutagenic events (we would not expect it to be because of multiple hits, revertants, hits within the same codon, code degeneracy, and chance coincidences), but it is not even a function only of the proportion of sites hit, X/L: for in the first two rows the numbers of amino acid differences are in a ratio of about 2, whereas X/L is the same ($\frac{1}{2}$) in both cases. This nonproportionality is even more dramatically illustrated by the second and third rows where the ratio of the X/L's is 4.5:1, whereas the number of amino acid differences is almost identical.

Five points need emphasizing.

(1) Exact calculations put very definite quantitative limits on the permissible values of the mutation rate, time of divergence, and the number and distribution of hyper- or hypovariable sites.

(2) Methods based on the proportion of sites hit or unhit are inherently incapable of yielding physically meaningful calculations in some cases. This includes methods in which the data is "normalized" to 100 residues.

(3) The "$N(d)/2$" approximation (Zuckerkandl and Pauling [28]) for X is sometimes a poor one; for example, in the first row of Table I, $N(d)/2 = 2.07$, while the true value of X is 9.

(4) The "negative log" approximation (Zuckerkandl and Pauling [29]) for X may be quite inaccurate; for example, for the first row of Table I the true value of X is 9, while the negative log estimate is 5.74.

(5) The parameter X, or its minimum or maximum value consistent with the experimental data, should be used in constructing phylogenetic trees, not the "minimum mutation distance."

3. Derivations for DNA

3.1. *General.* Amino acid differences among proteins arise from mutational changes that occur in the nucleic acid segments which code for these proteins. The number of amino acids which have mutated in the proteins may differ from the number of mutations which have affected the nucleic acid segment for several reasons: (a) multiple hits at the same site—several mutagenic events occur at a single nucleotide position in the segment instead of each event occurring at a different nucleotide along the segment; (b) back mutation—a *multiply* hit single nucleotide site may end up as the same nucleotide as it was originally; (c) since each amino acid in a protein is coded by a triplet of nucleotides, several of the mutagenic events may fall within the same triplet—this would give rise to only a single amino acid substitution; (d) degeneracy—some amino acids are coded for by more than one nucleotide triplet so that a mutagenic event occurring within this triplet need not lead to an amino acid substitution; (e) viability—if a particular nucleotide mutation leads to a nonviable organism, one will not be able to observe this mutation as an altered nucleotide base or as an amino acid substitution; a mutagenic event that resulted in the formation of one of the three chain terminating (nonsense) codons might in some cases fall in this category; and finally, (f) even though the rate of mutation may be accurately known over a certain region of space, what one must frequently examine is a particular subregion of this space; for example, if the mutation rate along a chromosome were known, the mutations themselves would show up as changes in the amino acid sequences of many nonhomologous proteins, only one, or even only a part of one of which is at hand for study. The *observed* mutations are therefore very much a function of the particular proteins or nucleic acids that are selected for analysis.

From the considerations of the preceding paragraph it is clear that no simple relationship exists between mutagenic events and observed protein mutations. As a consequence, a detailed understanding of paleogenetical studies requires the quantitative evaluation of each of the above factors so that their relative importance can be assessed.

3.2. *Definitions.* *Homologous* proteins are proteins that are *in fact* related to each other by point mutations in the common ancestral DNA coding for those proteins. It is possible for two proteins (or base sequences) to be identical or to bear any arbitrary degree of relatedness to each other without being homologous. This could occur by convergent evolution from two quite different ancestral DNA sequences. Such proteins are properly referred to as *analogous.* Experimentally it is usually difficult to distinguish between the two cases, but at times, when the mutagenic pathway is known, as with certain chemical mutagens, the distinction may be possible.

A *mutagenic event* is defined as a one step change of one nucleotide to a different nucleotide: that is, $\mathrm{C} \rightarrow \mathrm{T} = (\mathrm{C} \rightarrow \mathrm{T})$, $\mathrm{C} \rightarrow \mathrm{T}$ $(\mathrm{C} \rightarrow \mathrm{T} \rightarrow \mathrm{C} \rightarrow \mathrm{T})$, $\mathrm{C} \rightarrow \mathrm{T}$ (at one nucleotide site) and $\mathrm{A} \rightarrow \mathrm{G}$ (at another nucleotide site) represent, respectively one, three, and two mutagenic events, where C, T, A, and G are abbreviations for deoxycytidine, thymidine, deoxyadenosine, and deoxyguanosine, respectively. Such an event, of course, must be incorporated into the gene pool of the species if it is to be evolutionarily effective.

A nucleotide site is said to have been *hit* each time that a mutagenic event has occurred at that site.

A nucleotide site is said to have been *altered* if the nucleotide base occupying that site differs from the nucleotide base originally there before the mutagenic event occurred.

Other definitions will be introduced throughout the text as they are needed.

3.3. *Precise statement of problems to be solved in this paper.* (1) Consider a polynucleotide which contains L individual nucleotides. Let exactly X mutagenic events occur randomly along the length of this polynucleotide. After the X mutagenic events have occurred, in general, a number x, which is less than L, nucleotide sites will have been hit; for example, all X mutagenic events might occur at the same nucleotide site. Let $N(x)$ designate the average number of nucleotide sites which have been hit. An explicit formula for $N(x)$ is derived.

(2) The average number $N'(x)$ of nucleotide sites that have been altered will in general be less than $N(x)$ because of back mutations. An explicit expression for $N'(x)$ is given.

(3) An explicit formula for the average number of nucleotide base differences $N(D)$ between two homologous polynucleotides is derived, including correction for chance coincidences.

(4) Consider a protein of T amino acids which is coded by a polynucleotide of $L = 3T$ individual nucleotide bases. Let exactly X mutagenic events occur randomly along the length of this polynucleotide. After the X mutagenic events have occurred, a number A, less than T, amino acid sites will differ from the corresponding sites in the ancestral protein. An explicit formula for $N(A)$, the average number of amino acid substitutions that have occurred, is derived.

(5) Because of chance identities, the number of amino acid differences $N(d)$ between two homologous present day proteins will be less than $N_1(A)$ plus $N_2(A)$, where the subscripts refer to each homologue; a formula for $N(d)$ is derived.

(6) The limits of validity of the commonly used approximation $N(A) = N(d)/2$ are derived.

(7) Formulas are given which permit the proportion of amino acid substitutions which have occurred by one base, two base, and three base changes to be calculated.

3.4. *Calculation of* $N(x)$*: multiple hits.* Let us make the following definitions. An x part partition of X is a decomposition of X into a set of x (nonzero) positive integer summands $\{a_1, \cdots, a_x\}$, where $\sum_i a_i = X$. Partitions having the same a_i are considered to be identical even though the order of the a_i may differ in two such partitions. Let a particular x part partition of X be denoted by $(x, X)_j$, and let $n_{a_{ij}}(x)$ be the number of integers in this partition having the value a_i. Note that $\sum_{a_i \neq a_k} n_{a_{ij}}(x) = x$.

To make these abstract definitions more concrete, consider the following example. A particular 3 part partition of 6 is the set of integers $\{4, 1, 1\}$. Here, $a_1 = 4$, $a_2 = 1$, and $a_3 = 1$. We denote this particular partition by $(3, 6)_1$, where the subscript $j = 1$ is to remind us that this partition refers specifically to the set of integers $\{4, 1, 1\}$. For $(3, 6)_1 = \{4, 1, 1\}$, $n_{a_{11}} = 1$, $n_{a_{21}} = 2$, and $n_{a_{31}} = 2$. As stated in the definition, $a_1 + a_2 + a_3 = 4 + 1 + 1 = 6$; and $n_{a_{11}} + n_{a_{21}} = 3$. A different 3 part partition of 6 would be the set of integers $\{3, 2, 1\}$, and this partition could be labeled $(3, 6)_2$, for example.

Define N_{jx} as the number of ways of realizing $(x, X)_j$ along a polynucleotide which contains L individual nucleotides. This definition of N_{jx} requires that we associate the partitions $(x, X)_j$ in some well defined way with the polynucleotide of length L. We do this as follows. The partition (x, X) *means* that x nucleotide sites have been hit a total of X times; and the particular x part partition of X, $(x, X)_j = \{a_1, a_2, \cdots, a_x\}$, *means* that the first nucleotide site has been hit a total of a_1 times, the second site a_2 times, and the xth site a_x times. (The nucleotides in the polynucleotide can be numbered in any convenient manner: for example, the 3′ terminal nucleotide could be taken as the first nucleotide and the 5′ terminal nucleotide as the Lth nucleotide.) Now the first nucleotide site can be hit a_1 times in a number of ways: for example, if $X = 30$ and $a_1 = 3$, the first, second, and third mutagenic events could occur at the first site, or alternatively the second, fifth, and twenty seventh mutagenic events could occur there. (The mutagenic events are numbered in any convenient manner.) Similar considerations hold for the other nucleotide sites. The total number of ways in which the first site can be hit a_1 times, the second site a_2 times, the xth site a_x times, and the $(x + 1)$th, $(x + 2)$th, . . . , and Lth sites zero times is *by definition* N_{jx}.

Now the average number $N(x)$ of polynucleotide sites that have been hit is by definition

$$N(x) = \sum_x xP(x), \tag{3.1}$$

where $P(x)$ is the probability that *exactly* x sites have been hit. But $P(x)$ is by definition

$$P(x) = \frac{\sum_j N_{jx}}{\sum_{x \leqq L} \sum_j N_{jx}} = \frac{\sum_j N_{jx}}{L^X}. \tag{3.2}$$

The denominators in (3.2) are the total number of ways X mutagenic events can hit L nucleotide sites. Thus, if we can find an expression for N_{jx}, $N(x)$ will be given explicitly by (3.1). This expression for N_{jx} is derived in the following paragraph.

If x nucleotide sites have been hit in a polynucleotide of L nucleotides, then $L - x$ have not been hit. This can happen in

$$W_1 = \frac{L!}{(L - x)!x!} \tag{3.3}$$

ways. Now let us limit our consideration to those sites which have been hit at least once. In particular, let us assume these sites have been hit in the precise manner defined by the physical meaning attached to $(x, X)_j$. These x sites can be hit in a total of

$$W_2 = X!x! \tag{3.4}$$

ways, because $X!$ is the number of ways X mutagenic events can occur along the polynucleotide, and $x!$ is the number of ways that the x a_{ij} can be permuted among themselves. The factor $x!$ arises from the fact that in the *definition* of $(x, X)_j$ the order of the a_{ij} in the partition was irrelevant, while the *physical meaning* attached to $(x, X)_j$ was such that identical partitions in which the a_{ij} occur in different orders refer to different physical situations. Not all of these W_2 ways represent distinct physical situations, for the a_{ij} hits at the ith site can occur in $a_{ij}!$ ways and each of these ways leads to the same physical result. Similarly, if in the partition there are $n_{a_{ij}}$ integers having the value a_{ij}, these integers can be permuted among themselves in $n_{a_{ij}}!$ ways without altering the physical result. The total number of ways x sites can be hit by X mutagenic events is thus

$$W_3 = \frac{W_2}{\prod_{a_{ij} \neq a_{kj}} n_{a_{ij}}! \prod_{a_{ij}} a_{ij}!}. \tag{3.5}$$

Therefore,

$$N_{jx} = W_1 W_3 = \frac{X!}{\prod\limits_{a_{ij} \neq a_{kj}} n_{a_{ij}}! \prod\limits_{a_{ij}} a_{ij}!} \frac{L!}{(L-x)!}. \tag{3.6}$$

This completes the solution to our problem. *Tables of Factorials 0! – 9999!* [22] make it unnecessary to calculate the factorials in equation (3.6) by hand.

The method that has been given above for calculating $N(x)$ in terms of partitions illuminates the physical details of the mutation process. However, writing out the partitions that are needed in this method is frequently tedious. This is, if anything, an understatement. The basic difficulty is that no general formula exists for $n_{a_{ij}}$. In some applications, the calculation of the detailed structure for the probability for back mutation, for example, the individual a_{ij} and $n_{a_{ij}}$ of each partition must be known, and there is no way to get them except to write down the partitions one by one in some systematic manner that insures against leaving any partition out. In this respect, the *Tables of Partitions* [7] (see especially equation 1.1, p. ix, and equation 2.2a, p. xi) are very helpful, for they list the total number $P(X, x)$ of each possible (x, X) as well as the total number $p(X, X) = \sum_{x=1}^{X} P(X, x)$ of all possible partitions for a given X. Clearly, it would be desirable to have a formula for $N(x)$ that does not require the calculational labor of (3.6). Such a formula can be obtained as follows. Define $m(X, x)$ to be the number of ways X mutagenic events can hit x nucleotide sites, where each site is hit *at least* once. Mathematically, the number we have designated by $m(X, x)$ is the number of mappings of X *onto* x. Thus,

$$\sum_j N_{jx} = m(X, x) W_1, \tag{3.7}$$

where W_1 is given by (3.3).

We now calculate $m(X, x)$. The total number of ways X mutagenic events can hit x nucleotide sites is x^X. Therefore, $m(X, x)$ is given by x^X less those number of ways in which X mutagenic events can hit x nucleotide sites when k sites are not hit at all, where k takes on successively the values $0, 1, 2, \cdots, x - 1$. This follows from the fact that we *defined* $m(X, x)$ to include only those situations where *every* one of the x sites is hit *at least once*. Those situations in which some site or sites are not hit at all must be subtracted. But as in (3.3) the number of ways in which k sites can be hit and $x - k$ sites not hit is

$$W_4 = \frac{x!}{k!(x-k)!}. \tag{3.8}$$

Those k sites which have been hit can be hit in a total of $m(X, k)$ ways by definition. The total number of ways in which the X mutagenic events can hit x sites when some of the sites are not hit at all is thus

$$W_5 = \sum_{k=0}^{x-1} \frac{x!}{k!(x-k)!} m(X, k). \tag{3.9}$$

Therefore,

$$m(X, x) = x^X - W_5 = x^X - \sum_{k=0}^{x-1} \frac{x!}{k!(x-k)!} m(X, k). \tag{3.10}$$

Now $N(x)$ can be calculated from (3.10), (3.7), (3.2), and (3.1). It should be noted that because of (3.10) no knowledge whatsoever about partitions is required in calculating $N(x)$. Finally, we notice that (3.10) is a recursion formula for $m(X, x)$; that is, starting from $m(X, 0) = 0$, all the other $m(X, k)$ and finally $m(X, x)$ can be calculated from (3.10) alone. This important fact reduces the calculation of $N(x)$ to a simple iterative procedure which can readily be carried out by a computer. For those who are satisfied with the truth of (3.10) and are less interested in its physical derivation as given above, we comment here that it can be proved by mathematical induction on the positive integers x.

This completes the solution to the problem of multiple hits at the same nucleotide site:

$$N(x) = \sum_{x=1}^{L} xP(x), \tag{3.11}$$

$$P(x) = \frac{m(X, x)L!}{L^X x!(L-x)!}. \tag{3.12}$$

By maximizing $P(x)$ with respect to x, the *most probable* value of x can be determined. It is not obvious to the author whether this most probable value has an explicit mathematical formulation, or whether it must be determined by numerical calculation from (3.12).

3.5. *Calculation of $N'(x)$: back mutation.* After a_i mutagenic events have occurred at a given nucleotide site, the probability $P(a_i)$ that the final nucleotide is the same as the original nucleotide at that site depends only on a_i and on the number of nucleotides to which a given nucleotide can mutate. We give this probability for two models.

Case 1. Any nucleotide is free to mutate to any one of three other nucleotides.

Case 2. Any purine or pyrimidine nucleotide is free to mutate to only a purine or pyrimidine nucleotide, respectively.

Case 1 is given because amino acid substitutions are known which require the mutation of a purine to a pyrimidine or *vice versa.* Case 2 is given because of its possible usefulness with respect to chemical mutagens that are known to involve purine to purine or pyrimidine to pyrimidine transformations (C → U by nitrous acid, for example). As can be seen from Table II $P(a_i)$ is a *strong* function of the model selected and selection of the incorrect model in a particular case can completely invalidate the quantitative and qualitative topology of a paleogenetic analysis (such as the construction of a phylogenetic tree). Other models are possible, but the two given will suffice for present purposes.

The average probability of having the same nucleotide at any site after a_i mutagenic events have occurred there is

$$P = \sum_{a_i=1}^{X} P(a_i) \frac{\sum_{x \leqq L} \sum_{J} N_{jx} n_{a_{ij}}(x)}{\sum_{a_i=1}^{X} \sum_{x \leqq L} \sum_{j} N_{jx} n_{a_{ij}}(x)} \tag{3.13}$$

and the average number of altered nucleotides is

$$N'(x) = (1 - P)N(x). \tag{3.14}$$

TABLE II

$P(a_i)$ AS A FUNCTION OF a_i

The formula for $P(a_i)$ is proved by mathematical induction on the positive integers a_i.

	Case 1	Case 2
a_i	$P(a_i) = \frac{1}{4}\left[1 + \frac{(-1)^{a_i}}{3^{a_i-1}}\right]$	$P(a_i) = \begin{matrix} 1 \text{ if } a_i \text{ even} \\ 0 \text{ if } a_i \text{ odd} \end{matrix}$
0	1	1
1	0	0
2	1/3	1
3	2/9	0
4	7/27	1
5	20/81	0
.	.	.

Equation (3.13) appears more complicated than it really is (see Section A.2 of the Appendix, for a numerical calculation). However, its use does require a knowledge of the detailed structure of each partition $(x, X)_j$ for *all* possible x, and as stated on an earlier page, the only way to get this structure is by writing down *all* the partitions and counting the a_i in each to find the $n_{a_{ij}}$. If one is not interested in the *details* of the revertant process, then the net result, $N'(x)$, can be rapidly obtained for any X by repetitive application of the following recursion formulas which do not require a knowledge of the partition structure, and which may be proved by mathematical induction:

$$N'_0(x) = 0, \tag{3.15}$$

$$N'_{X+1}(x) = 1 + \left(1 - \frac{4}{3L}\right) N'_X(x). \tag{3.16}$$

Another very useful relation for reducing the calculational labor is given by

$$N'_{2X}(x) = 2N'_X(x) - \frac{4}{3} \frac{[N'_X(x)]^2}{L}. \tag{3.17}$$

Equations (3.15) through (3.17) are only valid for Case 1. For Case 2 the corresponding formulas are

$$N'_0(x) = 0, \tag{3.18}$$

$$N'_{X+1}(x) = 1 + \left(1 - \frac{2}{L}\right) N'_X(x), \tag{3.19}$$

$$N'_{2X}(x) = 2N'_X(x) - \frac{2[N'_X(x)]^2}{L}. \tag{3.20}$$

3.6. *Calculation of $N(D)$: nucleotide differences between homologous DNA's.* If the sequence of the common ancestor polynucleotide is known, then $N'(x)$ can be compared directly with the experimentally observed number of nucleotide substitutions for each homologue. In practice, the sequence of the ancestral polynucleotide is seldom known, and what one measures experimentally is the number and type of nucleotide differences $N(D)$ and coincidences $(L - N(D))$ between two homologous polynucleotides having the same common ancestor. A coincidence and a difference are defined, respectively, as the occurrence of identical or nonidentical nucleotide bases at the same position in the two homologous polynucleotides. The probabilities needed to calculate $N(D)$ are given in Table III.

TABLE III

PROBABILITIES NEEDED FOR CALCULATING $N(D)$

$p_i = [L - N'_i(x)]/L$, where the subscript i refers to homologue 1 or homologue 2.

	p_1p_2		$1 - p_1p_2$	
A	T	T	T	T
1	T	0	T	0
2	T	0	0	T
		$1 - p_1 - p_2 + p_1p_2$	$p_1(1 - p_2)$	$(1 - p_1)p_2$
		$1 - p_1 - p_2 + p_1p_2$		
1	0	G	G	0
2	0	G	0	G
Case 1:		⅓	⅓	⅓
Case 2:		1	0	0

The first column in this table indicates the ancestral polynucleotide A and the two homologues under consideration. The second column indicates the probability that both homologues have the same nucleotide base at a particular site as the ancestral polynucleotide. The last three columns indicate the probability that one or the other or both of the homologues has (have) a different nucleotide base at a particular site from the ancestral polynucleotide. For illustration, the ancestral base has been taken as thymidine. The last two columns clearly represent differences between the two homologues. Also, however, the third column, which represents those homologous bases that differ from the ancestral base at a given site, contains base pairs that may be the same (but not thymidine) or different. The probability for their being the same is given in the third column of the lower part of the table (guanosine has been chosen for illustration), and the probability for their being different is given in the last two columns of the lower part of the table. Probabilities are given for both Case 1 and Case 2 (see Section 3.5). The total probability that the two homologues will differ from *each other* at any single site is therefore

$$\begin{aligned} p &= p_1(1-p_2) + (1-p_1)p_2 + (\tfrac{2}{3})(1-p_1-p_2+p_1p_2) \\ &= (\tfrac{2}{3})(1+\tfrac{1}{2}(p_1+p_2)-2p_1p_2) \end{aligned} \tag{3.21}$$

for Case 1, and

$$\begin{aligned} p &= p_1(1-p_2) + (1-p_1)p_2 \\ &= p_1 + p_2 - 2p_1p_2. \end{aligned} \tag{3.22}$$

for Case 2.

The probability $P(D)$ that exactly D differences will be observed between the two homologues is then

$$P(D) = \frac{L!}{D!(L-D)!} p^D(1-p)^{L-D}. \tag{3.23}$$

The average value of D, $N(D)$ is thus pL, which becomes, upon substituting the values for p_i into (3.21) and (3.22),

$$N(D) = N_1'(x) + N_2'(x) - \frac{4}{3}\frac{N_1'(x)N_2'(x)}{L} \tag{3.24}$$

for Case 1,

$$N(D) = N_1'(x) + N_2'(x) - 2\frac{N_1'(x)N_2'(x)}{L} \tag{3.25}$$

for Case 2.

It should be noted that (3.24) and (3.25) are identical to (3.17) and (3.20), respectively, so that

$$N(D) = N_{2X}'(x). \tag{3.26}$$

Physically, this curious identity means it does not matter whether one considers each contemporary homologous DNA to have evolved from a common ancestral DNA over a time period such that each homologue has, on the average, received X hits, or whether one considers the homologues to have evolved one from the other during that same time period in such a way that the reference homologue has undergone $2X$ hits. The increased number of multiply hit sites and revertants in the latter case just equals the losses due to chance identity of base sites in the former. The two equivalent pathways are illustrated graphically in Figure 1.

3.7. *Approximations.* Under those conditions where the fraction to the right of $P(a_i)$ in (3.13) may be approximated by the Poisson distribution,

$$\frac{\sum_{x \leqq L}\sum_j N_{jx}n_{a_{ij}}(x)}{\sum_{a_i=1}^{X}\sum_{x\leqq L}\sum_j N_{jx}n_{a_{ij}}(x)} \approx P\left[\left(\frac{X}{L}\right), a_i\right] \equiv \frac{\exp\left\{-\frac{X}{L}\right\}\left[\frac{X}{L}\right]^{a_i}}{a_i!}. \tag{3.27}$$

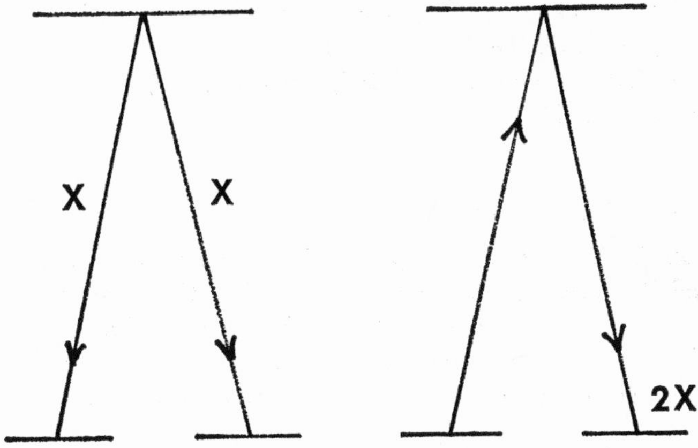

$$2N'(x) - \frac{4}{3}\frac{N'(x)^2}{L} = N(D) = N'_{2X}(x)$$

FIGURE 1

Equivalent mutational pathways.

Equation (3.13) reduces, to the simple form

$$P = \sum_{a_i=1}^{\infty} \frac{1}{4}\left[1 + \frac{(-1)^{a_i}}{3^{a_i-1}}\right] \frac{\exp\left\{-\frac{X}{L}\right\}\left[\frac{X}{L}\right]^{a_i}}{a_i!} = \frac{1}{4}\left[1 + 3\exp\left\{-\frac{4}{3}\frac{X}{L}\right\}\right] \tag{3.28}$$

for Case 1. Equation (3.14) thus becomes

$$N'(x) = \frac{3}{4}\left[1 - \exp\left\{-\frac{4}{3}\frac{X}{L}\right\}\right]L \tag{3.29}$$

for Case 1, and

$$N'(x) = \frac{1}{2}\left[1 - \exp\left\{-2\frac{X}{L}\right\}\right]L \tag{3.30}$$

for Case 2. Equations (3.24) and (3.25) then become

$$N(D) = \frac{3}{4}\left[1 - \exp\left\{-\frac{8}{3}\frac{X}{L}\right\}\right]L \tag{3.31}$$

for Case 1, and

$$N(D) = \frac{1}{2}\left[1 - \exp\left\{-4\frac{X}{L}\right\}\right]L \tag{3.32}$$

for Case 2.

Equations (3.29), (3.30), (3.31), and (3.32), aside from being useful in their own right, form a convenient, rapid check on the numerical accuracy of more exact calculations. The approximate equations are very handy computationally: a table of exponentials suffices to solve them.

In a somewhat different, but related, context, (3.29) has been independently derived by Neyman [19].

4. Derivations for proteins

4.1. *General.* The number of amino acid differences A between a present day protein and its ancestral homologue may be less than the average number of nucleotide base differences $N'(x)$ between the corresponding nucleic acids which code for these proteins for two reasons: first, amino acid codon degeneracy; and second, several of the differing nucleotide bases may fall within the same codon triplet.

4.2. *Amino acid codon degeneracy.* Some amino acids are coded for by more than one nucleotide triplet, so that a mutagenic event occurring within this triplet need not lead to an amino acid substitution. A recent tabulation of these codon triplets may be found in Watson [24]. If we consider a specific triplet coding for a particular amino acid, a *single* amino acid substitution will result at this position if either one, two, or three of the nucleotides of the triplet undergo mutation which results in the formation of a new triplet which codes for an amino acid differing from that in the ancestral protein. In Section A.1 of the Appendix the probability that a given amino acid will mutate to another amino acid if exactly one, two, or three nucleotide bases in the triplet coding for that amino acid are altered is calculated on the assumption that each of the sixty four codon triplets is equally probable. Although the three chain terminating triplets are necessarily less probable, that they sometimes do survive and find experimentally observable expression has been demonstrated by Weigert and Garen [25] (see also [11]). To the extent that these three codons cause deviations from randomness, this will show up in the viability parameter β (see equation (6.1)) or in the necessity to reduce X in order to obtain agreement with experiment. Experimental evidence that this assumption is approximately true has been provided by King and Jukes' analysis of known polypeptide sequences [13]. As in the preceding section, figures are quoted for both the case where any nucleotide base is free to mutate to any other (Case 1) and for the case where purines and pyrimidines may mutate only to a purine or pyrimidine (Case 2), respectively. These probabilities are when exactly one, two, or three nucleotides within a triplet have been altered, 0.7604, 0.9826, and 0.9931, respectively, for Case 1, and 0.6563, 0.9688, and 1.0000, respectively, for Case 2. Briefly, an amino acid substitution is almost certain when any combination of mutagenic events occurs within the amino acid codon, except for a single altered nucleotide in the third codon position. The above figures demonstrate that there is no large advantage with respect to conserving structure for an organism to develop a mechanism in which nucleotide base mutations are limited to purine to purine or pyrimidine to pyrimidine.

In order to complete the quantitative evaluation of the effect of codon degeneracy on the number of amino acid substitutions, we must calculate exactly how many codon triplets have sustained exactly one, exactly two, and exactly three altered (relative to the ancestral homologue) nucleotide bases. This problem is tackled in the next section.

4.3. *Multiple mutations within the same triplet.* If we consider a specific triplet coding for a particular amino acid position, a single amino acid substitution will result at this position if either one, two, or three of the nucleotides of the triplet undergo mutation, *provided* the number of single, double, and triple mutations are weighted by the appropriate probability for mutation from the preceding section. In particular, the average number of amino acid substitutions $N(A)$ is a function $F[N'(x), T]$, where T is the total number of amino acid residues in the homologue under consideration. At this point, it should be emphasized that it is *not* permissible to treat each nucleotide site as independent from the other sites. To do so will, in general, underestimate the number of amino acid substitutions. The reason for this is that $N'(x)$ already is corrected for multiple mutations at the same nucleotide site. The nature of the function F is complicated and best given by example in Section A.3 of the Appendix; its calculation is, however, straightforward. Some indication of its general properties may be had from the following considerations.

Represent the T sets of triplets from $L = 3T$ nucleotides in T rows as follows:

Row 1 _ _ _
Row 2 _ _ _
⋮
Row T _ _ _.

An amino acid substitution is represented by any row with one, two, or three altered nucleotides. For example, if $N'(x) = 7$, and $L = 18$,

Form 1	Form 2
x x x	x x _
x x _	x x _
x _ _	x x _
x _ _	x _ _
_ _ _	_ _ _
_ _ _	_ _ _

each represent four amino acid substitutions. We shall call such an array a *form*. A form is defined by stating the number of rows having zero, one, two, and three altered nucleotides, respectively, that is,

_ _ _	x _ _
_ x _	x _ x
x x x	_ x x
_ _ _	_ _ _
x _ _	_ _ _
_ x x	x x _

belong to Form 1 and Form 2, respectively. First we write down all possible forms. For a given $N'(x)$ and T, there can be A amino acid substitutions, where

$$I + \varepsilon \leqq A \leqq S, \tag{4.1}$$

$\varepsilon = \varepsilon(p)$: $\varepsilon(0) = 0$, $\varepsilon(1) = \varepsilon(2) = 1$, I and p are integers defined by the congruence

$$N'(x) \equiv p \pmod 3 \tag{4.2}$$

where $N'(x) = 3I + p$, and S is the smaller of $N'(x)$ and T. For example, if $N'(x) = 7$, $I = 2$, $p = 1$, $\varepsilon = 1$, and $S = 6$: $3 \leqq A \leqq 6$, that is, three, four, five, or six amino acid substitutions are possible. Before writing down any forms, A should be calculated since this will permit one to disregard those forms which are irrelevant to the problem, in this case any form with only one or two occupied rows. One must now count the number of ways W_i that the ith form can be realized. Let a_{0i}, a_{1i}, a_{2i} and a_{3i} be the number of rows in a given form which has zero, one, two and three altered nucleotides, respectively. Then, for that form,

$$\frac{N'(x)!T!3^{a_{1i}+a_{2i}}}{\prod_{j}^{3} a_{ji}!}. \tag{4.3}$$

A convenient check on the calculations is provided by the fact that

$$\sum_{\text{all } i} W_i = \frac{L!}{[L - N'(x)]!}. \tag{4.4}$$

In the absence of codon degeneracy, W_i would be the number of ways of realizing, for a given form, exactly A_i amino acid substitutions, where

$$A_i = a_{1i} + a_{2i} + a_{3i}. \tag{4.5}$$

However, because of codon degeneracy, W_i must be reduced by approximately the factor f_i, where

$$f_i = 0.7604 f_{1i} + 0.9826 f_{2i} + 0.9931 f_{3i} \tag{4.6}$$

for Case 1,

$$f_i = 0.6563 f_{1i} + 0.9688 f_{2i} + 1.0000 f_{3i} \tag{4.7}$$

for Case 2, and

$$f_{ji} = \frac{a_{ji}}{A_i}. \tag{4.8}$$

The numerical coefficients for the f_{ji} were taken from Section A.1 of the Appendix. To find the number of ways W_A in which exactly A amino acid substitutions can occur, one adds together the number of ways of realizing *all* those forms which have *exactly* A rows occupied by *at least* one altered nucleotide:

$$W_A = \sum_{i=i_A} f_i W_i, \tag{4.9}$$

where the subscript A on i_A is to remind us that the summation is over *only* those forms having exactly A rows occupied. The probability that exactly A

amino acid substitutions have occurred between the ancestral and present day homologue is thus about

$$P(A) = \frac{W_A}{\sum_{\text{all } i} W_i}, \tag{4.10}$$

and the average number of amino acid substitutions is

$$N(A) = F[N'(x), T] = \sum_A AP(A). \tag{4.11}$$

Before leaving this section, we note that another quantity that is sometimes of interest is the proportion of amino acid substitutions that have occurred by one base, two base, and three base changes. This proportion p_j, $j = 1, 2$, or 3, is approximately,

$$p_j = \frac{\sum_i f_{ji} c_j W_i}{\sum_i f_i W_i}, \tag{4.12}$$

where the c_j are the coefficients of the f_{ji} in (4.6) and (4.7).

For the ith form, (4.6) and (4.7) have the effect of excluding those ways which represent fewer than A_i amino acid substitutions because of degenerate rows having one altered nucleotide *only* or two altered nucleotides *only* or three altered nucleotides *only*. Equations (4.6) and (4.7) do not exclude those ways which represent fewer than A_i amino acid substitutions because of combinations or degenerate rows of mixed type. For example, ways in which two rows of a form are degenerate, one of the rows containing one altered nucleotide, the second row containing either two or three altered nucleotides are not excluded.

For this reason W_A and $P(A)$ in equations (4.9) and (4.10) should be overestimates. However, since those ways not excluded by (4.6) and (4.7) represent amino acid substitutions fewer than A_i, those $P(A)$ and W_A with A less than A_i will be underestimated. If the total number of forms being considered is reasonably large, these two opposite effects will partially offset one another. Thus, despite the above limitations $N(A)$ in (4.11) is reasonably accurate because the average is taken over all forms. In equations (4.9), (4.10), and (4.12) W_A, $P(A)$, and p_j are less accurate because the summations are over fewer forms.

We are currently attempting to find a method of treating multiple hits within the same codon that is both less cumbersome and more exact than the method given in this section. Nevertheless, the above treatment is an improvement over existing methods which virtually ignore the problem.

4.4. *Calculation of $N(d)$: the average number of amino acid differences between two present day homologues.* The number of amino acid differences d between two present day homologues may be less than $N(A)$ because though each of the homologues may differ at a particular site from the ancestral homologue, the two homologues themselves may have the same amino acid at that site. In Section 3.6 of this paper, it was shown that the general form of the equation necessary to correct for this accidental coincidence between two present day sites is

$$N(d) = N_1(A) + N_2(A) - \alpha \frac{N_1(A)N_2(A)}{T}, \tag{4.13}$$

where the subscripts refer to the two present day homologues in question, and α is a numerical constant that depends only on the structure of the genetic code. This constant can be evaluated by considering two homologues that have evolved at the same rate for a sufficiently long time so that all sequences have been randomized. Then, $N(d) = N_1(A) = N_2(A)$, and

$$\alpha = \frac{1}{[N(A)/T]_{\text{equil}}} = \frac{1024}{963} = 1.0633, \tag{4.14}$$

because the probability, after randomization, that two homologous proteins will have Arg, Ser, or Leu at the same site is $3(6/64)^2$; that both homologues will have Ala, Thr, Gly, Val, or Pro is $5(4/64)^2$; that both will have Ile or Term is $2(3/64)^2$; that both will have Lys, His, Cys, Glu, Gln, Asp, Asn, Tyr, or Phe is $9(2/64)^2$; and that both will have Trp or Met is $2(1/64)^2$. Adding these probabilities together gives the probability that two present day homologues will both have the same amino acid at a given site. This probability is $244/64^2$. The probability that these two homologues will differ at a given site is

$$\left[1 - \frac{244}{64^2}\right] = \frac{963}{1024} = \left[\frac{N(A)}{T}\right]_{\text{equil}}. \tag{4.15}$$

Finally,

$$N(d) = N_1(A) + N_2(A) - \frac{1024}{963}\frac{N_1(A)N_2(A)}{T}. \tag{4.16}$$

The probability that exactly d amino acid substitutions will be observed between two present day homologous proteins is

$$P(d) = \frac{T!}{d!(T-d)!}p^d(1-p)^{T-d}. \tag{4.17}$$

4.5. *The "$N(d)/2$" approximation.* The number of amino acid substitutions between an ancestral protein and each of the two present day homologues is sometimes estimated by assuming $N_1(A) = N_2(A)$ and by assuming that their common value $N(A) = N(d)/2$. The first assumption may be experimentally checked by seeing if the number of amino acid substitutions between each homologue and a third evolutionary distant present day homologue are approximately equal. The exact relation between $N(A)$ and $N(d)$ may be found from (4.16):

$$\begin{aligned} N(A) &= \frac{T}{\alpha}\left[1 - \left(1 - \frac{\alpha N(d)}{T}\right)^{1/2}\right] \\ &= \frac{1}{2}N(d)\left[\begin{aligned} &1 + \frac{1}{4}\frac{\alpha}{T}N(d) + \frac{1}{4}\cdot\frac{3}{6}\left(\frac{\alpha}{T}\right)^2 N(d)^2 \\ &+ \frac{1}{4}\cdot\frac{3}{6}\cdot\frac{5}{8}\left(\frac{\alpha}{T}\right)^3 N(d)^3 \\ &+ \cdots \end{aligned}\right]. \end{aligned} \tag{4.18}$$

Thus, if we are willing to accept an error in $N(A)$ no larger than ten per cent, the criterion for the validity of the "$N(d)/2$" approximation becomes

$$N(d) \leqq \frac{0.4T}{\alpha}. \tag{4.19}$$

5. Measures of error

5.1. *Measures of error in* (ND), $N'(x)$, $N(x)$, $N(A)$, *and* $N(d)$. The theoretically calculated value of $N(D)$ tends to be insensitive to small variations in X, the total number of mutagenic events, irrespective of their source (that is, whether they are statistical variations or nonstatistical ones). The reason for this is that an increase in X is partially offset by a compensating decrease due to a greater number of multiple hits, and increased back mutation. An analogous argument holds for small decreases in X. The same considerations hold for $N'(x)$ and $N(x)$. These considerations, of course, do not apply when $X \ll L$ or $X \gg L$.

The probability distributions that have been derived in this paper $P(x)$ (3.2) and (3.12), $P(a_i)$ (Table II), and $P(D)$ (3.23) are all well defined. Their frequency distributions therefore each possess a unique variance (second moment) that can, for a given X and L, be calculated in a straightforward manner by standard statistical methods (Hoel [8]), and this variance is a quantitative measure of the deviation from the average values $N(x)$, $N'(x)$, and $N(D)$ that one might expect to find in practice. This variance is particularly easy to calculate only for $P(D)$ and is given by

$$\sigma_D = \left(N(D)\left[1 - \frac{N(D)}{L}\right]\right)^{1/2}. \tag{5.1}$$

The true variance will be somewhat greater than this because the contribution from the variance of $N'(x)$ has been ignored in (5.1). Equation (5.1) shows that when $X \ll L$ or $X \gg L$ the error in $N(D)$ will be small. These are precisely the instances when we need an estimate of the error most badly, for in these instances the compensatory mechanisms that were discussed in the preceding paragraph do not apply. On the other hand, when X is of the order of L, these compensatory mechanisms do apply, so that (5.1) should still give a reasonably valid estimate of the error in $N(D)$, because of the insensitivity of the latter towards small variations.

Similarly, the standard deviation of the distribution of (4.17) is, closely,

$$\sigma_d = \left(N(d)\left[1 - \frac{N(d)}{T}\right]\right)^{1/2}. \tag{5.2}$$

The actual standard deviation of d will be somewhat greater than this because the variance of $N(A)$ has been ignored in (5.2).

6. Discussion

6.1. *General.* The only assumption made in deriving the statistics in this paper is that of *spatial* randomness along L. No assumptions about time are involved, and the statistics remain valid for *any* particular time dependence of X, linear or nonlinear, random or nonrandom. No assumptions have been made about the number of mutagenic events that have occurred in each of several homologues. They may be the same or different. The statistics can thus handle the case of homologous macromolecules which have evolved at different rates.

In Sections A.2 and A.3 of the Appendix and Table I, it is demonstrated that to neglect the phenomena of multiple hits, back mutation, hits within the same codon, the degeneracy of the genetic code, and accidental (chance) coincidence between two homologous sites, can, in actual experimental cases, lead to errors in the value of $N(d)$ of a factor of 3.5. Obviously, with errors of this magnitude, it is impossible to construct with confidence any meaningful phylogenetic trees, or to conclude that a series of sequence homologies did or did not arise by a stochastic pathway. The formulas in this paper permit one to quantitatively correct for the above phenomena so that unambiguous answers to the questions can be given.

If the observed number of differences between two homologous macromolecules differ from the calculated value of $N(D)$ or $N(d)$ by much more than twice the statistical error ((5.1) and (5.2)), the cause of such discrepancy may lie with one of the following factors.

(1) The spatial distribution of the mutagenic events is nonrandom along L. In fact, if the other factors below can be eliminated as causes of discrepancy, the statistics of this paper may be used as an algorithm to search for nonrandomness *within* a single molecule by considering subsegments of that molecule which contain $\ell < L$ nucleotide bases.

(2) $N(D)$ or $N(d)$ are not the appropriate statistic. These are an average value of D and d. Other values can and will occur with a relative frequency given by (3.23) and (4.17). If for some applications, an investigator wishes to utilize the most probable, rather than the average, values of x, x', D and d, these most probable values may be calculated from the equations given in this paper.

(3) The input data is incorrect, that is, one's estimate of X is wrong; this corresponds to incorrect assumptions, for example, linearity or randomness in time, about the mutation rate. If after all other causes have been eliminated, a discrepancy still remains, one should seriously consider revising the numerical value of the mutation rate, as well as assumptions about its temporal dependence.

(4) Viability—this is the least well known of all the quantities and may be one cause of any nonrandomness falling under category (1) above. Consider several homologous macromolecules and in particular that region of each for which one has calculated $N'(x)$. Call the number of sites in this region which for *all* the macromolecules contain the same nucleotide base at a given site. (Different sites may contain different nucleotide bases.) The viability β is defined as

$$\beta = 1 - \frac{f}{L}. \tag{6.1}$$

This may or may not be a good estimate of the viability depending on the number of homologues available. All calculations are then repeated starting with a revised estimate of X and L, namely, βX and βL to see if improved agreement results. This procedure should be followed only *after* the first three factors above have been adequately accounted for: otherwise β becomes no more than a "fudge factor."

6.2. *Minimum mutation distance.* The relationship between the *minimum mutation distance* [4] and the calculations in the present paper can be made clear by considering the proportion of amino acid substitutions that have occurred by actual one base, two base, and three base changes: p_1, p_2, and p_3 (Appendix, Section A.3). In the sense implied by the concept of minimum mutation distance, a three base change can occur in two ways: (a) for one present day homologue each base of the codon triplet differs from the corresponding base of the ancestral homologue, or (b) for one present day homologue two bases of the codon triplet differ from the corresponding ancestral bases, and for the second present day homologue the corresponding two bases are identical to those in the ancestral homologue while the remaining third base differs from the ancestral homologue. Thus, the proportion of amino acid substitutions that have a minimum mutation distance of 3 ($P_{\text{M.M.D.}=3}$) is very nearly,

$$(6.2) \qquad P_{\text{M.M.D.}=3} = 0.05(2)\left[p_3\left(1 - \frac{N(A)}{T}\right) + \frac{1}{3}\,p_1p_2\,\frac{N(A)}{T}\right].$$

The factor 0.05 arises from the fact that 95 per cent of actual three base changes are "silent" because the algorithm by which minimum mutation distance is computed counts these "silent" 3 base changes as either 1 or 2 base changes. If the numerical values of p_1, p_2, p_3, $N(A)$, and T, which are given in A.3 of the Appendix are substituted into (6.2), then $P_{\text{M.M.D.}=3} = 0.0078$. Equation (6.2) is an underestimate for it neglects all cases where the sum of the total number of base changes for a given codon is greater than three (the sum is taken over both homologues). Analogous calculations may be made to find $P_{\text{M.M.D.}=2}$ and $P_{\text{M.M.D.}=1}$.

Alternatively, one can utilize the principles embodied in (3.15) through (3.17) and calculate

$$(6.3) \qquad P_{X,\text{M.M.D.}=3} = 0.05p_{2X,3}.$$

In either case, it is clear that, except for very low mutation rates, the minimum mutation distance bears a minimum relationship to the true course of events.

6.3. *Application to experimental data.* In Table IV are shown the homologous sequence fragments of the A fibrinopeptides of the sheep, goat, ox, and reindeer, with which the numerical calculations will be compared. These short fragments were chosen for two reasons: first, they contain a constant region of three residues, and it is of interest whether the theoretical methods can detect this region; and second, for the variable region the minimum mutation rate estimated by Doolittle and Blombaeck [2] as 10^{-7} mutagenic events/year/codon may be as rapid as the rate of evolution of DNA itself (Kohne [14]), and thus the deficiencies of the "minimum mutation distance" show up most clearly. The fragments were kept short so as not to initially become bogged down in calculational irrele-

TABLE IV

ILLUSTRATIVE FIBRINOPEPTIDE A FRAGMENTS FOR COMPARISON WITH CALCULATED VALUES

See Section 6.3 for explanation.

Organism	Sequence	Amino acid differences	M.M.D.
Goat:	(Asp-Ser-Asp-Pro-Val-Gly)		
		0	0
Sheep:	(Asp-Ser-Asp-Pro-Val-Gly)		
		3	4
Ox:	(Gly-Ser-Asp-Pro-Pro-Ser)		
		2	2
Reindeer:	(Gly-Ser-Asp-Pro-Ala-Gly)		
Amino Acid Position Number:	17 16 15 14 13 12		

vancies. For computational purposes (See Appendix, Section A.2) it is assumed that these four artiodactyls diverged from their most recent common ancestor 15 million years ago giving a total of 9 ($10^{-7} \times 6 \times 15 \cdot 10^{6}$) primary mutagenic events randomly distributed over a region of 18 nucleotide bases or 6 amino acids. Thus, the ratio, hits: sites: : 9: 18 is ½, a value chosen to avoid the trivial cases where the sites are saturated with hits or hardly hit at all. The actual time of divergence of the most distantly related pair (sheep-reindeer) may be closer to 30 million years [2]. If the latter figure is used the general conclusions are only strengthened.

The first row of Table I demonstrates both that the expected number, 5.14, of amino acid differences between the homologous fibrinopeptide fragments being compared is less than the number of primary mutagenic events by a factor of 3.5 and that the experimentally observed number of differences (0–3) is inconsistent with a stochastic mechanism. The second row of Table I shows that when the constant region of three residues is taken into account, agreement with experiment is obtained. The third and fourth rows of Table I demonstrate that the observed differences can also be explained by assuming a mutation rate 2/9 of that in the first two rows, or, 2.2×10^{-8} mutagenic events/year/codon. The values in parentheses in Table I illustrate the fact that at the level of nucleotide base or amino acid differences, it is not statistically possible, in fragments as short as those being considered here, to distinguish between unrestricted mutation, where any base may mutate to any one of the other three bases, and restricted mutation, where purine ↔ pyrimidine mutations are forbidden.

The recursion formula for $m(X, x)$ (3.10) was developed in collaboration with Mr. Andrew Lebor during several evenings of discussion in the Spring of 1966. I wish to thank Dr. Thomas H. Jukes for pointing out that most three base changes are "silent" when the experimental data are analyzed with the concept of minimum mutation distance and for providing a quantitative measure of this "silence."

ADDENDUM

Since this paper was presented I have become aware, through the courtesy of Patricia Altham, University of Cambridge, Department of Pure Mathematics and Mathematical Statistics, that Equations (3.2) and (3.12) and (3.1) and (3.11) can be explicitly formulated as

$$P(x) = \frac{1}{L^X} \frac{L!}{x!(L-x)!} \sum_{\nu=0}^{x} (-1)^\nu \frac{x!}{\nu!(x-\nu)!} (x-\nu)^X \tag{7.1}$$

and

$$N(x) = L - L\left(1 - \frac{1}{L}\right)^X \tag{7.2}$$

with variance

$$\sigma^2(x) = L\left(\frac{L-1}{L}\right)^X \left[1 - \left(\frac{L-1}{L}\right)^X\right] + L(L-1)\left[\left(\frac{L-2}{L}\right)^X - \left(\frac{L-1}{L}\right)^{2X}\right]. \tag{7.3}$$

Similarly (3.14) can be explicitly written for Case 1

$$N'(x) = \frac{3}{4}L\left[1 - \left(1 - \frac{4}{3L}\right)^X\right] \tag{7.4}$$

with variance

$$\sigma^2(x) = \frac{3}{16}L\left[1 + 2\left(1 - \frac{4}{3L}\right)^X - 3\left(1 - \frac{4}{3L}\right)^{2X}\right] + \frac{9}{16}L(L-1)\left[\left(1 - \frac{8}{3L}\right)^X - \left(1 - \frac{4}{3L}\right)^{2X}\right], \tag{7.5}$$

and for Case 2

$$N'(x) = \frac{1}{2}L\left[1 - \left(1 - \frac{2}{L}\right)^X\right], \tag{7.6}$$

with variance

$$\sigma^2(x) = \frac{1}{4}L\left\{1 - \left(1 - \frac{2}{L}\right)^{2X} + (L-1)\left[\left(1 - \frac{4}{L}\right)^X - \left(1 - \frac{2}{L}\right)^{2X}\right]\right\}. \tag{7.7}$$

In a like manner, the exact variance, $\sigma^2(D)$, of D is given by (7.5) and (7.7) if X is replaced by $2X$ on the right side of these equations. The exact variance of D is somewhat less or greater than that given by the square of (5.1) depending on whether $L > 3$ or $L < 3$, respectively. The approximate nature of (5.1) results from the fact that (3.23) is an oversimplification of the true distribution of D because whether or not a difference occurs at a site is not really independent of whether differences exist at other sites so long as the total number of mutagenic events $2X$ is fixed. Analogous considerations apply to (4.17) and (5.2).

The above formulas are computationally more convenient than the ones in the text. In addition, the quantitative expressions for the variances (7.3), (7.5), and (7.7) supplement the discussion in Section 5.

A good reference for the mathematical techniques that are needed to solve the "occupancy problems" that arise in studies of molecular evolution is William Feller's book [3].

Note added in proof. The "derivations for proteins" in Section 4 and the discussion on minimum mutation distance in Section 6 have since been made quantitatively exact rather than approximate by Holmquist, Cantor, and Jukes [30]. The methods described here have also been applied to analyze evolutionary changes in the cytochrome c globins and immunoglobulins by Jukes and Holmquist [31].

APPENDIX

A.1. Probabilities for amino acid mutation

Probabilities for amino acid mutation are given in Tables AI, AII, and AIII. In Table AI, the calculation for the probability corresponding to Ser for Case 1 is

$$\frac{25}{27} = \frac{2}{3}\left[1 - \frac{1}{3}\cdot\frac{1}{3}\cdot\frac{1}{2}\right] + \frac{1}{3}\left[1 - \frac{1}{3}\cdot\frac{1}{3}(1)\right]. \tag{A.1.1}$$

The computation for the averaged probability is

$$0.9931 = \frac{1}{64}\left[(6)\frac{25}{27} + 58(1)\right]. \tag{A.1.2}$$

These examples indicate the general method of computation. In the remaining tables only the results are given.

A.2. Illustrative example: DNA

What follows is a detailed numerical analysis of the A fibrinopeptides of the ox, sheep, goat, and reindeer. These peptides contain 19 amino acids. Doolittle

TABLE AI

PROBABILITIES FOR AMINO ACID MUTATION
ALL 3 CODON POSITIONS ALTERED
$\underline{x}\ \underline{x}\ \underline{x}$

Amino acid	Probability Case 1	Case 2
Ser	25/27	1
All others	1	1
Average probability	0.9931	1

TABLE AII

PROBABILITIES FOR AMINO ACID MUTATION
ANY 2 CODON POSITIONS ALTERED
x x _, _ x x, x _ x

Term indicates chain terminating codon: UAA, UAG, or UGA.

Type of change	Amino acid	Probability Case 1	Case 2
x x _	Ser	25/27	1
	All others	1	1
	Average probability	0.9931	1
_ x x	Term	25/27	1/3
	All others	1	1
	Average probability	0.9965	0.9688
x _ x	Arg	7/9	1
	Leu	7/9	1/3
	All others	1	1
	Average probability	0.9583	0.9375
	Averaged average probability	0.9826	0.9688

TABLE AIII

PROBABILITIES FOR AMINO ACID MUTATION
ANY 1 CODON POSITION ALTERED
x _ _, _ x _, _ _ x

Type of change	Amino acid	Probability Case 1	Case 2
x _ _	Arg	7/9	1
	Leu	7/9	1/3
	All others	1	1
	Average probability	0.9583	0.9375
– x	Term	7/9	1/3
	All others	1	1
	Average probability	0.9896	0.9688
_ _ x	Met	1	1
	Trp	1	1
	Term	7/9	1/3
	Lys	2/3	0
	His	2/3	0
	Asp	2/3	0
	Asn	2/3	0
	Glu	2/3	0
	Gln	2/3	0
	Cys	2/3	0
	Tyr	2/3	0
	Phe	2/3	0
	Ile	1/3	1/3
	Arg	2/9	0
	Ser	2/9	0
	Leu	2/9	0
	All others	0	0
	Average probability	0.3333	0.0625
	Averaged average probability	0.7604	0.6563

and Blombaeck [2] have estimated a minimum mutation rate for these peptides of 10^{-7} mutations/year/amino acid. For simplicity, we shall also assume that 10^{-7} mutations/year is the rate at which mutagenic events occur per nucleotide base triplet; the actual rate, of course, will be slightly greater than this because of codon degeneracy. In particular, consider only that segment of six amino acids numbered 12 through 17 by the above authors. What is the number of differences to be expected between any two of the corresponding homologous DNA's which code for positions 12 through 17 after each homologue has had 15 million years to develop from a common ancestral DNA?

For the solution, first we have $L = 3 \times 6 = 18$ and $X = 10^{-7} \times 15 \cdot 10^{6} \times 6 = 9$. In general X will be nonintegral. In such a case one carries through the calculations for the integers on either side of X and at the end takes a weighted average.

Second, we need to list all $x \leqq 18$ part partitions of 9 as shown in Table AIV. We use the *Table of Partitions* [7] in order to find the number of partitions $P(9, x)$. The physical meaning of the particular partition 5, 4 is that two nucleotide bases and only two have been altered. One has been hit a total of five times, the other only four. The computation of the column N_{jx} is illustrated in the case when $j = 5, 2, 2$ and $x = 3$:

$$N_{jx} = \frac{9!}{(1!2!)(5!2!2!)} = \frac{18!}{15!} = 1{,}850{,}688. \tag{A.2.1}$$

We note that

$$p(9, 9) = \sum_{x=1}^{18} P(X, x) = 30, \tag{A.2.2}$$

so that we have left no partitions out, and that

$$L^X = 18^9 = \sum_{x \leqq 18} \sum_{j} N_{jx} = 1{,}980 \times 10^8, \tag{A.2.3}$$

so that our summation of the N_{jx} is correct. Thus, $P(X, x)$ and $p(X, X)$, from [7], and L^X provide independent checks on the accuracy of the calculations. The probability that exactly one, two, $\cdots$, nine sites have been hit is thus

$$\begin{aligned} P(1) &= 18/1{,}980 \times 10^8 = 0.000, \\ P(2) &= (3{,}054 + 11{,}016, +25{,}704 + 38{,}556)/1{,}980 \times 10^8 = 0.000, \\ P(3) &= 0.000, \\ P(4) &= 0.002, \\ P(5) &= 0.036, \\ P(6) &= 0.178, \\ P(7) &= 0.373, \\ P(8) &= 0.321, \\ P(9) &= 0.089. \end{aligned} \tag{A.2.4}$$

The *most probable* value of x is therefore 7 and the average value is

$$\begin{aligned} N(x) &= 9(0.089) + 8(0.321) + 7(0.373) \\ &\qquad + 6(0.178) + 5(0.036) + 4(0.002) \\ &= 7.24. \end{aligned} \tag{A.2.5}$$

TABLE AIV

PARTITIONS OF 9 FOR $x \leqq 18$ AND COMPUTATIONS FOR N_{jx}, THE NUMBER OF WAYS OF REALIZING $(x, X)_j$

	Partition	$P(9, x)$	N_{jx}
(1, 9):	9	1	18
(2, 9):	8, 1		3,054
	7, 2		11,016
	6, 3		25,704
	5, 4	4	38,556
(3, 9)	7, 1, 1		176,256
	6, 2, 1		1,233,792
	5, 3, 1		2,467,548
	5, 2, 2		1,850,688
	4, 4, 1		1,542,240
	4, 3, 2		6,168,960
	3, 3, 3	7	1,370,880
(4, 9)	6, 1, 1, 1		0×10^8
	5, 2, 1, 1		1
	4, 3, 1, 1		1
	4, 2, 2, 1		1
	3, 3, 2, 1		0
	3, 2, 2, 2	6	1
(5, 9)	5, 1, 1, 1, 1		1
	4, 2, 1, 1, 1		13
	3, 3, 1, 1, 1		9
	3, 2, 2, 1, 1		39
	2, 2, 2, 2, 1	5	10
(6, 9)	4, 1, 1, 1, 1, 1		17
	3, 2, 1, 1, 1, 1		168
	2, 2, 2, 1, 1, 1	3	168
(7, 9)	3, 1, 1, 1, 1, 1, 1		134
	2, 2, 1, 1, 1, 1, 1	2	605
(8, 9)	2, 1, 1, 1, 1, 1, 1, 1	1	635
(9, 9)	1, 1, 1, 1, 1, 1, 1, 1, 1	1	176

From this example, it is clear that one need only write out those partitions whose probabilities are high. In the present case, it is sufficient to calculate only the 5 through 9 part partitions of 9. This reduces the number of N_{jx} which must be calculated from 30 to 12, a considerable saving in time and effort. Those partitions which *do* have high probability may be found *prior* to writing out any of the partitions by using (3.10) and (3.12). We have

$$m(9, 1) = 1$$

$$m(9, 2) = 2^9 - \frac{2!}{1!1!}(1) = 510 \tag{A.2.6}$$

$$\vdots \qquad \vdots \qquad \vdots\,.$$

If these values of $m(X, x)$ are substituted into (3.12), we get

$$\begin{aligned} P(1) &= 1(18!)/18^9(1!)(17!) = 0.000 \\ P(2) &= 510(18!)/18^9(2!)(16!) = 0.000 \\ &\vdots \qquad\qquad \vdots \qquad\qquad \vdots \; . \end{aligned} \tag{A.2.7}$$

These values agree exactly with those calculated by the longer method.

TABLE AV

COMPUTATION OF THE PROBABILITY OF EXACTLY a_i HITS AND OF THE PROBABILITY $P(a_i)$ THAT THE FINAL NUCLEOTIDE IS THE SAME AS THE ORIGINAL NUCLEOTIDE

a_i	$\sum_{x \leq L} \sum_j N_{ij} n_{a_{ij}}(x)$	Probability of exactly a_i hits	$P(a_i)$ Case 1	$P(a_i)$ Case 2
1	11.11×10^{11}	.784	.000	0
2	2.66	.188	.333	1
3	.36	.025	.222	0
4	.03	.002	.259	1
5	.00	.000	.247	0
6	.00	.000	.250	1
7	.00	.000	.250	0
8	.00	.000	.250	1
9	.00	.000	.250	0
	14.16×10^{11}			

Third, to find the effect of back mutation, we construct Table AV. As a sample calculation for column 2 let $a_i = 1$. Then we have

$$\begin{aligned} N_{1j}n_{1j}(x) = {} & 0(18) + 1(3{,}054) + [2(176{,}256) \\ & + 1(1{,}233{,}792) + 1(2{,}467{,}548) + 1(1{,}542{,}240)] \\ & + 10^8[3(0) + 2(1) + 2(1) + 1(1) + 1(0)] \\ & + 10^8[4(1) + 3(13) + 3(9) + 2(39) + 1(10)] \\ & + 10^8[5(17) + 4(168) + 3(168)] \\ & + 10^8[6(134) + 5(605)] \\ & + 10^8 \cdot 7(635) + 10^8 \cdot 8(176) \\ = {} & 11.11 \times 10^{11}. \end{aligned} \tag{A.2.8}$$

To calculate the probability that a site has been hit exactly a_i times for $a_i = 1$, we have $0.784 = 11.11 \times 10^{11}/(14.16 \times 10^{11})$.

The *most probable* probability for back mutation at a site is therefore,

$$P(2) = 0.188 \times 0.333 = 0.0626 \tag{A.2.9}$$

for Case 1, and

$$P(2) = 0.188 \times 1 = 0.188 \tag{A.2.10}$$

for Case 2, and the average probability for back mutation at a site is

$$\text{(A.2.11)} \quad P = 0(0.784) + 0.188(0.333) + 0.025(0.222) + 0.002(0.259) = 0.0687$$

for Case 1, and

$$\text{(A.2.12)} \quad P = 0(0.784) + 0.188(1) + 0.025(0) + 0.002(1) = 0.1884.$$

Fourth, since in the present instance there is no large difference between the most probable values and the average values, we shall continue the calculations with the average values only. The average number of altered nucleotides in each homologue will be,

$$\text{(A.2.13)} \qquad N'(x) = (1 - 0.0687)7.24 = 6.74$$

for Case 1, and

$$\text{(A.2.14)} \qquad N'(x) = (1 - 0.1884)7.24 = 5.88$$

for Case 2.

Fifth, now we can calculate the average number of nucleotide base differences between the two homologues. Assuming for simplicity that each has evolved from the ancestral polynucleotide at roughly equal rates, we have

$$\text{(A.2.15)} \quad N(D) = 6.74 + 6.74 - \frac{4}{3}\frac{(6.74)(6.74)}{18} = 10.12 \pm 1.80(\sigma)$$

for Case 1, and

$$\text{(A.2.16)} \quad N(D) = 5.88 + 5.88 - 2\frac{(5.88)(5.88)}{18} = 7.91 \pm 2.04(\sigma)$$

for Case 2.

This completes the numerical calculations. Had no corrections of any sort been made, the incorrectly calculated value of $N(D)$ would have been $9 + 9 = 18$, an error of the order of 103 per cent. Had corrections been made for multiple hits at the same nucleotide site alone, or for chance coincidence between homologous sites alone, the errors would have been of the order of 63 per cent and 16 per cent, respectively. Taking into account multiple hits and chance coincidence, but not back mutation, reduces the error to about 7 per cent. Attention should also be directed to the magnitude of the statistical error. The number of observed differences between the two homologues in this particular example could lie anywhere between 5 to 13 and still justify the statistical conclusion that the two homologues are each derived from a common ancestral polynucleotide. In view of this wide range within which it is not possible to reasonably conclude that the hypothesis of common ancestry is false, it is all the more important that all possible corrections be made before conclusions are drawn about the phylogeny of two homologues.

A.3. Illustrative example: proteins

In Section A.2 the number of expected differences between any single present day homologue and the ancestral DNA coding for these positions was found to be $N'(x) = 6.74$ and 5.88 for unrestricted (Case 1) and restricted (Case 2) mutation, respectively. Because the DNA sequences are not experimentally available for these peptides, we here continue the calculations by using the methods described in the main body of this paper (Section 4) to find $N(d)$, the number of amino acid differences expected between two present day homologues. In what follows, to conserve space, we shall show the calculations for Case 1 and quote them for Case 2.

Because $N'(x)$ is not integral (6.74), we must calculate $N(A)$ for both $N'(x) = 6$ and $N'(x) = 7$. Detailed calculations are given only for $N'(x) = 7$

TABLE AVI

CALCULATIONS FOR $N(A)$

Case: $N'(x) = 7; I = 2, p = 1; \varepsilon = 1; 3 \leqq A \leqq 6; T = 6.$
Details of computations: $W_i = 907{,}200 = (7!6!3^{1+0})/(3!1!0!2!)$;
$f_i = 0.9155 = (1/3)(0.9155) + (0)(0.9826) + (2/3)(0.9931)$;
$f_iW_i = F_iW_i = 830{,}542 = 0.9155(907{,}200)$.

	$A = 3$		$A = 4$		$A = 5$		$A = 6$
	x x x x x x x – – – – – – – – – – –	x x x x x – x x – – – – – – – – – –	x x x x x – x – – x – – – – – – – –	x x – x x – x x – x – – – – – – – –	x x x x – – x – – x – – x – – – – –	x x – x x – x – – x – – x – – – – –	x x – x – – x – – x – – x – – x – –
W_i	907,200	2,721,600	24,494,400	24,494,400	12,247,200	73,483,200	22,044,960
f_{1i}	1/3	0	1/2	1/4	4/5	3/5	5/6
f_{2i}	0	2/3	1/4	3/4	0	2/5	1/6
f_{3i}	2/3	1/3	1/4	0	1/5	0	0
f_i	0.9155	0.9861	0.8741	0.9270	0.8069	0.8492	0.7974
f_iW_i	830,542	2,683,770	21,410,555	22,706,309	9,882,266	61,938,989	17,578,651
$W_{A=3}$	= 830,542 + 2,683,770 = 3,514,312						
$W_{A=4}$	= 44,116,864						
$W_{A=5}$	= 71,821,255						
$W_{A=6}$	= 17,578,651						
W_A	= 137,031,082						
W_i	= 160,392,960						

in Table AVI. The probabilities that exactly A amino acid substitutions have occurred between the ancestral and present day homologue are, from equation (4.10),

$$P(3) = \frac{3{,}514{,}312}{160{,}392{,}960} = 0.0219, \quad P(4) = 0.2750, \quad P(5) = 0.4477, \quad P(6) = 0.1096. \tag{A.3.1}$$

The average number of amino acid substitutions is thus,

$$N(A) = 3(0.0219) + 4(0.2750) + 5(0.4477) + 6(0.1096) = 4.06. \tag{A.3.2}$$

The proportions of these substitutions that have occurred by one base, two base, and three base changes can be obtained using (4.12) and are

$$p_1 = 0.503, \quad p_2 = 0.424, \quad p_3 = 0.073, \tag{A.3.3}$$

where the calculation for p_1 is

$$\begin{aligned} p_1 = \frac{0.7604}{137{,}031{,}082}[&\tfrac{1}{3}(907{,}200) + 0(2{,}721{,}600) + \tfrac{1}{2}(24{,}494{,}400) \\ &+ \tfrac{1}{4}(24{,}494{,}400) + \tfrac{4}{5}(12{,}247{,}200) + \tfrac{3}{5}(73{,}483{,}200) + \tfrac{5}{6}(22{,}044{,}960)] \\ = 0.503.& \end{aligned} \tag{A.3.4}$$

When similar calculations are made for $N'(x) = 6$, we find

$$N(A) = 3.66, \quad p_1 = 0.573, \quad p_2 = 0.377, \quad p_3 = 0.050. \tag{A.3.5}$$

Since $N'(x) = 6.74$ is 74/100 of the way between 6 and 7, the number of amino acid substitutions is

$$N(A) = 3.66 + 0.74(4.06 - 3.66) = 3.96, \tag{A.3.6}$$

and

$$p_1 = 0.525, \quad p_2 = 0.412, \quad p_3 = 0.066. \tag{A.3.7}$$

A corresponding result for Case 2, where $N'(x) = 5.88$ is $N(A) = 3.28$. The average number of amino acid substitutions between two present day homologues is, from equation (4.16),

$$\begin{aligned} N(d) &= 3.96 + 3.96 - \tfrac{1}{6}(3.96)(3.96)(1.0633) \\ &= 5.14 \pm 0.86 \end{aligned} \tag{A.3.8}$$

for Case 1, and

$$N(d) = 4.65 \pm 1.02 \tag{A.3.9}$$

for Case 2. This completes the calculations.

The observed number of amino acid substitutions between any pair of the fibrinopeptides A under discussion varies from zero (sheep-goat) to three (ox-sheep, ox-goat); the pairs ox-reindeer, sheep-reindeer, and goat-reindeer each have two substitutions. The calculated value of 5.14 appears to be considerably too high, and its error ($\sigma = 0.86$) is too small to allow the difference to be explained as statistical fluctuation. One might be tempted to conclude that the correct value of the mutation rate is nearer 0.39×10^{-7} mutagenic events/year/codon rather than 1.0×10^{-7}, and indeed, this is a valid possibility (see Table I rows three and four; also, Section 6.3). However, the fact that the figure 1.0×10^{-7} represents a minimum estimate argues against this interpretation. A second reasonable interpretation emerges if we consider the possibility that the individual mutagenic events are spatially nonrandom along the nucleic acid segment coding for these peptides. Examining the actual amino acid sequence in positions 12 through 17, we find that positions 14 through 16 contain the same sequence in all these mammals, namely, H-Ser-Asp-Pro-Oh. Another possibility is that organisms which sustained mutations in these positions did not survive. In either case we can estimate the *viability* β from (6.1), $\beta = 1 - \frac{3}{6} = 0.5$. We now repeat the calculations of Sections 3 and 4 using the revised values of X and L, namely,

$$X' = \beta X = 0.5(9) = 4.5, \quad L' = \beta L = 0.5(18) = 9. \tag{A.3.10}$$

When this is done we find that

$$N(d) = 2.62 \pm 1.51 \tag{A.3.11}$$

for Case 1, and

$$N(d) = 2.40 \pm 1.38 \tag{A.3.12}$$

for Case 2, in agreement with the experimental values found above.

REFERENCES

[1] M. O. DAYHOFF, *Atlas of Protein Sequence and Structure*, Vol. 4, Silver Spring, Md., National Biomedical Research Foundation, 1969.

[2] R. F. DOOLITTLE and R. BLOMBAECK, "Amino-acid sequence investigations of fibrinopeptides from various mammals: evolutionary implications," *Nature*, Vol. 202 (1964), pp. 147–152.

[3] W. FELLER, *An Introduction to Probability Theory and Its Applications*, Vol. 1, New York, Wiley, 1968, (3rd ed.).

[4] W. M. FITCH and E. MARGOLIASH, "Construction of phylogenetic trees: A method based on mutation distances as estimated from cytochrome *c* sequences of general applicability," *Science*, Vol. 155 (1967), pp. 279–284.

[5] ———, "An improved method for determining codon variability in a gene and its application to the rate of fixation of mutations in evolution," *Biochem. Genet.*, Vol. 4 (1970), p. 5797.

[6] L. GATLIN, "The information content of DNA II," *J. Theor. Biol.*, Vol. 18 (1968), pp. 181–194.

[7] H. GUPTA, C. E. GWYTHER, and J. C. P. MILLER, *Tables of Partitions, Royal Society Mathematical Tables*, Vol. 4, Cambridge, University Press, 1958.

[8] P. G. HOEL, *Introduction to Mathematical Statistics*, New York, Wiley, 1954, (2nd ed.). (See pp. 61, 74, 101.)
[9] W. R. HOLMQUIST, "The origin, partial structure and properties of Hemoglobins A_{Ic}," Ph.D. Thesis, California Institute of Technology, Pasadena, 1966. (See pp. 249–258.)
[10] T. H. JUKES, "Some recent advances in studies of the transcription of the genetic message," *Adv. Biol. Med. Phys.*, Vol. 9 (1963), pp. 1–41.
[11] ———, *Molecules and Evolution*, New York, Columbia University Press, 1966. (See Chapter 4.)
[12] M. KIMURA and T. OHTA, *On the rate of molecular evolution*, Mishima, Japan, National Institute of Genetics, 1970.
[13] J. L. KING and T. H. JUKES, "Non-Darwinian evolution," *Science*, Vol. 164 (1969), pp. 788–798.
[14] D. E. KOHNE, "Evolution of higher organism DNA," *Quart. Rev. Biophys.*, Vol. 33 (1970), pp. 327–375.
[15] E. MARGOLIASH, "The amino acid sequence of cytochrome *c* in relation to its function and evolution," *Can. J. Biochem.*, Vol. 42, No. 5 (1964), pp. 745–753.
[16] M. MARTIN and B. H. HOYER, "Adenine plus thymine and guanine plus cytosine enriched fractions of animal DNA's as indicators of polynucleotide homologies," *J. Molec. Biol.*, Vol. 27 (1967), pp. 113–129.
[17] H. MATSUBARA, T. H. JUKES, and C. R. CANTOR, "Structural and evolutionary relationship of ferredoxins," *Brookhaven Symp.*, Vol. 21 (1968), pp. 201–216.
[18] S. B. NEEDLEMAN and C. D. WUNSCH, "A general method applicable to the search for similarities in the amino acid sequence of two proteins," *J. Mol. Biol.*, Vol. 48 (1970), pp. 443–453.
[19] J. NEYMAN, "Molecular studies of evolution: A source of novel statistical problems," *Proceedings of the Purdue Symposium on Statistical Decision Theory*, Lafayette, Indiana, Purdue University Press, 1970.
[20] L. PAULING and E. ZUCKERKANDL, "Chemical paleogenetics, molecular "restoration studies" of extinct forms of life," *Acta Chem. Scan.*, Vol. 17, Suppl. #1 (1963), pp. S9–S16.
[21] T. A. REICHERT and A. K. C. WONG, "An application of information theory to genetic mutations & the matching of polypeptide sequences," *Biotechnology Program*, Pittsburgh, Carnegie-Mellon University, 1970, Personal communication.
[22] J. B. REID and G. MONTPETIT, *Tables of Factorials 0:–9999!*, Washington, D.C., National Academy of Sciences, National Council Publications 1039, 1962.
[23] C. E. SHANNON and W. WEAVER, *The mathematical theory of communication*, Urbana, The University of Illinois Press, 1949.
[24] J. D. WATSON, *Molecular Biology of the Gene*, New York, W. A. Benjamin, 1970, (2nd ed.). (See Chapter 13.)
[25] M. G. WEIGERT and A. GAREN, "Base composition of nonsense codons in *E. coli;* Evidence from amino-acid substitutions at a tryptophan site in Alkaline phosphatase," *Nature*, Vol. 206 (1965), pp. 992–994.
[26] A. C. WILSON and V. M. SARICH, "A molecular time scale for human evolution," *Proc. Nat. Acad. Sci. USA*, Vol. 63 (1969), pp. 1088–1093.
[27] E. ZUCKERKANDL, "The evolution of hemoglobin," *Sci. Amer.*, Vol. 189, May (1965), pp. 110–118.
[28] E. ZUCKERKANDL and L. PAULING, "Molecular disease, evolution, and genic heterogeneity," *Horizons in Biochemistry* (edited by M. Kasha and B. Pullman), New York, Academic Press, 1962, pp. 189–225.
[29] ———, "Evolutionary divergence and convergence in proteins," *Evolving Genes and Proteins* (edited by V. Bryson and H. J. Vogel), New York, Academic Press, 1965, pp. 97–166.

Added in proof.

[30] R. HOLMQUIST, C. CANTOR, and T. H. JUKES, "Improved procedures for comparing homologous sequences in molecules of proteins and nucleic acids," *J. Molec. Biol.*, Vol. 64 (1972), pp. 145–161.

[31] T. H. JUKES and R. HOLMQUIST, "Estimation of evolutionary changes in certain homologous polypeptide changes," *J. Molec. Biol.*, Vol. 64 (1972), pp. 163–179.

STATISTICAL ASPECTS OF THE NON-DARWINIAN THEORY

W. J. EWENS
UNIVERSITY OF TEXAS AT AUSTIN

1. Introduction

Quite apart from its possible biological relevance, the non-Darwinian theory of evolution currently under discussion is of considerable interest to statisticians. This is so, of course, because any mathematical formulation of the proposition that a considerable proportion of observed allelic substitutions have no selective significance and have occurred purely by chance, and, hence, any quantitative testing of this theory, must be conducted by statistical methods. The ramifications of the statistical testing which will be required to discuss the half dozen or so major supporting arguments for the theory can hardly be supposed yet to have been analyzed, even superficially; in particular, this is is true for those arguments relying on protein sequence data, amino acid frequencies, and the genetic code. It may be that novel forms of statistical tests will be required for these analyses. The main aim of this paper, while going in this direction, though rather restricted, is to devise a statistical test of the non-Darwinian theory based on a form of data currently being obtained in large volume by biologists, namely, the number and frequencies of different alleles at a locus provided by a sample of individuals from one generation of a population. A subsidiary aim is to show that quite simple statistical arguments can cast some doubt on the usefulness of one support for the non-Darwinian theory, namely, the support arising from the principle of substitutional loads. It will be convenient to consider this subsidiary aim first.

2. The substitutional load

Our aim is not to question the validity of the concept of substitutional load itself, but rather its usefulness as a support for non-Darwinian evolution. To do this we trace the main outline of the way the substitutional load is calculated.

Consider a diploid population of fixed size N and suppose that at a certain locus two alleles, A_1 and A_2, are possible. The three genotypes are supposed subject to differential selection and it is assumed that this selection acts entirely

On leave from the Department of Mathematics, LaTrobe University, Bundoora, 3038, Victoria, Australia.

Written under support of USPHS Grant GM-15769.

through differential preadult mortality; specifically, that the probabilities that newborn individuals survive to reach the age of sexual reproduction are 1 for A_1A_1, $1 - sh$ for A_1A_2 and $1 - s$ for A_2A_2. Here s is assumed small and positive, $0 \leqq h \leqq 1$. Suppose in any generation that the frequency of A_1 is x. Then a proportion $1 - s(1 - x)\{1 + x(2h - 1)\}$ of all offspring survive and, hence, each parent generation is required to produce $N[1 - s(1 - x)\{1 + x(2h - 1)\}]^{-1}$ offspring so that, after the preadult selection has occurred, exactly N offspring survive to reach sexual maturity. The number of "selective deaths" is thus

$$(1)\quad N[1 - s(1 - x)\{1 + x(2h - 1)\}]^{-1} - N \cong Ns(1 - x)\{1 + x(2h - 1)\} = N\ell,$$

say. The substitutional load L is conventionally defined as the sum of the values of ℓ, whereby x increases from $x_0(\approx 0)$ to $x_1(\approx 1)$, and is

$$(2)\quad L = \int_0^\infty \ell\, dt = \int_{x_0}^{x_1} \ell \frac{dt}{dx}\, dx = \int_{x_0}^{x_1} [s(1 - x)\{1 + x(1 - 2h)\}][sx(1 - x)\{1 - h + x(2h - 1)\}]^{-1}\, dx.$$

For $h = \frac{1}{2}$, $x_0 = 10^{-4}$, we have $L = 18.4$, while for other values of h, L is sometimes greater than, sometimes less than, this value. A "representative" value for L is conventionally taken as $L = 30$.

We now turn to the biological interpretation of L. If the replacement process of A_1 for A_2 requires T generations, then clearly from (1) and (2) $NT + NL$ total offspring are required during this process; in other words, on average each individual must produce $1 + L/T$ offspring to face the possible forces of preadult selection. Alternatively, we may say that the optimal genotype leaves an average of $1 + L/T$ offspring who reach sexual maturity and that *all* genotypes must produce this number of offspring.

The crux of the argument arises when many loci are considered simultaneously. If we assume the same fitnesses at all loci as those given above, and if the substitutions at the different loci start, on average, n generations apart, there will be T/n substitutions in progress at any one time. The mean number of offspring of the optimal genotype (that is, the genotype having the configuration "A_1A_1" at each locus) is

$$(3)\quad \left(1 + \frac{L}{T}\right)^{T/n} \cong \exp\left\{\frac{L}{n}\right\},$$

and using the value $L = 30$ and the estimate $n = \frac{1}{2}$ quoted as deriving from protein sequence data, this gives $e^{60} \cong 10^{26}$ offspring. In other words, the argument suggests that in order for the substitutions at all the loci to proceed at the required rate, each individual must leave 10^{26} offspring and that these survive differentially in such a way that N offspring eventually reach sexual maturity. The substitutional load argument centers on the impossibility that each individual can leave such a large number of offspring.

It is at this point that elementary statistical arguments suggest that this reasoning be reviewed. We first consider the probability that an individual chosen at random is of the optimal genotype. If $h = \frac{1}{2}$, $x_0 = 10^{-4}$, $x_1 = 1 - 10^{-4}$, $s = 0.01$, $n = \frac{1}{2}$, there are 7,360 substitutions in progress at any one time and the probability that any randomly chosen individual has the optimal genotype A_1A_1 at all 7,360 substituting loci is about $10^{-15,000}$. It seems of dubious value to base our considerations on such individuals. Rather, statistical considerations suggest that in any real population, no individual will have a genetic configuration which is too "extreme" (note that this in no way slows down the substitution rates), and, in particular, the mean number of surviving offspring required of the optimal genotype *we can expect to appear* in the population will not differ much from unity. This requirement can be calculated using the statistics of extreme values (of a sample of N individuals) and effectively assuming "independence" of loci. (This assumption, which, for example, ignores linkage between loci, will provide an upper bound to the offspring requirement, which should in fact be rather lower if linkage were taken into account.) For $N = 10^5$, $s = 0.01$, $h = \frac{1}{2}$, we find that this offspring requirement is about 1.5. In other words, substitutions at the required rate can occur if all individuals leave 1.5 offspring and these survive differentially according to their genotype (with the optimal *existing* genotype all surviving) in such a way that N offspring altogether survive to sexual maturity. When the effects of linkage, extensive linkage disequilibrium, and the possibility that considerable selection actually occurs through fertility difference, are all taken into account, the value 1.5 is probably reduced to about 1.2. There appears little difficulty for a population actually to achieve this value: that is to say, the statistical form of reasoning we have adopted suggests that there is little difficulty in ascribing the observed substitutions to selective forces with selective differentials of order one per cent. (Of course, this does not mean selection must be the responsible agency; there appears to be no reason, however, for us to say selection *cannot* be responsible.)

A further statistical point relating to the above argument is as follows. In a typical size population of, say, 10^6, no individual can leave more than 10^6 offspring who survive to maturity. The calculation that 10^{26} offspring of certain genotype do so survive must then result from inexact modelling; in this case, from the implicit assumption that mean numbers are "multiplicative over loci." In other words, not only have the load arguments supporting the non-Darwinian theory been carried out entirely with reference to individuals whose probability of occurrence is of order $10^{-15,000}$, they have also ascribed mean numbers of viable offspring to such (essentially nonexistent) individuals about 10^{20} times in excess of the maximum possible value.

3. The sampling theory of neutral alleles

A considerable amount of data is being obtained currently of the following form: a sample of n individuals ($2n$ genes) is taken from a population of unknown size N. In the sample it is observed that k different alleles occur, with numbers $n_1, n_2, \cdots, n_k$ ($\sum n_i = 2n$). We now ask, can such data be used to test the hypothesis that the alleles are selectively neutral with respect to each other?

To do this, is it assumed that new alleles are formed in the following fashion: each gene will mutate with fixed (but unknown) probability u, to an allelic type not currently existing, nor previously existing, in the population. (Note that this assumption is inspired by protein sequence data, where our detailed knowledge of the sequence of amino acids determined by any gene makes this assumption reasonable. The assumption is perhaps less valid if our mode of differentiating alleles is by electrophoresis, since the theoretical effects of the nonidentification seemingly unavoidable in this procedure are not yet known. It is, therefore, possible that application of the following techniques to electrophoretically obtained data should be viewed with extreme caution.) More explicitly, we assume a multinomial mode of sampling the genes of a daughter generation from the genes of the parent generation. This implies that if we fix attention on some allele A_k, and suppose that i genes of this allele exist in any generation, then the probability $p_{i,j}$ that there exist j genes of this allele in the next generation is

$$p_{i,j} = \binom{2N}{j}\left\{\frac{i}{2N}(1-u)\right\}^j\left\{1 - \frac{i}{2N}(1-u)\right\}^{2N-j}. \tag{4}$$

Note that we are assuming that no selective differences exist between alleles; thus, our aim is to develop distribution theory under the neutral hypothesis, with a view to subsequent testing of it using real data.

If it is supposed that sufficient time has elapsed for a stationary situation to be reached, then passing to the diffusion approximation to (4), standard theory (Ewens [2], Chapters 5 and 6) shows that the probability that, in any randomly chosen generation, there exists an allele in the population with frequency in the range $(p, p + \delta p)$ is

$$f(p)\,\delta p = \theta p^{-1}(1-p)^{\theta-1}\,\delta p, \tag{5}$$

where $\theta = 4Nu$. It will turn out that all of our subsequent theory can be carried out by using this "frequency spectrum" $f(p)$. In particular, we note that if $p_1, p_2, \cdots$ are the (unknown) frequencies of the various alleles present in any generation, and if $\phi(p)$ is any function of p that is $O(p)$ near $p = 0$, then

$$E \sum \phi(p_i) = \theta \int_0^1 \phi(p)p^{-1}(1-p)^{\theta-1}\,dp, \tag{6}$$

to a sufficiently close approximation. For example, if $\phi(p) = p$, use of (6) yields the trivial identity $E \sum p_i = 1$, while if $\phi(p) = p^2$, we find

$$E \sum p_i^2 = \frac{1}{1+\theta}. \tag{7}$$

Note that the left side in (7) is the probability that two genes drawn at random are of the same allelic type. It is, thus, a measure of genetic variability in the population and the quantity $1/(1 + \theta)$ is, consequently, of some interest to geneticists. We shall return later to the question of estimating this quantity from experimental data.

Now suppose a sample of $2n$ genes is drawn one by one from the population. We suppose $N \gg n$ so that binomial approximations are adequate. Then if we know the frequencies $p_1, p_2, \cdots$ of the various alleles in the population in the generation from which the sample was drawn, the probability that a previously unseen allele appears for the first time on the $(j + 1)$th draw is

$$\sum_i (1 - p_i)^j p_i. \tag{8}$$

Also the probability that the first j draws all yield the same allele is

$$\sum_i p_i^j. \tag{9}$$

Equation (6) shows that the unconditional probabilities to be attached to these events are

$$\theta \int_0^1 (1 - p)^j p[p^{-1}(1 - p)^{\theta-1}]\, dp = \frac{\theta}{\theta + j}, \tag{10}$$

and

$$\theta \int_0^1 p^j[p^{-1}(p - 1)^{\theta-1}]\, dp = \theta(j - 1)!\, \frac{\Gamma(\theta)}{\Gamma(\theta + j)}, \tag{11}$$

respectively. It follows from (11) that the probability that the first $(j + 1)$ draws all yield one allelic type, given that this is true of the first j draws, is

$$\frac{\theta j! \Gamma(\theta)\Gamma(\theta + j)}{\theta(j - 1)! \Gamma(\theta)\Gamma(\theta + j + 1)} = \frac{j}{j + \theta}. \tag{12}$$

From this it follows that the probability that a new allele appears on the $(j + 1)$th draw, given that the first j draws yield only one allelic type, is $\theta/(\theta + j)$. Note that this is identical to (10). We now argue more generally that the probability that a new allele appears on the $(j + 1)$th draw is $\theta/(\theta + j)$ *whatever* the allelic composition was of the first j draws. A formal proof of this proposition has been given by Karlin and McGregor [6]. Intuitively, this (unusual) result can be seen as follows. If we label the new allele seen on the $(j + 1)$th draw as A_k, then this gene is descended from some original mutant A_k allele. The stochastic behavior of the line of descendants of this mutant is independent of the allelic composition of the rest of the population, and so far as A_k is concerned the allelic forms of the non-A_k genes are just irrelevant labels. In particular, the probability that A_k appears for the first time at the $(j + 1)$th draw is independent of this irrelevant labelling, that is, of the numbers and frequencies of the alleles which appeared on the first j draws.

It follows from this that if we write $\pi_{j,i}$ for the probability that the first j draws yield exactly i different alleles, we have

$$\begin{aligned} \pi_{j,1} &= (j - 1)![(\theta + 1)(\theta + 2)\cdots(\theta + j - 1)]^{-1}, \\ \pi_{j,j} &= \theta^{j-1}[(\theta + 1)(\theta + 2)\cdots(\theta + j - 1)]^{-1}, \end{aligned} \tag{13}$$

and the recurrence relation

$$\pi_{j+1,i} = \pi_{j,i}\left\{\frac{j}{\theta + j}\right\} + \pi_{j,i-1}\frac{\theta}{\theta + j}. \tag{14}$$

We are particularly interested in the values $\pi_k = \pi_{2n,k}$. Solution of (14) shows that

$$\pi_k = \frac{\ell_k \theta^k}{L(\theta)}, \tag{15}$$

where

$$\begin{aligned} L(\theta) &= \theta(\theta + 1)\cdots(\theta + 2n - 1) \\ &= \ell_1\theta + \ell_2\theta^2 + \cdots + \ell_{2n}\theta^{2n} \end{aligned} \tag{16}$$

and the ℓ_k are moduli of Stirling numbers of the first kind. Equation (15) gives the probability distribution of the number k of different alleles in the sample. We have in particular

$$\begin{aligned} E(k) &= \frac{\theta}{\theta} + \frac{\theta}{\theta + 1} + \cdots + \frac{\theta}{\theta + 2n - 1}, \\ \operatorname{Var}(k) &= \frac{\theta}{\theta} + \frac{\theta}{\theta + 1} + \cdots + \frac{\theta}{\theta + 2n - 1} \\ &\quad - \left[\frac{\theta^2}{\theta^2} + \frac{\theta^2}{(\theta + 1)^2} + \cdots + \frac{\theta^2}{(\theta + 2n - 1)^2}\right], \end{aligned} \tag{17}$$

and, further, that for $2n$ large, k had an approximate normal distribution with this mean and this variance. Note also that the distribution (15) is complete.

We turn now to a more complex problem, namely, the distribution of the vector $\{k; n_1, \cdots, n_k\}$, where $n_1 \cdots n_k$ are the numbers of genes of the k alleles in the sample. We find (Ewens [3]) that this distribution is of the form

$$\frac{g(n_1, \cdots, n_k)\theta^k}{L(\theta)}, \tag{18}$$

where $g(n_1, \cdots, n_k)$ does not depend on θ. Comparison of (15) and (18) shows that k is sufficient for θ. Hence, any estimable function of θ is best estimated (in the sense of minimum variance unbiased) by some (unique) function of k. In particular, this is true of the (estimable) function $1/(1 + \theta)$, whose optimal estimator is

$$c(k) = \frac{\text{coeff } \theta^k \text{ in } \theta(\theta + 2)(\theta + 3)\cdots(\theta + 2n - 1)}{\text{coeff } \theta^k \text{ in } \theta(\theta + 1)(\theta + 2)\cdots(\theta + 2n - 1)}. \tag{19}$$

Curiously, because of the genetical interpretation of $1/(1 + \theta)$, (see the discussion following equation (7)), this function has traditionally been estimated by

$$\frac{n_1^2 + \cdots + n_k^2}{(2n)^2}. \tag{20}$$

This estimator has been found in Monte Carlo simulations to have very large variance (see, for example, Bodmer and Cavalli-Sforza [1]) and the present theory indicates why this is the case, namely, that (20) uses precisely the inappropriate part of the vector $\{k; n_1, \cdots, n_k\}$. Further, Monte Carlo runs (Guess and Ewens [4]) suggest that the variance of $c(k)$ is only about 35 per cent that of (20). It is clear that some standard statistical theory is of some aid here in estimating an important genetical parameter.

Our main use, however, of the sufficiency of k for θ will be to provide a test for the hypothesis of selective neutrality, for under this theory the distribution of $n_1, \cdots, n_k$, given k, must be independent of θ and, hence, the same for all models of the form (4) (whatever the values of N and u are). It is found that this conditional distribution is [3]

$$f(n_1 \cdots n_k | k, 2n, \text{neutrality}) = \frac{(2n)!}{k! \ell_k n_1 n_2 \cdots n_k}. \tag{21}$$

More precisely, we assume the k alleles in the sample have been labelled in some conventional fashion $A_1 \cdots A_k$: equation (21) gives the probability that there are n_1 genes of the allele labelled $A_1, \cdots$, and n_k genes of the allele labelled A_k. Note that the fact that (21) sums to unity is established by an identity for Stirling numbers going back at least to Cauchy (see, for example, Jordan [5], p. 146, equation (5)).

The distribution (under the hypothesis of selective neutrality) of any test statistic can be found from (21). The actual choice of test statistic is not easy since the alternative hypothesis (that selection exists) does not seem sufficiently precise to yield an unambiguous statistic. Here we shall be content with using (more or less arbitrarily) the (information) statistic

$$B = -\sum x_i \log x_i, \tag{22}$$

where $x_i = n_i/2n$. The mean and variance of B can readily be calculated from (21) and by noting that marginal distributions from (21) are calculated almost immediately. Hence, if we write $E(B)$ and $\sigma(B)$ for the mean and standard deviation of B under the hypothesis of selective neutrality, given the appropriate value of k, it is possible for any set of data to evaluate

$$L = \frac{B - E(B)}{\sigma(B)}, \tag{23}$$

which under selective neutrality is a random variable having mean zero and variance one. Evaluation of L seems particularly useful when sets of interrelated data (that is, from each of a number of species in each of a number of locations) are available, since the patterns of the values for L for such data are often revealing.

If, on the other hand, it is desired to carry out a test of hypothesis, an approximate but reasonable procedure is to suppose $B/\log k$ has a beta distribution (whose parameters are known from $E(B)$ and $\sigma(B)$). A standard transformation yields a variable having an F distribution under the neutrality hypothesis. A FORTRAN program carrying out all the required computations is provided in [3].

TABLE I

EXAMPLE USING $n = 350$ AND $k = 4$

Sample	Allele frequencies				L	F	$(d.f.)$
1	.35	.30	.20	.15	2.42	34.44	(3.5, 4.7)
2	.83	.11	.04	.02	0.02	1.02	(3.5, 4.7)
3	.99	.005	.0025	.0025	−1.74	0.07	(3.5, 4.7)

In Table I we indicate the sort of result obtained by an example. Suppose $n = 350$, $k = 4$, and consider three different sets of values of $x_1, \cdots, x_4$. The first sample yields significant evidence of selection (of some form) holding all alleles at high frequency, while the third sample yields evidence of selection favoring one allele. The second sample has almost a "perfect" set of "neutral" frequencies.

It is proper to conclude on a note of caution. The present lack of theoretical knowledge on the effect of nonidentification may be sufficiently strong to vitiate application of the above to electrophoretically derived data. (The above theory assumes total ability to differentiate different alleles.) Our model also ignores the effects of linkage, possible fluctuation in population size, and so forth. (On the other hand, it is conjectured that the distribution (21) applies for a wide range of "neutral" models, not just the model (4).) Finally, the test does not appear to be particularly powerful (in a statistical sense) and using it one may often maintain the hypothesis of neutrality when, in fact, mild selection does occur. Altogether, it appears that the test of hypothesis, and use of the index function L, may best serve as a cautiously used adjunct to other and independent methods of testing the non-Darwinian theory.

REFERENCES

[1] W. F. BODMER and L. L. CAVALLI-SFORZA, "Variation in fitness and molecular evolution," *Proceedings of the Sixth Berkeley Symposium on Mathematical Statistics and Probability*, Berkeley and Los Angeles, University of California Press, 1972, Vol. 5, pp. 255–275.

[2] W. J. EWENS, *Population Genetics*, London, Methuen, 1969.

[3] ———, "The sampling theory of selectively neutral alleles," *Theor. Pop. Biol.*, Vol. 3 (1972), pp. 87–112.

[4] H. GUESS and W. J. EWENS, "Theoretical and simulation results relating to the neutral allele theory," *Theor. Pop. Biol.*, to appear.

[5] C. JORDAN, *Calculus of Finite Differences*, New York, Chelsea, 1950.

[6] S. KARLIN and J. L. MCGREGOR, *Addendum* to "The sampling theory of selectively neutral alleles" [3], *Theor. Pop. Biol.*, Vol. 3 (1972), pp. 113–116.

AUTHOR REFERENCE INDEX
VOLUMES I, II, III, IV, V, and VI